Occupational Health Guide to Violence in the Workplace

Occupational Safety and Health Guide Series

Series Editor

Thomas D. Schneid
Eastern Kentucky University
Richmond, Kentucky

Titles in the Series

Managing Workers' Compensation: A Guide to Injury Reduction
by Keith Wertz and Brad Layton

Motor Carrier Regulation Compliance Guide
by Erik Scott Dunlap

Physical Hazards of the Workplace
by Larry R. Collins

Occupational Health Guide to Violence in the Workplace

Thomas D. Schneid
Eastern Kentucky University
Richmond, Kentucky

CRC Press
Taylor & Francis Group
Boca Raton London New York

CRC Press is an imprint of the
Taylor & Francis Group, an **informa** business

Disclaimer

Although the author has taken great pains to ensure that the information included in this text is accurate and up-to-date, prudent professionals are advised to research the specific issue to ensure complete accuracy. As we are all aware, the law changes with every court decision and governmental standards are being modified on a daily basis. The reader should also be aware that not all areas of potential liability are covered in this text. The author has attempted to identify the areas that have the greatest frequency or carry the greatest potential liability in terms of frequency and severity. The author provides no warranty, either expressed or implied, as to the accuracy of the law, standards, or other information contained in this text. Although suggestions are offered, the authors do not intend this text to provide specific legal counsel with regard to individual circumstances. Competent local legal counsel should be acquired to assist in specific circumstances and situations.

CRC Press
Taylor & Francis Group
6000 Broken Sound Parkway NW, Suite 300
Boca Raton, FL 33487-2742

First issued in paperback 2019

© 1999 by Taylor & Francis Group, LLC
CRC Press is an imprint of Taylor & Francis Group, an Informa business

No claim to original U.S. Government works

ISBN-13: 978-1-56670-322-2 (hbk)
ISBN-13: 978-0-367-40011-8 (pbk)

Library of Congress Card Number 98-38656

Library of Congress Cataloging-in-Publication Data

Schneid, Thomas D.
    Occupational health guide to violence in the workplace / by Thomas D. Schneid.
      p.   cm.
    Includes bibliographical references and index.
    ISBN 1-56670-322-0 (alk. paper)
    1. Violence in the workplace–United States–Prevention.
  2. Industrial buildings–Security measures–United States.
  3. Industrial safety–Law and legislation–United States.
  4. Industrial hygiene–Law and legislation–United States.
  I. Title
HF5549.5.E43S36   1998
658.3'82--dc21

          98-38656
          CIP

Visit the Taylor & Francis Web site at
http://www.taylorandfrancis.com

and the CRC Press Web site at
http://www.crcpress.com

About the author

Dr. Thomas D. Schneid is Professor in the Department of Loss Prevention and Safety and is also serving as the coordinator for the Fire and Safety Engineering program at Eastern Kentucky University. He has earned a B.A. in Education, M.S. and C.A.S. in Safety, M.S. in International Business, J.D. (law), L.LM. (Graduate law in labor/employment), and a Ph.D. in Environmental Engineering. Dr. Schneid is the founding member of the law firm of Schumann and Associates, PLLC located in Richmond, Kentucky.

Dr. Schneid has authored or co-authored numerous texts including *Americans With Disabilities Act: A Compliance Manual* (Van Nostrand Reinhold, 1994); *Fire Law* (Van Nostrand Reinhold, 1995); *Fire and Emergency Law Casebook* (Del Mar Publishing Co., 1996); *Food Safety Law* (co-authored with M. Fagel, B.R. Schumann, and M. Schumann, Van Nostrand Reinhold, 1997); *Legal Liabilities for Safety and Loss Prevention Professionals* (co-authored with M. Schumann, Aspen Publishing Company, 1997); *Creative Safety Solutions*, CRC/Lewis, 1998). Additionally, he has authored over 75 published articles on various safety, environmental, and legal topics.

Dr. Schneid was selected as the "Rising Star in Safety" in 1997 by *Occupational Hazards* magazine and is listed in *Who's Who in American Law and Professional Safety.*

Acknowledgments

Special thanks to my wife, Jani, and my children, Shelby and Madison, for motivating me and permitting the time to research and complete this text.

My thanks to Shiella Patterson and Amy Eades for all of their assistance in assembling and editing this text. And thanks to the faculty, staff and students in the Department of Loss Prevention and Safety at Eastern Kentucky University.

And, as always, my thanks to my parents for all their hard work and support in providing me the ability to acquire the education and skills necessary to write this book.

Contents

Foreword ix

Analyzing case law and briefing a case xi

Introduction xix

Chapter 1   Overview and history 1

Chapter 2   Individual profiles and warning signs 15

Chapter 3   Workplace violence prevention and response program 23

Chapter 4   OSHA and other potential liability 37

Chapter 5   Workers' compensation liability 73

Chapter 6   Negligent hiring 113

Chapter 7   Negligent retention 127

Chapter 8   Negligent supervision 171

Chapter 9   Negligent training 199

Chapter 10   Negligent security 221

Chapter 11   Wrongful death actions 241

Chapter 12   Other legal considerations 295

Chapter 13   What to expect after an incident 373

Chapter 14   Other selected readings and studies 381

Appendix A   Table of cases and cites 421

Appendix B   NIOSH information...429

Appendix C   OSHA information 437

Appendix D   OSHA guidelines..441

Appendix E   Employee workplace rights 451

Appendix F   OPM guidelines and selected web cites................................461

Index 479

Foreword

The word is getting out. Violence in the workplace is not only real, it's growing. It is happening not only in the U.S. Post Office, but where you live and work! In fact, the National Institute for Occupational Safety and Health (NIOSH) has found that an average of 20 workers are **murdered each week** in the United States. This is not the whole story. We generally hear about the victims in highly publicized workplace murders. Unfortunately, there are many more victims of workplace violence. It is estimated that 1 million workers (roughly 18,000 per week) are victims of workplace **assaults** each year.

For employers, there are many concerns and risks associated with these workplace issues and the potential for violence in the workplace. Employers must face competing interests of employee privacy and worker protection. Should employers make their workplaces so guarded and rigid that they have metal detectors in every hallway and security checkpoints at every door? Should employers search all bags, coats, and luggage that enter the workplace? At what point has an employer gone too far in the interest of preventing workplace violence, while creating an atmosphere of suspicion and mistrust, all in the name of employee safety?

On the other hand, what effect does a lone gunman with an automatic weapon and murder on his mind do to the workplace morale of an organization? How long will valued employees stay with your organization? How productive will they be? What effect will this have on your employees' families? Can you take this risk? What is your true liability? Can it be measured only in dollars and cents?

The answers to these questions are easier than the solutions to the problem. Employers can go too far in the interest of employee safety, and security is not cheap. Then what is the happy medium? This author believes that the answer is within your own organization. The key to discovering the answer is a full self evaluation. Discover what your risks are in your organization, plan a course of action, be prepared to modify and revise your plan of action, involve employees in the decision-making process, and review your program on a regular basis.

The key to planning is knowledge. Know what your risks and liabilities are for workplace violence. Learn from the mistakes and experiences of others. Can the same situation happen in your organization? What are your hiring practices? Do you conduct meaningful reference checks? Do you have an outlet for disgruntled employees? How do you discipline employees? Do you conduct fair and consistent employment terminations? Do you conduct exit interviews?

Is your organization liable only for employees getting injured or killed on the job through workplace violence? If you answered yes to this question, think again. Employers have a legal obligation to protect **third parties** (such as visitors, clients, vendors, and patrons) from violence committed by **non-employees**.

What types of businesses or industries need to quickly take stock and review their potential for workplace violence? There are a number of businesses and industries that face a higher potential for workplace violence. Jobs that place workers at risk for workplace violence include delivering services or goods, positions requiring the exchanging of money, interacting with the public, working late at night or in the early morning hours, working alone, and guarding valuable goods or property.

What approach should the average employer take? Employers should take a proactive approach to this problem, just like any other business issue. Don't wait until a former employee walks into your facility with a loaded weapon and asks to see his former boss. Get a game plan. Review your security. Review your policies and contingency plans. Get started now if you haven't done anything. If you've started, finish and review. Keep current and protect yourself and your workforce. This is a real problem and prevention is the key to success. Knowledge is the key; action is the lock. Get started now and secure your success.

Michael S. Schumann, Esq.
Schumann and Associates, PLLC
Richmond, Kentucky

Analyzing case law and briefing a case

Case law is the accumulation of court decisions which in essence, shapes and develops new law and clarifies old law. In each chapter throughout this text, you will find several cases that will exemplify the particular point of law addressed in the chapter. It is important that you read, evaluate, and brief each of these cases in order to acquire a complete understanding of the law in the particular area. As the basic rule of thumb, when analyzing and/or briefing a case you should read through the case in detail at least three times. On the first reading, identify the type of case, the court, and acquire a basic understanding of the case type. In the second reading, you should identify the basic issues, facts, and holding of the decision as well as any dissenting opinion. In the third reading of the case, you should take notes with regard to the actual brief of the case which you will be referring to later. It is essential that you take good notes and brief your case extremely well so you can refer to the brief and refresh your memory at some later point in your studies.

Finding the case

Finding the case in the library is often one of the most difficult parts of your total analysis. As can be seen from the various cases provided in this text, other cases are referenced and numerous "cites" are provided throughout this text. All reported decisions of cases in the United States judicial system are listed according to the publication in which the case appears (called a reporter), the volume in which the case is located, and the initial page number. For example, 36 S.E. 2d 924. If you were searching for this case, you would go to the *South Eastern Reporter*, second edition, look for volume number 36 and find the case at page 924. In the federal judicial system, the publications tend to be in accordance with the region of the country, and in most state judicial systems the location will be in a state reporter. It should be noted that not all cases are reported and published. Generally, trial court decisions are not published because these decisions do not serve as mandatory precedent for other courts to follow. Usually only decisions of federal and state appellate courts are published.

Statutes, regulations, and standards tend to be within other publications, such as the *Code of Federal Regulations*. This system is set up utilizing the same publication, volume, and page number as with the judicial court decisions. However, the series numbers may reflect the particular regulation or standard. For example, 29 C.F.R. 1910.120 is the Occupational Safety and Health Administration's Standard with regard to hazardous waste operation and emergency response. If you were searching for this particular standard, you would go to the *Code of Federal Regulations*, which is signified by the C.F.R. designation, volume 29, and Section 1910.120.

In addition to the normal procedure in locating a case, some law offices and law libraries also provide an electronic database through which to locate cases. The two major databases are WESTLAW and LEXIS. Each of these databases normally provides training and publications to assist you in locating a particular case. As a general rule, these databases provide a basic menu of the various areas where the law is located and numerous sub-databases that guide your search. For example, if you were searching for a federal decision, you may enter the federal database and then narrow your search to the particular area of law which you are seeking. If you have the case name, it can normally be utilized to pull up the case. If you are searching for a particular issue of the law, these databases will provide the case cites for your review and evaluation. It is highly recommended that you acquire the particular training or assistance from the librarian at the law library prior to conducting any search on WESTLAW or LEXIS.

It is highly recommended that you become familiar with the particular library you will be using. Take a few minutes and walk through the library to locate the different publications that are available and note the location of each of these publications. Thumb through a few of the publications and test your skill at locating particular cases so you are able to find cases in a timely manner in the future.

Briefing the case

The basic purpose for briefing a case is to help you understand the particular legal issues of the case and their significance. There are various methods of briefing a case, and the following format is only meant to be an example of one of the methods. Your instructor may suggest an additional format for you or you may devise your own system which helps you analyze these cases. No matter what method you should adopt, ensure that you read the case thoroughly prior to beginning to take notes for your brief.

The basic framework in which we would recommend you brief a case includes the following:

A. Issues
B. Facts
C. Holding (Decision)

D. Dissent
E. Your Opinion
F. Underlying Policy Reason

Issues

Identify the basic issue or issues that are in question before the court. In order to find the basic issue or issues involved, you have to identify the rule of law that governs the dispute and ask how it applies to the particular facts of the case. In most circumstances, you will be writing the issue for your case brief as a question that combines the rule of law with the material facts of the case. For example, does the arson statute in the state of Kentucky apply to a minor child?

Facts

The facts of the particular case describe the events between the conflicting parties that led to this particular litigation and tells how the case came before the court that is now deciding it. Often included in the facts are the relevance to the issue the court must decide and the basic reasons for its decision. When you first read through the case, you will not know which facts are relevant until you know what the issues are in the particular case. Thus, it is vitally important that you read the case thoroughly prior to beginning to summarize the facts of the particular case.

In addition to the specific facts of the situation, it is important to see what court decisions have come prior to the case you are reviewing. Often, the decisions which are published are appellate decisions and, thus, a district court or circuit court has decided the matter previously and now the matter is on appeal. If the particular facts of the situation in an appellate case are not provided in detail, you may want to research and review the district or circuit court decisions in order to acquire the particular facts in your case.

In this section, you should also include the relevant background for this case. You should identify who the plaintiff and the defendant are, the basis of the plaintiff's suit, and the relief the plaintiff is seeking. You may also want to include the procedural history of the case, such as Motions to Dismiss and other motions that are relevant to the case. In an appeals case, the decisions of the lower courts, grounds for those decisions, and the parties who appealed should also be noted.

Within this facts section, you should be as brief as possible. However, all pertinent points should be noted. Although this is a judgment call, most statements of fact in a brief should not be more than two or three paragraphs long. Given the fact that you would have read the case at least three times while briefing the case, the facts provided in your brief of the case should be the major points used to refresh your memory.

Holding

The holding is the court's decision on the question that was actually presented before it by the parties. The court, in a split decision, may also provide the minority's decision or dissent. The holding can normally be identified by the statement the court has decided or the majority decision. A holding, in essence, provides the answer to the question you were asking in your issue statement. If there is more than one issue involved in the case, there may be more than one holding in any given case.

Dissent

Often with U.S. Supreme Court cases and appellate cases, the majority decision is the decision of the court. However, the minority position is also provided an opportunity to give their reasoning as to their dissent in the decision-making process. Although the dissenting opinion is not law and has no bearing on the case, the dissent provides another point of view on the particular issue of the case and also may be referred to in some later case.

Opinion

After you have reviewed the case thoroughly and have analyzed the court's decision and briefed your case, you should have a good idea whether you agree or disagree with the court's opinion. In this section, you should provide your personal opinion as to whether you agree or disagree with the court and the reasons you agree or disagree with the decision.

Policy

In many cases, there is an underlying social policy or particular social goal which the court wishes to further. When a court explicitly refers to those policies in a particular case, this information should be included in your brief, since it will provide a better understanding of the court's decision. For example, in the historic case of *Brown v. Board of Education* the decision of the court was formed through an underlying social policy to eliminate discrimination in our school system. This underlying social policy is often very important in Appellate and Supreme Court decisions. Attached is an example brief for your review and evaluation. It is highly recommended that you test your skills by briefing several of the cases within this text or other cases prior to your initial briefing of a case for a class. In addition to the above stated information, below are several other helpful hints which may assist you in briefing the case:

1. It is best to try and confine your brief to a maximum of one page. If your brief is over two pages, you have probably provided too much

information. Remember, a brief is to refresh your memory at the time you need to recall this information for class or other purposes.

2. The cases printed in this textbook have been edited by the author for the purposes of this book. In many cases, the full court opinion may run 20 or more pages. If you find that you are having difficulty understanding the case in the edited format, you may want to go to the library and read the full text opinion.

3. During your first couple of attempts at briefing a case, it is often difficult to extract the important elements and issues of a case. Please keep in mind that not all judges are expert writers, so the opinions may often be confusing or difficult to understand. Additionally, you should realize that all courts do not follow the same format in writing opinions; thus you may find some decisions more difficult to understand than others. So, you may find that judges sometimes go off on a tangent and discuss other rules and points of law that are not essential to the determination of the particular case. It is your job to be able to filter through this information to identify the particular issues and laws that are applicable to the case.

4. You may often run across Latin or "legal" terms with which you are not familiar. Since you will need to have a clear understanding of the terminology utilized in the particular case, it is advisable to look up the term in a legal dictionary. A good idea is to have a *Black's Law Dictionary, Ballantine Law Dictionary, Gilmer's Law Dictionary,* or other law dictionary available while you are reading and briefing the case. Standard dictionaries often do not provide these terms, or the explanation provided may be incomplete.

5. When reading the cases provided in this text, you may want to look at the particular chapter and section headings of the textbook in which the case appears. If you are having difficulty identifying the particular issue of the case, the issue is normally related to the topic discussed in the chapter or section heading. The cases in this text have been inserted to illustrate the subject matter being discussed in each of the chapters.

6. Remember, the issue or issues in the particular case should always be stated in the form of a question. You should never begin your issue with the words, "whether or not" because this will not frame the question properly. Also, the terminology, "should plaintiff win" or "is there a contract" are not correct forms of stating the particular issue.

7. In determining the particular rule of law, ask yourself, "If I had to tell someone who knew nothing about this case what this case is about or what it stands for in one sentence, what would I say?" Often, the rule of law can be determined by taking the issue and putting it in the form of a declaration. For example, in the case of *Miranda v. Arizona*, 384 U.S. 436 (1966), the issue and rule may be as follows:

Issue: When a person is taken into police custody, or otherwise deprived of his freedom of action in a significant way, must his constitutional rights to remain silent and to have an attorney present be explained to him prior to questioning?

Rule: When a person is taken into police custody, the following warnings must be given prior to questioning:

1. That he has the right to remain silent.
2. That any statement he makes may be used against him as evidence.
3. That he has the right to have an attorney present.
4. If he cannot afford an attorney, one will be appointed for him.

8. Last, do not use other people's briefs. Without having read a particular case and analyzed the court decision yourself, use of another individual's brief of a case is essentially worthless. A brief is simply the codification of your thoughts and work that you will refer to in the future in order to refresh your memory.

Example case brief

Case Name: *Marshall v. Barlow's, Inc.*, 436 U.S. 307 (1978)

Issue: Is Section 8(a) of the Occupational Safety and Health Act unconstitutional in that it violates the Fourth Amendment?

Facts: Appellee (Barlow's, Inc.) initially brought this action to obtain injunctive relief against a warrantless inspection of its business premises by Appellant (Secretary of Labor Marshall). The inspection was permitted under Section 8(a) of the Occupational Safety and Health Act of 1970, which authorizes agents of the Secretary of Labor to search the work area of any employment facility within OSHA's jurisdiction for safety hazards and OSHA violations without obtaining a search warrant or other process. A three-judge Idaho District Court ruled in favor of Barlow's and concluded that the Fourth Amendment required a warrant for this type of search and that the statutory authorization for warrantless inspections was unconstitutional. This appeal resulted.

Holding: Yes, Section 8(a) of the Occupational Safety and Health Act of 1970 was unconstitutional in that it violated the Fourth Amendment. The U.S. Supreme Court affirmed the decision of the Idaho District Court and granted Barlow's an injunction enjoining the enforcement of the act to that extent.

The court states that the rule against warrantless searches applies to commercial premises as well as private homes. Although an exception to this rule is applied to certain "carefully defined classes of cases," including closely regulated businesses such as the firearm and liquor industries, this exception does not automatically apply to all businesses engaged in interstate commerce, as the Secretary alleges.

Opinion: I agree with the court in this case. I feel that requiring search warrants ensures that the search is a reasonable one and that the particular business being inspected is not merely being singled out (for one reason or another). I agree with the court in that requiring search warrants will not make inspections less effective nor will it prevent necessary inspections, but rather will serve to ensure fairness in inspections.

Policy: Although no specific public policy was mentioned in the case, the implied policy was that of pro-business, anti-regulation.

Introduction

"Violence is the last refuge of the incompetent."

Isaac Asimov

"Nothing good ever comes of violence."

Martin Luther

Work can be dangerous to the life of your employees, your co-workers, and even to yourself! The American workplace has recently become an even more dangerous place, with the dramatic increase in incidents of workplace violence. In fact, in the most recent study conducted in 1996, violence in the workplace resulted in 1144 deaths and several thousand workplace injuries.* The potential for violence in our workplaces, from internal or external sources, is a risk we must recognize and attack through a proactive effort and be prepared to address the aftermath. Safety and loss prevention professionals, human resource and personnel professionals, and management in general must recognize this very real potential risk of harm which can result from a workplace violence incident and prepare to eliminate or minimize this potential risk in their workplaces.

So what is the probability of a workplace violence incident in my company? Here are several basic questions you may want to ask yourself:

- Do I have employees? How many?
- Have I ever terminated an employee? Layoff?
- Do my employees have spouses or "significant others"?
- Does my company produce products which may be controversial?
- Is your workplace open to the public?
- Do I have employees on workers' compensation?

In most companies, the answer to most of these questions will be "yes." In essence, virtually every workplace in America, from the front seat of a New York cab driver to your local fast food restaurant, is at risk. The risk

---

* *Census of Fatal Occupational Injuries*, 1996, Bureau of Labor Statistics, Washington, D.C.

Causes of Workplace Death in 1996
Total fatalities 6,112

Work-related deaths dropped to a five-year low in 1996, but
there were still 6,112 occupational fatalities during the year.
Here is the breakdown:

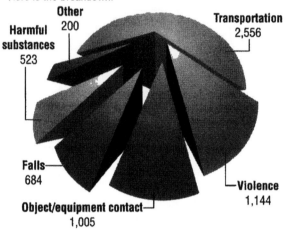

Other
200

Harmful substances
523

Transportation
2,556

Falls
684

Object/equipment contact
1,005

Violence
1,144

From *Census of Fatal Occupational Injuries,* 1996, Bureau of Labor
Statistics, Washington, D.C.

may be higher in some workplaces than others, but virtually every workplace
has some degree of risk which should be evaluated.

What type of risks should I be evaluating? As can be seen from the
various cases, the weapon of choice in a large number of workplace violence
incidents is a firearm. Given the number of handguns, rifles, and shotguns
in circulation in the United States, virtually anyone can acquire access to
these types of weapons. But how about other "weapons" such as explosive
devices (as seen in Oklahoma City), automatic military-style firearms (such
as the AK–47), sabotage of an in-plant fuel storage tank, setting fire to the
operation, contamination of drinking water or food stocks for employees,
contamination of products used by employees, or even using a company
vehicle as a weapon? The range of potential weapons through which the
risk can arise can be diverse; however, the type of workplace and monetary
expenditures to reduce the risk are often limiting factors.

If the risk cannot be eliminated or minimized and I have a workplace
violence incident, what are the costs? To start with, what is a life worth?
What will be the impact on my other employees? What is the cost of the
negative media exposure? What is the cost of the loss of public trust? Will
the public purchase my products? What is the loss of potential future talent?
Will individuals want to work for my company? Will my employees quit
and look for other employment? What impact will this incident have on my
company's profit and loss statement? On my company's stock price?

And how about the legal actions? Are the injured employees compensated under workers' compensation? Can my company be sued for negligent security? Can others injured or killed during the incident sue my company? Will my company be visited by OSHA? Other federal, state, and local agencies? Can my company be fined?

The focus of this text is threefold: (1) to identify areas of potential risk for employers through analysis of prior incidents of workplace violence; (2) to provide proactive methods to employers to eliminate or minimize incidents of workplace violence; and (3) to identify areas and theories of potential legal liability and defenses following a workplace violence incident. The goal is to identify the potential risk and take a proactive, systematic approach to eliminate the risks associated with workplace violence. If these goals are achieved and maintained, there is no need for the legal considerations. However, for those who do experience a workplace violence incident, your world will change immediately and dramatically. We live in a litigious society! Inspection by various governmental agencies are mandated. Lawsuits of varying nature will be prosecuted. The media will camp on your doorstep. Preparation is the key to keep a bad situation from becoming worse!

And remember, workplace violence doesn't always happen to the other guy. It can happen to you. Be prepared. Take proactive action to prevent workplace violence in your workplace. And prepare for after an incident. The few hours spent in preparation will pay exceptional dividends in preventing or minimizing a workplace violence incident.

chapter one

Overview and history

> *"The history of the past interests us only in so far as it illuminates the history of the present."*
>
> Ernest Dimnet

> *"History is a race between education and catastrophe."*
>
> H.G. Wells

Workplace violence has fast become the leading cause of work-related deaths in the United States and has opened an expanding area of potential liability against employers who failed to safeguard their workers. According to statistics from the National Institute of Occupational Safety and Health (NIOSH), over 750 workplace killings a year have been reported in the 1980s.[1] Additionally, according to the National Safe Workplace Institute, there were approximately 110,000 incidents of workplace violence in the United States in 1992.[2] A common misconception is that violent incidents are a fairly new phenomenon, however, workplace violence has been happening for a substantial period of time. The primary reason for the emphasis in this area at this time is because of the increased frequency and severity of the incidents of workplace violence.

According to the U.S. Bureau of Labor Statistics, there were 1,063 homicides on the job in 1993 and of these deaths 59 were killed by co-workers or by disgruntled ex-employees.[3] By 1996, 1,144 of the 6,112 on-the-job fatalities were caused by workplace violence.[4] The 1993 report also noted that there were 22,396 violent physical acts which occurred on the job in 1993, and approximately 6% of these incidents were committed by present or former co-workers. This percentage has steadily increased each year.

---

[1] Bensimon, Helen Frank, Violence in the workplace, *Training and Development* at 27 (January 1994).
[2] Id.
[3] *Census of Fatal Occupational Injuries*, Bureau of Labor Statistics, U.S. Department of Labor, August 1994.
[4] *Census of Fatal Occupational Injuries*, Bureau of Labor Statistics, U.S. Department of Labor, 1996.

A man and woman walked out of the main post office in Milwaukee trying to comfort each other yesterday after the shooting.

MORRY GASH/
ASSOCIATED PRESS

Postal worker wounds two, kills one, himself

ASSOCIATED PRESS

MILWAUKEE — Turned down for promotion to a day job and written up for sleeping at work, a postal clerk pulled a pistol yesterday and killed one co-worker, wounded another and a supervisor.

Anthony Deculit, 37, then put the 9mm handgun into his mouth and killed himself.

There were 1,500 workers on duty at the city s main post office when the shots rang out about 12:45 a.m. Workers ducked for cover and scrambled to get out.

Postal inspector Ida Gillis said no one knew how a gun was smuggled inside.

As many as a dozen shots were fired. Deculit's supervisor, Joan Chitwood, 55, was hit in the right eye. She was hospitalized yesterday in satisfactory condition. Last month, she sent Deculit a warning letter after catching him sleeping on the job.

Deculit, placed on five months' probation, filed a grievance against Chitwood and asked for transfer to the day shift, but was turned down.

Co-workers said Deculit and Russell "Dan" Smith, 42, the slain man, disliked each other so much they didn't even speak.

Smith was shot in the head, back and arm. Another postal worker was slightly injured by a shot to the foot.

Co-worker Michael Witkowski

4 killed on the job

ASSOCIATED PRESS

ORANGE, Calif. — A disgruntled state transportation worker shot and killed four men at a maintenance yard before he was shot to death by police in a gun battle that sent his former co-workers hopping fences and diving for cover.

An officer and an additional transportation department employee were hospitalized.

More than 60 people were at the maintenance yard run by the state Transportation Department when Arturo Reyes Torres stormed in Thursday afternoon, armed with an assault rifle, a shotgun and a handgun, authorities said.

Reyes, 41, who had worked at the yard for about 15 years, was fired last month after he was videotaped selling scrap aluminum from the yard. The Orange County Register reported. Darrell Henson, who used to work with Reyes, said his friend was distraught about the firing.

tried to talk to Deculit.

"He was standing there pointing the gun at me," Witkowski said. "He said, 'Mike, you don't want to be here.' And I told him, 'Tony, don't do this.'"

Deculit asked Witkowski to call his wife and gave him the phone number. Seconds later, he turned the gun on himself.

"He blew his brains out," Witkowski said. "I don't get it."

Attacks Up Against Judges, Lawyers

GENEVA—Attacks on judges and lawyers worldwide increased to 572 reported instances in 1996—a 25-percent rise from 1995—according to a report by an international group of jurists.

Last year, in the 49 countries covered by the report, 26 judges and lawyers were killed; two were "disappeared"; 97 were arrested, detained or tortured; 32 were physically attacked; 91 were threatened with physical violence; and hundreds were sanctioned or prevented from carrying out their duties. The report gave no reason for the rise in attacks.

The findings are in the eighth annual report by the Geneva-based Center for the Independence of Judges and Lawyers, or CIJL: "Attacks on Justice: The Harassment and Persecution of Judges and Lawyers." The CIJL released the report in October.

The report described "several patterns that constitute a threat to the judiciary," such as the creation of special tribunals, the removal of judicial discretion in sentencing and the elimination of judicial tenure. Pervasive corruption, public denunciations of judges and intimidation of defense lawyers also threatened the legal profession's independence, the report said.

In the United States, it said, election of judges casts serious doubt on their ability to be independent, particularly when it comes to the death penalty.

In Colombia and Peru, the practice by judges of concealing their identities in court undermines justice, the report said. In Ethiopia, the dismissal of 77 judges made those who lack tenure sensitive to political reactions to their rulings.

Corruption within the Mexican police force and the judiciary prevents "citizens from reporting crime and human rights violations" there.

Human rights lawyers are particularly targeted in countries such as Sudan, Pakistan, the Philippines and Nigeria, among others, according to the report.

The CIJL is a part of the Geneva-based International Commission of Jurists, a nongovernmental body of 45 eminent jurists devoted to promoting the rule of law.

—JOHN ZAROCOSTAS

Compiled from National Law Journal staff, correspondent and Associated Press reports.

(Reprinted with permission.)

Workers remember postal shooting

by LIBBY QUAID
Associated Press Writer

EDMOND, Okla. (AP) — Postal clerk Vesta McNulty has worked every Aug. 20 for the past 10 years. She makes a point of it

She was there on the summer morning in 1986 when a part-time mail carrier walked into Edmond's main post office and opened fire — killing 14 co-workers and wounding six others before shooting himself in the head.

Returning the day after Patrick Henry Sherrill's rampage may have been the most difficult of all. "If I hadn't come back then, I probably would never have come back," she says.

Many survivors make it a tradition to work on the anniversary of the siege, which until the Oklahoma City bombing was the deadliest one-day attack in state history.

Families of some victims plan to lay a wreath at the post office's memorial fountain on Tuesday. Other workers will pause for a moment, as they have in years past, to remember the friends they lost.

Ernie Bingham has also worked every Aug. 20 since the shooting. He had been sorting letters for less than an hour when Sherrill walked in that morning. The attack shattered his sense of security.

"What you think is a safe place isn't," said Bingham, who now manages the city's other post office.

Police handling of the case drew criticism and some lawsuits, which were later dismissed. Officers were reluctant to storm the building because they thought they had a hostage situation. When they went in about 1 1/2 hours later, they found 15 bodies, including Sherrill.

Some blamed the U.S. Postal Service for promoting a working environment in which tension escalated with mail volume. Managers were also blamed for increasing pressure on workers.

Sherrill was upset that day about the most recent evaluation in his troubled career with the Postal Service.

In 1982, he had quit rather than be fired from his job as a clerk in the Oklahoma City office. His record went undetected when he was hired three years later in Edmond, where supervisors also tried to fire him for continual infractions.

"They were putting pressure on him to improve his performance, to be the best carrier he could be. Those are worthy goals, but the methods they used, belittling him, I just perceive it was mishandled," said former letter carrier Mike Bigler.

He said Sherrill made "normal mistakes," like taking too long for lunch. "I wasn't perfect when I started, either," he said.

Bigler, 46, was shot in the back that morning but was able to return to work the next day. The bullet pierced the flesh on his right shoulder blade and exited about eight inches to the left.

He now works for General Motors and runs a prison ministry with his wife.

Bigler says he missed his chance to help Sherrill before he snapped.

"Patrick asked me, 'Can I talk to you?' I said, 'Yeah, just a minute,' but by then he was already on his route. I didn't get there soon enough."

Survivors who remained at the Edmond Post Office have grown a little more secure as the Postal Service insisted on new safety regulations and union officials and management began new training for supervisors. Automation has assumed some of the work.

"We've tried to make it a friendly workplace," said Valerie Welsch, a spokeswoman for the district office in Dallas.

Sweeping reviews followed the massacre when a General Accounting Office investigation revealed 30,000 improperly locked or blocked exits in postal buildings across the country.

Several other postal shootings, including the 1991 killings of four supervisors in Royal Oak, Mich., and the shootings of four in Montclair, N.J., in 1995, kept the issue of safety at the forefront in the past decade.

Edmond Postmaster Ted Tyler said the Postal Service works to perform in-depth screening and background checks of potential employees and has a no tolerance policy for anyone who makes threats, outbursts or brings weapons to work.

"Self-control is now a condition for employment," Tyler said.

Emergency drills now are routine. Still, the postal employees know it would be impossible to stop another Patrick Henry Sherrill.

"I feel safer, but I think you have to go on," Ms. McNulty said. "Not that I'm not cautious, I just don't look over my shoulder every minute."

The Richmond Register • Page B3

Monday, August 19, 1996

Man killed at Ford plant

PLYMOUTH TOWNSHIP, Mich. (AP) — A 30-year Ford Motor Co. employee walked into work and gunned down a security supervisor before killing himself — the second killing in two years at the plant, police said.

Authorities refused to discuss a motive for the shootings early Sunday at the Ford Motor Climate Control Plant, in Wayne County outside Detroit.

The gunman, 51-year-old Stephen Cox of Ypsilanti, and the 57-year-old victim, Donald Jagod, were acquaintances, Police Chief Carl Berry said.

Cox, who worked as a cleaner at the plant, reported to work about 5:30 a.m., Berry said. Soon after, he entered the security office with a handgun and opened fire on Jagod, a security supervisor.

Jagod, a retired Dearborn police office who had worked at the plant for four years, was unarmed, Berry said

(Reprinted with permission.)

Table 1.1 Workplace Homicides in the United States, 1980–92[a]

Year	Number	Rate
1980	929	.96
1981	944	.94
1982	859	.86
1983	721	.72
1984	660	.63
1985	751	.70
1986	672	.61
1987	649	.58
1988	699	.61
1989	696	.59
1990	725	.61
1991	875	.75
1992	757	.64
Total	9,937	.70

[a] NIOSH (1995); Data not available for New York City and Connecticut; data calculated per 100,000 workers.

According to the data from the National Traumatic Occupational Fatalities (NTOF) Surveillance System during the 13-year period from 1980 through 1992, 9,937 workplace homicides occurred, with an average workplace homicide rate of 0.70/100,000 workers (see Table 1.1). Over the course of the 1980s, workplace homicides decreased; but in the 1990s, the numbers began to increase, surpassing machine-related deaths and approaching the

Occupational health guide to violence in the workplace

Table 1.2 Workplace Homicides by Industry
and Sex — United States, 1980–1992[a]

| Industry | Homicides (% of total) | |
	Male workers	Female workers
Retail trade	36.1	45.5
Services	16.0	22.2
Public administration	10.5	2.9
Transportation/communication/public utilities	10.6	3.8
Manufacturing	7.0	4.9
Construction	4.1	.6
Agriculture/forestry/fishing	2.7	.6
Finance/insurance/real estate	2.4	6.8
Wholesale trade	1.7	1.1
Mining	.6	.1
Not classified	8.5	11.7

[a] NIOSH (1995); Data for New York City and Connecticut were not available for 1992; percentages add to more than 100% because of rounding.

number of workplace motor-vehicle-related deaths.[5] NTOF is an ongoing, death-certificate-based census of traumatic occupational fatalities in the United States, with data from all 50 states and the District of Columbia.

In this study, the majority (80%) of workplace homicides during the study period of 1980 through 1992 occurred among male workers. The leading cause of occupational injury death varied by sex, with homicides accounting for 11% of all occupational injury deaths among male workers and 42% among female workers.[6] The majority of female homicide victims were employed in retail trade (46%) and service (22%) industries. A large number of male homicide victims were employed not only in retail trade (36%) and service (16%) industries but in public administration (11%) and transportation/communication/public utilities (11%) (see Table 1.2). Although homicide is the leading cause of occupational injury death among female workers, male workers have more than three times the risk of work-related homicide (see Table 1.3).

The age of workplace homicide victims ranged from 16 to 93 during the study period of 1980 through 1992. The largest number of workplace homicides occurred among workers aged 25 to 34, whereas the rate of workplace homicide increased with age. The highest rates of workplace homicide occurred among workers aged 65 and older; the rates for these workers were more than twice those for workers aged 55 to 64. This pattern held true for both male and female workers.[7]

[5] NTOF includes information for all workers aged 16 or older who died from an injury or poisoning and for whom the certifier noted a positive response to the *injury at work?* item on the death certificate. For additional discussion of the NTOF system and the limitations of death certificates for the study of workplace homicide, see Castillo and Jenkins (1994).
[6] NIOSH (1995); Data not available for New York City and Connecticut; data calculated per 100,000 workers.
[7] NIOSH 1995.

Table 1.3 Workplace Homicides by Age Group
and Sex — United States, 1980–1992[a]

Age group	Male workers Number	Rate	Female workers Number	Rate	All workers Number	Rate
16–19	242	.55	102	.25	344	.41
20–24	796	.87	285	.35	1,081	.62
25–34	2,020	.89	591	.33	2,611	.65
35–44	1,841	.99	423	.28	2,265	.68
45–54	1,344	1.04	293	.29	1,637	.71
55–64	1,055	1.22	191	.31	1,246	.84
65+	620	2.59	115	.71	735	1.83
Total	**7,935**	—	**2,001**	—	**9,937**	—
Average	—	**1.01**	—	**.32**	—	**.70**

[a] NIOSH (1995); Data from New York City and Connecticut were not available for 1992; rates are per 100,000 workers; totals include victims for whom age data were missing (17 male workers) and 1 worker whose sex was not reported.

Table 1.4 Workplace Homicides by
Race — United States, 1980–1992[a]

Race/ethnicity of victims	Number	% of total	Rate
White (includes Hispanic)	7,239	72.8	.59
Black	1,938	19.5	1.39
Other	760	7.6	1.87

[a] Id.

Table 1.5 Workplace Homicides by Bureau of the Census by
Geographic Region[a]

Region	Number	% of total	Rate
North Central	1,797	18.1	.50
North East	1,043	10.5	.35
South	4,819	48.5	1.02
West	2,278	22.9	.79

[a] Id.

This study also noted that although the majority of workplace homicide victims were white (73%), black workers (1.39/100,000) and workers of other races (1.87/100,000) had the highest rates of work-related homicide (see Table 1.4).

During the study period, the largest number of homicides and the highest rates per 100,000 workers occurred in the South (N = 4,819; rate = 1.02/100,000) and the West (N = 2,278; rate = 0.79/100,000) (see Table 1.5). Four states, namely Louisiana, Nebraska, Oklahoma, and New York, were unable to provide data on work-related homicides. In addition, data for 1992 were unavailable from New York City and Connecticut.

Table 1.6 Workplace Homicides by Method[a]

Method	Number	% of total
Firearm	7,590	76.4
Cutting or piercing instrument	1,231	12.4
Strangulation	185	1.9
All other methods	931	9.4

[a] Id.

Table 1.7 Workplace Homicides by Industry[a]

Industry	Number	% of total	Rate
Retail trade	3,774	38.0	1.60
Public administration	889	8.9	1.30
Transportation/communication/public utilities	917	9.2	.94
Agriculture/forestry/fishing	222	2.2	.50
Mining	45	0.5	.40
Service	1,713	17.2	.38
Construction	335	3.4	.37
Finance/insurance/real estate	327	3.3	.35
Wholesale trade	155	1.6	.27
Manufacturing	650	6.5	.24
Not classified	910	9.1	—

[a] Id.

In this study, the preferred method of committing these workplace homicides was a firearm. Between 1980 and 1992, 76% of work-related homicides were committed with firearms, and another 12% resulted from wounds inflicted by cutting or piercing instruments (see Table 1.6). During this period, the number of firearm-related homicides declined, then gradually increased, with the number of firearm-related workplace homicides in 1991 exceeding that in 1980.

During the study period, the greatest number of deaths occurred in the retail trade (3,774) and service (1,713) industries, whereas the highest rates per 100,000 workers occurred in retail trades (1.6), public administration (1.3), and transportation/communication/public utilities (0.94) (see Table 1.7).

As can be seen from Table 1.8, the largest number of deaths occurred in grocery stores, eating and drinking places, taxicab services, and justice/public order establishments. Taxicab services had the highest rate of work-related homicide during the 3-year period 1990–1992 (41.4/100,000). This rate was nearly 60 times the national average rate of work-related homicides (0.70/100,000). This figure was followed by rates for liquor stores (7.5), detective/protective services (7.0), gas service stations (4.8), and jewelry stores (4.7) (Table 1.8). The rates show an increase from the previously published rates for 1980–1989 for taxicab services, detective/protective services, grocery stores, and jewelry stores. Rates decreased in liquor stores, gasoline

Table 1.8   Workplace Homicides in High-Risk Industries[a]

Industry	Number	Rate	Number	Rate
Taxicab services	287	26.9	138	41.4
Liquor stores	115	8.0	30	7.5
Gas service stations	304	5.6	68	4.8
Detective/protective services	152	5.0	86	7.0
Justice/public order establishments	640	3.4	137	2.2
Grocery stores	806	3.2	330	3.8
Jewelry stores	56	3.2	26	4.7
Hotels/motels	153	1.5	33	0.8
Barber shops	14	1.5	4	—
Eating/drinking places	734	1.5	262	1.5

[a] Id.

service stations, justice/public order establishments, and hotels/motels; they remained the same in eating and drinking places.

According to the Bureau of Labor Statistics data, 73 to 82% of the homicides occurred during a robbery or other crime, whereas only 9 to 10% were attributed to business disputes, and only 4 to 6% were attributed specifically to co-workers or former employees. The percentage of work-related homicides attributed to firearms (76%) is slightly higher than that found in the general population, where 71% of the 1993 murders with victims aged 18 or older were committed with firearms.[8] Changes in the risk of workplace homicide in specific industry and occupation groups may be attributable to a number of factors, including increased recognition and recording of cases as work-related, changes in training or other work practices, increased levels of crime in certain settings, and the distribution of resources in response to perceived levels of crime.

This study found that the circumstances of workplace homicides differ substantially from those portrayed by the media and from homicides in the general population. For the most part, workplace homicides are not the result of disgruntled workers who take out their frustrations on co-workers or supervisors, or of intimate partners and other relatives who kill loved ones in the course of a dispute; rather, they are mostly robbery-related crimes.[9]

In addition to the incidents of workplace violence resulting in death, incidents of workplace violence resulting in assaults has also increased. Although limited information is available in the criminal justice and public health literature regarding the nature and magnitude of nonfatal workplace violence, a few victimization studies include designation of victimizations that occurred at work.[10] These studies indicate that factors related to an increased risk for workplace victimization included routine face-to-face contact with

---

[8] FBI, 1994.
[9] NIOSH, 1995.
[10] NIOSH, 1995: Also see, Collins and Cox, 1987; 1982 Victim Risk Supplement to the National Crime Victimization Survey, Lynch (1987).

large numbers of people, the handling of money, and jobs that required routine travel or that did not have a single work site.

A number of recent estimates have been made of the current magnitude of nonfatal assaults in U.S. workplaces. The first comes from the BLS Annual Survey of Occupational Injuries and Illnesses (ASOII). The ASOII, an annual survey of approximately 250,000 private establishments, indicated that 22,400 workplace assaults occurred in 1992, representing 1% of all cases involving days away from work. Unlike homicides, nonfatal workplace assaults are distributed almost equally between men (44%) and women (56%). The majority of the nonfatal assaults reported occurred in the service (64%) and retail trade (21%) industries.[11]

Another estimate of the magnitude of nonfatal workplace assaults comes from a survey by the Northwestern National Life Insurance Company, which indicates that 2.2 million workplace assaults (defined as physical attacks) occurred between July 1992 and July 1993.

Another estimate of assaults in the workplace comes from the National Crime Victimization Survey (NCVS), which indicates that each year nearly 1 million persons were assaulted while at work or on duty. The Bureau of Justice Statistics also analyzed workplace victimizations by type of work setting and found that 61% occurred in private companies, 30% occurred among government employees, and 8% of the victims were self-employed.[12]

When individuals in the NCVS were asked whether this workplace victimization was reported to the police, 56% indicated that it was not. For 40% of respondents, the reason cited for not reporting to the police was that the event was believed to be a minor or private matter. Another 27% did not report to the police because the incident was reported to another official such as a company security guard.[13]

The NCVS also solicits information about days away from work and lost wages due to the victimization. As a result of workplace victimizations, approximately half a million workers lost 1.75 million days of work annually (an average of 3.5 days per crime) and victims lost more than $55 million in wages, not including days covered by sick or annual leave. As a result of the 16% of victimizations in which injuries were incurred, 876,800 workdays were lost annually and $16 million were lost in wages, not including days covered by sick or annual leave.[14]

An alarming area which should also be considered within the realm of workplace violence are incidents of sabotage and violence directed at the company by outside individuals or organizations such as the bombing of the World Trade Center, the Unabomber, and the tragedy in Oklahoma City. Although these incidents of workplace violence are on a larger scale and utilize a random-type weapon, individuals on the job are at risk, and employers incur liability.

---

[11] BLS, 1994.
[12] Bachman, 1994.
[13] Id.
[14] Id.

Incidents of workplace violence have been highly publicized. The most visible organization with a substantial number of workplace violence incidents is the U.S. Postal Service, which recorded some 500 cases of workplace violence toward supervisors in an 18-month period in 1992 and 1993.[15] Additionally, the U.S. Postal Service also recorded 200 incidents of violence from supervisors toward employees.[16] It should be noted, however, that the U.S. Postal Service was one of the first organizations to take steps to address the workplace violence issue, and the recent hostage situation in Denver, Colorado, reflected success in addressing the hostage situation.

Below are just a few of the other highly publicized incidents which resulted in injury or death to individuals on the job:

- The shooting spree at the Chuck E. Cheese restaurant in Denver in which a kitchen worker killed four employees and wounded a fifth
- The ex-employee of the Fireman's Fund Insurance Company who killed three individuals, wounded two others, and killed himself in Tampa, Florida
- The 1986 Edmond, Oklahoma, shooting where a letter carrier killed fourteen and wounded six others
- The disgruntled postal worker in Dearborn, Michigan, who shot another employee in May 1993
- The former postal worker who killed four employees and injured another in the Montclair, New Jersey, post office
- The postal worker in Milwaukee who killed one co-worker, wounded another employee and a supervisor, and killed himself
- The disgruntled state transportation worker who shot and killed four employees in Orange, California
- The ex-postal worker held seven employees hostage on Christmas eve in Denver, Colorado

So what exactly is "workplace violence"? Generally, workplace violence is defined as "physical assaults, threatening behavior, or verbal abuse occurring in the work setting."[17] Although incidents of threatening behavior, such as bomb threats or threats of revenge, are not statistically available, there is a substantial likelihood that these types of incidents are also on the upswing.

Many companies and organizations in the United States have taken steps to safeguard their employees in the workplace through a myriad of security measures, policy changes, and other methods. The potential legal liabilities in this particular area have drastically increased for employers. In most circumstances, the employer would be responsible for any costs incurred by the employee through the individual state workers' compensation system.

---

[15] Kurlan, Warren M., Workplace violence, *Risk Management*, at 76 (June 1993).
[16] Id.
[17] Physical Assault, from our research, has run the gamut from an employee shoving or punching an employee through the use of a weapon or explosive to kill the individual.

Now, however, new and novel theories such as negligent retention, negligent hiring, negligent training, as well as the potential of governmental monetary fines, such as through OSHA, have also merged to increase the potential risk.

Most experts concede that there are no magic answers when it comes to addressing problems in the area of work-related violence. Given the fact that the potential of violence exists on a daily basis and the method in which the violence can be precipitated can come from a wide variety of areas, the intangibles lend themselves to the fact that workplace violence is a very complicated issue. Is this a new issue? Absolutely not. Incidents of workplace violence have been occurring since the Industrial Revolution in the United States. The frequency (the number of incidents) has substantially increased as well as the severity of these types of workplace incidents. This may correlate to a variety of reasons, including, but not limited to, the increased violence in our society, the availabilty of weapons, the down-sizing of the workplace, the management style, and numerous of other reasons. Additionally, when you include the different types of workplaces in America as well as the variety of management approaches, there is no one simple answer to this multifaceted question.

The Occupational Safety and Health Administration has provided guidelines for specific industries, such as the retail industry and health care operations.[18] Many employers have taken proactive steps to develop a general strategy in order to protect their employees and thus reduce the potential legal risks as well as providing ancillary efficacy benefits to employees and management. In addition to the proactive strategy, many employers have developed a reactive plan and implemented stringent employee screening and monitoring processes to identify and address potential incidents of workplace violence in order to minimize potential risks.

In most circumstances, employers are better able to combat the potential risk of workplace violence when the threat is initiated by an employee rather than an ex-employee or outside individual. Researchers have provided a general profile of individuals with a propensity toward workplace violence, (See Chapter 3) which include such areas as employees with depression, suicidal threats, poor health, and other traits. Incidents precipitated by ex-employees, spouses of employees, and individuals outside the organization are substantially harder for the employer to address, given the lack of control which is present in the workplace.

As employers attempt to address the potential risks of workplace violence and the correlating legal risks and costs, the employers must be very cautious so as not to trample upon the individual's rights and freedoms. As employers develop and implement more stringent activities and programs to curtail or minimize the potential risks of workplace violence, they must be extremely careful not to create additional legal risks through their actions.

---

[18] Workplace Violence: OSHA says guidelines will target the retail and health care sectors, 1995 DLR 16 (BNA, January 23, 1995).

Privacy laws, acquisition of information laws, and discrimination laws provide avenues of potential redress for employees and individuals in these areas.

Companies now walk a legal tightrope because of the expanding emphasis on workplace violence. To a great extent, this area of law is still expanding, thus employers should attempt to maintain an approach that provides the maximum protection to employees while not affecting employees' privacy rights or other individual rights. This is a difficult endeavor but one which is becoming a necessity in the American workplace.

chapter two

Individual profiles and warning signs

> *"The whole of science is nothing more than a refinement of everyday thinking."*
>
> Albert Einstein

> *"Science is the knowledge of consequences, and dependence of one fact upon another."*
>
> Thomas Hobbes

An individual who may commit an act of workplace violence does not have a sign on his\her back that states, "I will commit a violent act in your workplace....Employer do something!" However, from the studies conducted on the workplace violence incidents which have occurred, there has been compiled a profile identifying certain warning signs and groups of individuals most susceptible or "at risk" for workplace violence.

It should be noted, however, that an individual who may commit a violent act in the workplace does not fit an exact profile and there are no clear-cut, absolutely specific categories which each and every individual or workplace violence incident fits exactly. However, the limited research conducted in this area has identified several broad categories common among violent employees. In general, the potentially violent employee will have worked at the place of employment for a substantial period of time (in comparison to a newly hired employee) and have achieved a small level of success in their job. The potentially violent employee may be considered by his/her supervisor or team leader or other members of management to be a "dedicated employee" who goes over and above the expected work duties, coming to work on Saturday or staying late to finish a project. The potentially violent employee, in essence, has adopted the work and workplace as the center of his/her life. These individuals view their self-worth, their status, and self esteem in accordance with their performance on the job. Thus, the removal of the job, demotion, failure to acquire a promotion, or other actions by management strikes at the emotional heart of the individual's self-worth and can be paralleled to a direct threat to the very essence of the individual.

The potentially violent employee may be under the perception that he/she earned a promotion or pay raise through his/her work activities and was passed over for a promotion or pay raise that he/she earned or deserves. This results in the individual holding a grudge against a certain person (or the company itself) whom the individual perceives as the reason or catalyst for this decision. Another common scenario for potentially violent employees is during a period of transition, such as when the company is downsizing and employees are to be laid off or when the company has been purchased by another and employees anticipate downsizing. The potentially violent employee may have a paranoid state of mind, believing that managers or the company itself are "out to get him/her." Potentially violent employees may also have problems outside the workplace that they are trying to address, which compounds or escalates the anxiety at work. Outside problems could range from a dispute with a spouse or family member to simply an annoying neighbor. Virtually every employee, including ourselves, has problems that need to be addressed on a daily basis. However, the potentially violent employee "always has a problem" of one type or another and usually talks about the problems constantly in the workplace.

Potentially violent employees may have a history of violence. This violence may have manifested itself in terms of a fight they had in a bar, child abuse, domestic abuse, or violence at previous places of work. These potentially violent employees often possess a fascination with guns and may own several guns and may carry a concealed weapon, legally or illegally.

Several studies have identified the Caucasian male between the age of 25 and 40 as the primary group involved in workplace violence based upon past incidents of workplace violence. Within the base level profile, the employee usually worked for his company for a substantial period of time, has a history of violence, and owns one or more firearms. The individual is often considered a "loner" and may or may not have a family. If the individual has a family, the family situation has substantial problems. This individual often intimidates other employees on the job or makes veiled threats or even stalks other employees.

Conversely, this individual may show no outward visible signs of hostility to other employees such as exemplified in the movie "Falling Down," and the stressors continue to escalate until violence occurs. Take the case of Larry, a technician for an electronics manufacturer in California. Larry was one of several employees laid off due to the recession in 1991. After he was laid off he kept returning to his former place of employment, seeking advice about finding another job or just visiting with people. Then one day he walked in with a bandolier across his chest and a shotgun in his hand. He shot out the switchboard before the receptionist could call for help. He set small fires and several pipe bombs; he then calmly walked upstairs and shot and killed a vice president and a regional sales manager. Another executive escaped with his life by hiding under a desk. When he was done, he walked out and got on his bicycle and rode away.[1]

There is no distinct category of individuals who commit violent acts. However, careful analysis of the previous incidents of workplace violence has identified several warning signs for employers. However, it should be noted that an employee simply possessing some of the warning signs does not automatically mean the person will perform a violent act. Prudent employers should establish a proactive prevention program so that when the signs are recognized the program can be initiated to assess and assist the employee if necessary.

The following are some of the warning signs that may be present for an individual having the potential of committing violence in the workplace:

- Holds grudges against co-employees or the company itself. The individual may not have received the promotion or raise he/she felt was deserved.
- Exhibits paranoid behavior. Individuals may feel that the whole world is out to get them.
- Regularly has problems. It may relate to home, work, or outside friends, but they are always talking about the problem they are currently dealing with.
- Is fascinated with firearms. May carry one on person. May have a collection and frequently talks about them.
- Knows that they are going to be fired, laid off, or feels like they will be laid off.
- Is a loner at work. Does not mix with co-employees, except maybe for a romantic interest with a co-employee. That may lead to the co-employee feeling threatened by the person.
- Intimidates others by verbal or physical means. Anonymous notes that threaten someone or harassing phone calls.
- Has difficulty accepting criticism and/or authority.
- Likes to push authority to its limit, to see how far they can go.
- Has a history of violent behavior, either before or after employment with the company.
- Has a history of substance abuse, drugs and/or alcohol.
- Is fascinated by acts of workplace violence, likes to talk about them with other co-workers.
- Focuses on any type of workplace event that may generate stress — downsizing, corporate takeover, or a recession.
- Engages in activities that expose employees to outside factors that may cause an act of violence.
- May come from an unstable or dysfunctional family.[2]

---

[1] Barrier, M., The enemy within, *Nation's Business*, Feb. 1995 at 18-22.
[2] Id. and Johnson, D., Kiehlbaugh, J., Kurutz, J., Workplace violence scenario for supervisors, *HR Magazine* at 63-67.

It should be noted that simply exhibiting the potential for workplace violence does not automatically place the person within the "at risk" category for workplace violence. Employers should exercise extreme caution when evaluating the warning signs in order to not violate individuals' rights while trying to protect the workplace. Additionally, there are several common myths about violent employees which should be addressed:

1. *"Violent employees just snap, without any warnings or clues."* This statement, for the most part, is not true. Employees with violent tendencies usually provide multiple clues or warnings, and these are usually provided to a large number of people or other employees. If co-workers, supervisors, and management have been trained in and are paying attention to these signs, the employee can often be provided assistance prior to the violent act.
2. *"If clues are provided, then there should be no incidents of workplace violence."* If co-workers, supervisors, and management recognize and report the warning signs and the behavior of other employees, and management takes the appropriate action to respond to the problem, this statement may sound true. But in most circumstances, employees who are threatened often make light of the threat or simply turn away. It is usually not the lack of signs by the potentially violent employee that is the problem but the unwillingness of employees, supervisors, and management to recognize these signs and address the potential situation, which often leads ultimately to workplace violence incidents.
3. *"Employees who commit workplace violence have nothing to live for and/or have everything to lose."* This is a common myth. In analyzing the workplace violence incidents to date, employees who commit workplace violence almost always have some degree of success in the workplace and in their personal lives. These individuals, however, are of the perception that something very important to them has been taken away (i.e., job promotion, raise, etc.) through an unfair decision-making process. These individuals often feel that the company has a moral or personal obligation to them to provide the job promotion, raise, etc. When the organization does not provide this important job promotion, raise, etc. to the individual, he/she is willing to risk everything, including job, home, family, etc. to strike back at the company, even if this results in injury or harm to fellow workers or themselves. [3]

Recent cases have added additional warning signs which may be also considered in the evaluation, including the following:

1. A disregard for the safety and health of fellow workers.
2. A fascination with workplace violence incidents and a verbal expression of approval for the use of violence to achieve a purpose.

---

[3] Workplace Violence: The First Line of Defense. Kenwood Group, 1994.

3. An extreme interest of automatic weapons, bombs, or other destructive devices.
4. Extreme desperation over recent financial, personal, or family problems. This is often expressed directly to co-workers or supervisors.
5. Extreme moral righteousness. The employee often expresses his/her belief that the company is not following its own rules or procedures or laws.
6. Other warning signs that may be present in individuals include psychosis, chemical dependency, and severe depression.
7. Last, the employee has an obsessive involvement with his/her job and expresses no outside interest. When the employee is provided constructive criticism on job performance, the employee often holds a grudge against the supervisor or team leader. This grudge is often verbalized to other employees and the employee often hopes that negative things will happen to the supervisor or team leader or their families.

As noted above, in attempting to acquire a profile of a potentially violent employee, the analysis of past workplace violence incidents have identified the male as being at a greater risk than the female, and the average age is between 25 and 40 years old. This general profile utilizes an analysis of workplace violence incidents which have already happened. Thus, professionals should be aware that virtually any employee of any age, sex, race, etc. can become an "at risk" employee under the right circumstances. Additionally, within the profile, three basic levels of job-related actions were also identified:

- At the initial level, the employee may be spreading gossip and rumors within the organization, be uncooperative with fellow workers and supervisors, argue with fellow workers and management, be belligerent to clients, make lewd comments, and use profanity. These actions or activities should also be included within the warning signs given the fact that if these activities or actions are unchecked, the employee often progresses to the second level of at-risk activities.
- At the second level, the hostilities identified in the initial level often increase. The employee's actions, such as violating company policy, sabotaging equipment, and theft are often common warning signs. Again, if left unchecked, the employee often progresses to the third level, where the threats become serious, the employee becomes involved in violent altercations with fellow workers and management, and the employee often begins carrying a weapon.
- The third level involves serious threats, fighting, weapons carrying and brandishing, and other escalation in the area of violence. [4]

---

[4] Filippi, S.T., *Violence in the Workplace: Containing the Problem.* Professional Safety, June, 1996, p. 37.

It is important that the professional design a system through which to identify and address the warning signs early in the initial stage and does not allow the hostilities to escalate. Any type of change in the working environment, such as a job layoff or termination, may be particularly traumatic to an "at risk" employee and should be a period of vigilance. "At risk" employees involved in such workplace modifications should be monitored, and warning signs addressed through specific procedures or programs through which the employee is provided personal notification, out-placement assistance, clear explanation of the reasoning, and possibly consultation.

In order to be able to identify the warning signs within employees in the workplace, it is essential that the company implement specific policies and training programs for employees and all levels of management to ensure that the warning signs are identified appropriately. Additionally, prudent companies may also wish to implement drug and alcohol programs and employee assistance programs to provide an avenue for employees to seek assistance prior to a workplace violence incident.

"At risk" employees are often chemically dependent on controlled substances and alcohol. Drug abuse and alcoholism are common methods of self medication when the employee may be over stressed. Chemical dependency in employees often reduces their emotional control and may ultimately increase the stress or perceived stress in the workplace. Additionally, utilizing prescription medications, when these are combined with controlled substances or alcohol, may ultimately have a counteracting effect on the prescribed medication. A simple, but often costly method of addressing this potential risk is a formal policy and procedure creating a drug-free workplace. This program should include, but not be limited to, a formal written policy, pre-employment drug testing, testing for cause or random testing, employee assistance programs, and additional components addressing the specific circumstances and legal guidelines. And last, management must listen to their employees. This is especially important for those individuals who exhibit one or more of the warning signs and can be considered "at risk" employees. The key person within this chain of identification is the supervisor or team leader who has daily interaction with the employee. Specific training and education with regard to identification of the warning signs, appropriate documentation, and available avenues of assistance should be provided to the supervisors or team leaders. Additionally, many companies also provide training to all employees in order that they may recognize warning signs of impending violence which may not be readily identifiable to the supervisor, team leader, or management.

Prudent professionals should exercise extreme caution when addressing the issues of warning signs with the specific individuals. Although the employer has a legal obligation to create and maintain a safe and healthy work environment, over exuberance in addressing the potential warning signs with an individual may also tread upon their individual legal rights. As can be seen in the following chapters, employers have a "narrow tunnel" of acceptable conduct. On one side of the legal tunnel, if employers do not address workplace violence issues, the company may be liable for such categories as negligent hiring, negligent retention, negligent supervision, negligent security, and workers' compensation. On the other side, if employers tread upon individual rights, such actions as defamation, Americans with Disabilities Act claims, and libel/slander may be available to the employee. Thus, it is extremely important that prudent employers develop a specific company policy and plan to address workplace violence issues (see Chapter 3).

In summation, employers should utilize the information provided in the profile in early warning signs to develop programs to address potential workplace violence in order to eliminate and/or minimize the potential risk in the workplace. The profile and warning signs are not absolutes and should only be used as guidance. Prudent professionals should be aware that virtually any employee under the right circumstances can potentially be an "at risk" employee. Additionally, an "at risk" employee may exhibit one or many of the warning signs or none at all. An "at risk" employee can be any sex, race, age, etc., and thus, although the profile was developed from prior incidents, virtually any employee can be "at risk."

chapter three

Workplace violence prevention and response program

> *"The method of the enterprising is to plan with audacity and execute with vigor."*
>
> Christian Bovee

> *"The man who is prepared has his battle half fought."*
>
> Miguel De Cervantes

Preparedness is the key to eliminating or minimizing workplace violence incidents. Prudent employers should identify the potential risks of workplace violence in the specific operations and prepare beforehand to address specific needs during crisis times. Without a preplan and preparation for a workplace violence incident, when and if a situation arises, management and employees will have virtually no knowledge or ability to properly react to the situation and will simply be "running around like a chicken with its head cut off," which can result in additional harm or damage.

Preplanning is simply assessing the potential risks and preparing to properly address the situation if it happens. Preplanning is most probably already in effect in your operations in the areas of plant evacuations or disaster preparedness planning. In essence, the management team and employees sit down and assess the potential risks beforehand and prepare personnel, equipment, services, and other components to address the situation in an emergency. Thus, when a workplace violence situation happens, the management team and employees know their specific responsibilities, evacuation routes, equipment locations, emergency telephone numbers, and all other necessary information in order to react properly.

Workplace violence programs can be stand-alone programs or incorporated in the companies' emergency or disaster preparedness programs. However, given the potential conflicts with individual rights and personnel issues, a workplace violence program may be more extensive than a basic disaster preparedness program. The following are basic components to consider when developing a workplace violence prevention program:

- Written policy
- Written program with individual responsibilities
- Threat assessment procedure
- Background check procedure
- Controlled substance testing program procedures and drug-free workplace polices and procedures
- Nonharassing policy and report procedures
- Management and employee training
- Alarm and evacuation procedures
- Employee assistance programs
- Physical security
- Central management control area
- Operation assessment procedures
- Investigation and documentation procedures
- Law enforcement and agency liaisons
- Emergency telephone listing or communications procedures
- After-incident critical stress debriefing
- Litigation and agency control
- Media control
- Injury or fatality family notification procedures
- Insurance and corporate notification procedures

The initial step for most companies is the development of a written policy statement. Among the advantages of issuing a statement are:

- It informs employees that the violence policy covers intimidation, harassment, and other inappropriate behavior that threatens or frightens them.
- It encourages employees to report incidents.
- It tells employees whom to call.
- It demonstrates senior management's commitment to dealing with reported incidents.

A workplace violence policy statement should include, but not be limited to, the following areas:

- All employees are responsible for maintaining a safe work environment.
- The policy covers not only acts of physical violence, but harassment, intimidation, and other disruptive behavior.

- The policy covers incidents involving co-workers and incidents involving individuals from outside the agency perpetrating violence against agency employees.
- The company or management will respond appropriately to all reported incidents.
- The company will act to stop inappropriate behavior.
- Supervisors and employees involved in responding to incidents will be supported by upper management in their efforts to deal with violent and potentially violent situations.[1] Some companies also include a reference to the disciplinary policy for noncompliance.

Additionally, the following should also be considered in developing a written policy statement:

1. A written policy statement should be brief and simple. Implementation details can be provided in training and in more detailed backup documents.
2. There are disadvantages to using definitions of terms such as violence, threats, and harassment in your written policy statement. Definitions can discourage employees from reporting incidents that they do not believe fall within the definition. The reporting system should not deter employees from reporting situations that frighten them. An employee knows a threat or intimidation or other disruptive behavior when he or she experiences it. Definitions are not necessary. If you want to clarify the scope of your organization's concept of one or more of the terms in the policy, you could use examples. For example, you may want to give examples of verbal and nonverbal intimidating behavior. Another consideration is that definitions are often restrictive and may create legal problems in the future when you are taking disciplinary actions against the perpetrators of workplace violence. Use of definitions can make it more difficult to defend a case on appeal.

   Avoid using such terms as "zero tolerance" or other terminology which can cause difficulties in interpretation. And always have your policy reviewed and approved by corporate and legal counsel prior to implementation.

---

[1] *Dealing with Workplace Violence: A Guide for Agency Planners*, Office of Personnel Management, 1997.

Sample Written Policy Statement

Workplace Violence Prevention Policy

It is the company's policy to promote a safe environment for its employees. The company is committed to working with its employees to maintain a work environment free from violence, threats of violence, harassment, intimidation, and other disruptive behavior. While this kind of conduct is not pervasive at our agency, no agency is immune. Every agency will be affected by disruptive behavior at one time or another.

Violence, threats, harassment, intimidation, and other disruptive behavior in our workplace will not be tolerated; that is, all reports of incidents will be taken seriously and will be dealt with appropriately. Such behavior can include oral or written statements, gestures, or expressions that communicate a direct or indirect threat of physical harm. Individuals who commit such acts may be removed from the premises and may be subject to disciplinary action, criminal penalties, or both.

We need your cooperation to implement this policy effectively and maintain a safe working environment. Do not ignore violent, threatening, harassing, intimidating, or other disruptive behavior. If you observe or experience such behavior by anyone on agency premises, whether he or she is an agency employee or not, report it immediately to a supervisor or manager. Supervisors and managers who receive such reports should seek advice from the Employee Relations Office at the main building regarding investigating the incident and initiating appropriate action. [**PLEASE NOTE: Threats or assaults that require immediate attention by security or police should be reported first to security at 111-1111 or to police at 911.**]

I will support all efforts made by supervisors and agency specialists in dealing with violent, threatening, harassing, intimidating, or other disruptive behavior in our workplace and will monitor whether this policy is being implemented effectively. If you have any questions about this policy statement, please contact the Human Resources office.

Thank you for your support and cooperation.

Herman President
President of the Company

The second component for the workplace violence prevention program is a careful assessment of the potential risks and the development of a written program. Although there is no requirement that employers assess their workplace or develop a written program, these activities are essential in order to properly identify the areas of risk and to prepare management team members and employees to react properly when necessary.

In assessing the potential risks of workplace violence, prudent professionals should assess any and all past incidents of workplace violence. These incidents should be scrutinized to identify what "went right" and what "went wrong" in these situations. Additionally, utilizing "what if" scenarios, each and every identifiable risk area should be evaluated. After identifying all of the potential risk areas, the risk areas should be prioritized as to probability.

After all potential risks are evaluated and prioritized, assess each potential risk for methods of preventing or minimizing the potential risk. Every potential solution should be evaluated from physical security to administrative controls.

Every aspect of a workplace violence situation should also be included in the written plan, including what to do if workplace violence happens after incident-critical stress debriefing for victims. The program should "leave no stone unturned" and every aspect and issue should be reduced to writing within the program. Areas to consider include, but are not limited to, the following:

- Posted evacuation routes
- Evacuation alarms and lighting
- Command centers
- Delineated responsibilities
- Emergency notification telephone numbers
- Corporate telephone numbers
- Methods of communication
- Management and employee training program curricula
- Psychologists or other qualified personnel for critical stress debriefing
- Chain of command
- Accounting for personnel after an evacuation
- Physical security assessments/inspection procedures

Screening applicants for employment is an important component in order to identify and potentially avoid individuals who may pose a greater risk of workplace violence. However, in many companies, this function belongs solely to the human resource or personnel departments. Thus, it is important that the human resource professional be involved in the development of the workplace violence prevention and response program.

In selecting new candidates for employment, screening for criminal history, controlled substance problems, and other potential signs of violence is essential (see Chapter 2). Management personnel responsible for the screening and hiring functions should be trained in identifying the warning signs and acquisition of related information.

An important component of the workplace violence prevention and response program is the development and issuance of training to all levels of management and all employees. Training should address the basic procedures for reporting, evacuation signals and routes, EAP options, danger signs, and other basic components of the program. Additionally, specialized training in conflict resolution, effective communications, effective methods of managing problem employees, and related areas should be offered to human resource personnel, supervisors, and other employees with specific program responsibilities.

For smaller companies or companies with limited resources, professionals may wish to call upon all applicable inside departments as well as outside agencies and sources to provide assistance. For example, for a situation that poses an immediate threat of workplace violence, all legal, human resource, employee assistance, community mental health, and law enforcement resources could be used to develop a response.

Although it is hoped that the procedure is never utilized, prudent professionals should prepare for after a workplace violence incident happens. Specific responsibilities as to notification of appropriate governmental agencies (such as OSHA, if there is a fatality), investigation and documentation of evidence, notification of families, notification of insurers, and a list of referral programs or other local mental health services for stress debriefing sessions should be carefully evaluated before critical incidents. Written workplace violence prevention and response programs should be carefully reviewed by company officials and legal counsel prior to implementation. Additionally, prudent professionals should write this important program on an educational level appropriate for their workforce and write the program in a defendable manner. If a workplace violence incident should occur followed by subsequent litigation, this program will be closely examined for deficiencies. Thus, the program should be written from a defensive posture as if a jury would be scrutinizing this document after an incident.

In developing the written program, the first rule of any threat is to take it seriously, as if each could end up being a violent act. Investigating and recording incidents is a key in preventing a violent act in the workplace. In recording the investigation, the four W's are important in starting the investigation off.

Who: Who made the actual complaint or threat? Who witnessed the actual threat?

When: When did the threat take place? Are there any events leading up to the actual threat? If so, when did they take place?

What: What exactly happened? This includes all facts that pertain to the actual threat and the investigators' assessment of what occurred.

Where: Where did the threat take place? In an office or outside the company? This is an important part which easily gets left out.

An incident report form should already be prepared by the supervisor, security, human resources, or other responsible management team member and is the easiest way to handle the investigation. Some sample questions that may be included are:

1. Who made the threat?
2. Who was the threat made against?
3. What were the actual words said in the threat?
4. Was there any threat of physical harm? If so what?
5. Where did the threat take place? (actual physical location)
6. What time did the threat take place? (be specific)
7. Who witnessed the actual threat?
8. What did the witnesses say occurred? (in their own words)
9. Ask the witnesses to write down their statements and attach these to the document.
10. Any other information that might pertain to the incident.[2]

These questions are not inclusive and a company needs to make its own incident form specific to its organization.

It is important when interviewing the person who made the threat to do so in a nonthreatening environment. Having a strategy to deal with the person's anger is an important part. Focus only on facts, not personal opinions or personality traits, this will help the person deal with the situation in a calm, rational manner and control emotional reactions.

In addition, there are signals that the interviewer can give off that will let the person know he or she is interested in what is being said. Some of these are: making eye contact; giving the person your full attention (do not work on notes or answer the phone; this can make the situation worse); ask open-ended questions; do not interrupt the person, let them finish what they are saying; speak in a calm rational voice; repeat what the person says in your own words to clarify the situation, so the interviewer understands fully.

Conduct any action necessary that would prevent an incident from occurring

This may involve disciplinary action; thus human resources or other member of the management team in charge should be involved. Human resources usually can refer the individual to the company's employee assistance program or other program for counseling. The investigator interviewing the person will come to a set of conclusions. The action that is taken should be based on the conclusions drawn from the interview. In some cases, the person who made the threats may feel better just after talking to the interviewer. This may resolve the issue, but in most cases it won't, and further action

[2] Id.

will be necessary. It is important to take a "zero tolerance" stance toward any threat.

Develop or improve security measures

Technology has advanced security measures. Gone are the days of the security guard walking a solitary route around a building. Present are enhanced security cameras and laser beams to detect intruders. Electronic systems can provide or deny access to persons entering a workplace. Every employer needs to assess their needs and provide appropriate security measures. There are several security companies that will provide a review and recommend security measures. There are some that will handle every aspect of the security measures. But providing security is a key to protecting employees from the threat of a violent employee and from other types of incidents that could threaten a company. Immediate improvements will provide a measure of safety, such as outdoor lighting, closed-circuit television, an intercom system, and an alarm system. These are just a couple of examples of what can de done, each company should survey its needs and provide appropriate security measures.

In addition to physical security measures, policies concerning security measures need to be included in the corporate policies and procedures manual. When an employee would like to work late, what does the policy state about that? Does another employee need to stay also? Are they even allowed to stay after hours? The policy also needs to include statements regarding security when a threat has been made. Is security tightened? If it is, how so? These issues need to be addressed in the corporate policy statements, and employees need to be aware of them.

In addition to the above measures, the local law enforcement department needs to be contacted. They can provide valuable information regarding threats occurring to other companies, mistakes that were made, and lessons learned from these mistakes.

Develop a crises plan in the event of actual workplace violence

The chain of command in the management team needs to be detailed so that every person knows his individual responsibilities and there is no confusion when an incident occurs. When an incident occurs, the responsibility will shift from the office manager or supervisor to a management team member. These responsibilities need to be detailed in writing, and training provided.

When are other agencies notified? The local police should be the first agency notified, so they can provide protection. A person from the employee assistance program needs to be notified and on hand. In an appropriate order, other agencies should be notified, such as the fire department and local emergency services.

In the crisis plan other resources should be made available, such as a trauma consultant, either a counselor or therapist to assist employees in dealing with the crisis, a security consultant, an in-house legal representative and a medical physician.

What procedures should be immediately instituted to determine how safe the workplace is? Is there an internal security team that could isolate the area until the police department arrives? An immediate assessment needs to be included to determine the effect of the incident on the workplace.

An immediate investigation of the incident needs to be conducted. What are the facts of the case? Who were the witnesses? What is the relationship between the investigation the employer is conducting and the investigation the police department will conduct?

What limits does the employer place on the release of information to other employees? Procedures need to be instituted about releasing information to the other employees. There needs to be regulations about what can be said during emergency situations.

What are the immediate public relations concerns? How much information will be released to the public and who will release the information?

How does a company preserve control of the day-to-day running of the business while an incident is occurring?

These are just a few ideas that can be a part of a crisis plan. Each company needs to develop a plan specific for the organization. After a plan has been written, a hypothetical situation should be conducted. This will let the employees know that a plan exists and the company is serious about dealing with workplace violence.

Use the courts to deal with threats of violence

The company should adopt a zero-tolerance policy toward threats of workplace violence or an actual incident of violence. When an employee has been discharged or no longer is employed by the company and makes a threat toward the company or a representative of the company, the use of the courts in dealing with violence is an effective and acceptable method.

In the area of physical security, prudent professionals should take advantage of all possible resources. Law enforcement and security officers should be involved in all stages of the planning process in an effective workplace violence prevention program. They can play an active role in prevention, intervention, and response to threatening situations, in addition to their traditional role of responding to actual incidents of physical violence. During the planning phase, law enforcement/security officers can:

- Identify types of workplace violence they can address and when and how they should be notified of a potential or actual incident
- Indicate whether their officers have arrest authority
- Identify their jurisdictional restrictions and alternative law enforcement agencies who may be able to provide assistance

- Identify threat assessment professionals
- Advise on what evidence is necessary and how it can be collected or recorded, so that law enforcement can assess the information and decide what action to take
- Explain anti-stalking laws applicable in the agency's jurisdiction and how and when to obtain restraining orders
- Suggest security measures to be taken for specific situations, such as in cases where employee assistance program counselors or other mental health professionals warn the agency that an individual has made a threat against an agency employee
- Arrange for supervisor/employee briefings or training on specific workplace violence issues such as:
  - Personal safety and security measures
  - Types of incidents to report to law enforcement/security
  - Types of measures law enforcement/security may take to protect employees during a violent incident, e.g., explanations of what it means to "secure the area," "secure the perimeter," and "preserve evidence"
  - Suggestions on how to react to an armed attacker
  - Suggestions for dealing with angry customers or clients
  - Suspicious packages
  - Bomb threats
  - Hostage situations
  - Telephone harassment and threats

When potentially violent situations arise, law enforcement/security officers can work with the incident response team to:

- Provide an assessment of the information available to determine whether law enforcement intervention is immediately necessary; for example, whether a criminal investigation is appropriate and whether a threat assessment professional should be consulted
- Identify what plan of action they deem appropriate
- Determine who will gather what types of evidence

Prudent professionals should not overlook the potential of utilizing security technology as part of the overall prevention effort. Given the advances in technology and the current pricing structure, physical security measures often can be a cost-effective alternative or supplement to prevention efforts. Law enforcement and security personnel can provide assistance in this area.

The following are some examples of ways to improve physical security:

- Post a security guard at the main building entrance or at entrances to specific offices.
- Install a metal detector or CCTV (closed-circuit television) camera or other device to monitor people coming in all building entrances.

- Issue all employees photo identification cards and assign temporary passes to visitors, who should be required to sign in and out of the building.
- Brief employees on steps to take if a threatening or violent incident occurs. Establish code words to alert co-workers and supervisors that immediate help is needed.
- Install silent, concealed alarms at reception desks.
- Ensure that officers (or guards) should have a clear view of the customer service area at all times.
- Arrange office furniture and partitions so that frontline employees in daily contact with the public are surrounded by natural barriers (desks, countertops, partitions) to separate employees from customers and visitors.
- Provide an under-the-counter duress alarm system to signal a supervisor or security officer if a customer becomes threatening or violent.
- Establish an area in the office for employees and/or customers to escape to if they are confronted with violent or threatening people.
- Provide an access-control combination lock on access doors.
- Mount closed-circuit television cameras for monitoring customer service activity from a central security office for the building.[3]

Companies should also consider addressing methods to safeguard computer systems. There have been cases where employees have sabotaged computer equipment, computer systems, and computer records. Therefore, whenever a threat of sabotage is suspected, procedures should be initiated to prevent the person from having access to the facility's computer system.

An important area of consideration is the use of employee assistance programs (EAPs). The major concern for most companies is the cost in terms of initial expenditures and ongoing costs. However, EAPs can play a major role in providing alternative assistance to employees. If a company cannot afford to provide employee assistance programs directly, consideration should be provided to assemble a listing of available social services or other outside agencies for referral.

One of, if not *the*, most important issue in successful EAP programs is confidentiality. Employees using company-provided EAP services are granted considerable privacy and confidentiality. Prudent professionals should also note that the confidentiality of employee records is also provided under various federal and state privacy laws.

In summation, proper planning can avoid potential incidents of workplace violence and minimize the impact of this type of divesting incident if it should take place in your operation. Like the old axiom, "plan your work and work your plan," the time and effort spent in preparation and planning will pay handsome dividends if only one potential workplace violence incident is avoided.

---

[3] Id.

Examples of Useful Handouts for Employees

The attached desk card summarizes the actions you should (or should not) take in a hostile or threatening situation. Print out and detach the card, tear or cut along the dotted lines, fold the card into a "tent," and tape the ends together underneath so the card will stand up on your desk with the text facing you. Review the card often. That way, if you are confronted by an angry, hostile, or threatening customer or co-worker, you will know what you should do. Everyone in your office, including supervisors and managers, should follow these same procedures. You can make copies of this card so that everyone has his or her own card.

Coping with Threats and Violence

For an angry or hostile customer or co-worker

- Stay calm. Listen attentively.
- Maintain eye contact.
- Be courteous. Be patient.
- Keep the situation in your control.

For a person shouting, swearing, and threatening

- Signal a co-worker, or supervisor, that you need help. (Use a duress alarm system or prearranged code words.)
- Do not make any calls yourself.
- Have someone call security or the local police.

For someone threatening you with a gun, knife, or other weapon

- Stay calm. Quietly signal for help. (Use a duress alarm or code words.)
- Maintain eye contact.
- Stall for time.
- Keep talking — but follow instructions from the person who has the weapon.
- Don't risk harm to yourself or others.
- Never try to grab a weapon.

Telephoned Threats

- Keep calm. Keep talking.
- Don't hang up.
- Signal a co-worker to get on an extension.
- Ask the caller to repeat the message and write it down.
- Repeat questions, if necessary.
- For a bomb threat, ask where the bomb is and when it is set to go off.
- Listen for background noises and write down a description.
- Write down whether it's a man or a woman; pitch of voice, accent, anything else you hear.
- Try to get the person's name, exact location, telephone number.
- Signal a co-worker to immediately call security or the local police.
- Notify your immediate supervisor.

Emergency Phone Numbers

Carefully tear out the "Emergency Phone Numbers" card at the dotted lines. Write in all the emergency numbers for your building. Tape this card to your desk by your phone or somewhere else close to your phone for handy reference. (Copies of this card also can be made.)

Supervisor/Human Resources _____
Security _____
Police/Sheriff _____
Fire Department _____
Ambulance _____
Medical Unit _____

chapter four

OSHA and other potential liabilities

> "We will never have real safety and security for the wage earners unless we provide for safety and security for the wage payers and wage savers, investors, and then, by all means, protection for both against the reckless wasters and wage spenders."
>
> William J.H. Boetcker

> "Security is the mother of danger and the grandmother of destruction."
>
> Thomas Fuller

Workplace violence is often perceived as a security issue rather than a safety issue in many companies. However, the primary federal group addressing workplace violence issues has been the United States Department of Labor through the Occupational Safety and Health Administration (OSHA) and the National Institute of Safety and Health (NIOSH). At this point in time, NIOSH has conducted extensive research into the issue of workplace violence and has published numerous documents on this subject (see Appendix B). Although OSHA has no specific regulations to address workplace violence or preventive measures, OSHA has published guidance documents for the health care and retail industries and has the catch-all "General Duty Clause," or Section 5(a)(1), which mandates employers create and maintain a safe and healthful workplace.

From a potential liability point of view, most employers are fearful of OSHA because of the compliance inspections that can result in monetary and criminal penalties for noncompliance with the prescribed standards. Given the fact that OSHA does not have specific standards addressing workplace violence at this time, OSHA would ascribe liability to the employer under the "General Duty" clause usually following a workplace violence incident. In essence, the employer is required to provide a workplace free from recognized hazards that are causing or likely to cause serious physical harm or death. If a workplace violence incident occurs and if the employer was aware of threats or otherwise "placed on notice" of the risk, the employer would possess an affirmative duty to address the hazard through increased security, a prevention program, or other measures. If the employer was on notice and a workplace violence incident happens, the employer would be in violation of the General Duty clause and thus could be cited. As of this writing, OSHA has cited several employers for violations under the General Duty clause, including a hospital, a rental company, and retail stores.

OSHA has published guidelines to assist employers in the area of workplace violence prevention and control. Although the guidelines are industry specific, these guidelines can be utilized as a model for all industries. As can be seen from Appendix C, OSHA is viewing this area with extreme interest, given the substantial number of fatalities and injuries. In 1996, OSHA held several meetings and developed the Workplace Violence Initiative. The guidelines for the health care and social service industries and night retail industries have been developed and published in the last three years. It should be noted that OSHA has additionally developed workplace violence prevention programs and outreach materials available to the public (see Appendix C).

Several state programs (known as State Plan States) have addressed the issue of workplace violence. The most prominent is California Occupational Safety and Health Administration (Cal-OSHA), which has taken an aggressive, proactive position in this area. Cal-OSHA has developed specific regulations for the health care industry and has published several documents providing guidance to employers in developing and implementing proactive programs.

In order to ensure that professionals are aware of the potential liability that can be incurred through OSHA or a state plan program, below is a brief overview of the agency structure, legislative history, and potential penalties.

Overview and history

The Federal Occupational Safety and Health Act of 1970

Before the Federal Occupational Safety and Health Act (hereafter referred to as the OSH Act) of 1970 was enacted,[1] safety and health issues were limited to specific industry safety and health laws and laws that governed federal contractors. It was during this period prior to the enactment of the OSH Act in 1970 that Congress gradually began to regulate safety and health in the American workplace through such laws as the Walsh–Healey Public Contracts Act of 1936, the Labor Management Relations Act (Taft–Hartley Act)

---

[1] 29 U.S.C. §63 et seq.

of 1947, the Coal Mine Safety Act of 1952, and the McNamara–O'Hara Public Service Contract Act of 1965.

With the passage of the controversial OSH Act in 1970, federal and state governmental agencies became actively involved in managing health and safety in the private sector workplace. Employers were placed on notice that unsafe and unhealthful conditions/acts would no longer be permitted to endanger the health, and often the lives, of American workers. In many circles, the Occupational Safety and Health Administration (OSHA) became synonymous with the "safety police," and employers were often forced, under penalty of law, to address safety and health issues in their workplaces.

Today, the OSH Act itself is virtually unchanged since its 1970 roots, and the basic methods for enforcement, standards development and promulgation, and adjudication remain intact. OSHA has, however, added many new standards in the past 26 years based primarily on the research conducted by the National Institute for Occupational Safety and Health (NIOSH) and recommendations from labor and industry and has revisited several of the original standards in order to update and/or modify the particular standard. The Occupational Safety and Health Review Commission (OSHRC) and the courts have been active in resolving many disputed issues and clarifying the law as it stands.

Legislative history

Throughout the history of the United States, the potential for the American worker to be injured or killed on the job was a brutal reality. Many disasters, such as that at Gauley Bridge, West Virginia,[2] fueled the call for laws and regulations to protect the American worker. As early as the 1920s, many states recognized the safety and health needs of the industrial worker and began to enact workers' compensation and industrial safety laws. The first significant federal legislation was the Walsh–Healey Public Contracts Act of 1936, which limited working hours and the use of child and convict labor. This law also required that contracts entered into by any federal agency for over $10,000 contain the stipulation that the contractor would not permit conditions that were unsanitary, hazardous, or dangerous to employees' health or safety.

In the 1940s, the federally enacted Labor Management Relations Act (Taft–Hartley Act) provided workers with the right to walk off a job if it was "abnormally dangerous." Additionally, in 1947, President Harry S. Truman created the first Presidential Conference on Industrial Safety.

In the 1950s and 1960s, the federal government continued to enact specialized safety and health laws to address particular circumstances. The Coal Mine Safety Act of 1952, the Maritime Safety Act, the McNamara–O'Hara Public Service Contract Act (protecting employees of contractors performing

---

[2] Page J., M. O'Brien, *Bitter Wages* (1973). Published in *Employment Law*, Rothstein, Knapp, and Liebman, eds., Foundation Press, 1987. During the construction of a tunnel in 1930–1931 476 workers died, and approximately 1,500 were disabled, primarily by silicosis.

maintenance work for federal agencies), and the National Foundation on the Arts and Humanities Act (requiring recipients of federal grants to maintain safe and healthful working conditions) were all passed during this time.

The federal government's first significant step in developing coverage for workplace safety and health was passage of the Metal and Nonmetallic Mine Safety Act of 1966. Following passage of this act, President Lyndon B. Johnson, in 1968, called for the first comprehensive occupational safety and health program as part of his Great Society program. Although this proposed plan never made it to a vote in Congress, the seed was planted for future legislation.

One particular incident shocked the American public and federal government into action. In 1968, a coal mine fire and explosion in Farmington, West Virginia, killed 78 miners. Congress reacted swiftly by passing a number of safety and health laws, including the Coal Mine Health and Safety Act of 1969, the Contract Work Hours and Safety Standards Act of 1969, (Construction Safety Act), and the Federal Railway Safety Act.

In 1970, fueled by the new interest in workplace health and safety, Congress pushed for more comprehensive laws to regulate the conditions of the American workplace. To this end, the OSH Act of 1970[3] was enacted. The overriding purpose and intent of the OSH Act was "to assure so far as possible every working man and woman in the Nation safe and healthful working conditions and to preserve our human resources."[4]

Coverage and jurisdiction of the OSH Act

The OSH Act covers virtually every American workplace that employs one or more employees and engages in a business that in any way affects interstate commerce.[5] The OSH Act covers employment in every state, the District of Columbia, Puerto Rico, Guam, the Virgin Islands, American Samoa, and the Trust Territory of the Pacific Islands.[6] The OSH Act does not, however, cover employees in situations where other state or federal agencies have jurisdiction which requires the agencies to prescribe or enforce their own safety and health regulations.[7] Additionally, the OSH Act exempts residential owners who employ people for ordinary domestic tasks, such as cooking, cleaning, and child care.[8] It also does not cover federal,[9] state, and local governments[10] or Native American reservations.[11]

---

[3] 29 U.S.C. §651 et seq.
[4] 29 C.F.R. §651(b).
[5] Id. at §1975.3(d).
[6] Id. §652(7).
[7] *See e.g.*, Atomic Energy Act of 1954, 42 U.S.C. §2021.
[8] 29 C.F.R. §1975(6).
[9] 29 U.S.C.A. §652(5) (no coverage under OSH Act when U.S. government acts as employer.)
[10] Id.
[11] *See e.g., Navajo Forest Prods. Indus.*, 8 O.S.H. Cases 2694 (OSH Rev.Comm'n. 1980), aff'd., 692 F.2d 709, 10 O.S.H. Cases 2159.

The OSH Act does require every employer engaged in interstate commerce to furnish employees "a place of employment ... free from recognized hazards that are causing, or are likely to cause, death or serious harm."[12] To help employers create and maintain safe working environments and to enforce laws and regulations that ensure safe and healthful work environments, Congress provided for the creation of the Occupational Safety and Health Administration (OSHA), to be a new agency under the direction of the Department of Labor.

Today, OSHA is one of the most widely known and powerful enforcement agencies. It has been granted broad regulatory powers to promulgate regulations and standards, investigate and inspect, issue citations, and propose penalties for safety violations in the workplace.

The OSH Act also established an independent agency to review OSHA citations and decisions, the Occupational Safety and Health Review Commission (OSHRC). The OSHRC is a quasijudicial and independent administrative agency composed of three commissioners appointed by the president who serve staggered six-year terms. The OSHRC has the power to issue orders, uphold, vacate, or modify OSHA citations and penalties, and direct other appropriate relief and penalties.

The educational arm of the OSH Act is the National Institute for Occupational Safety and Health (NIOSH), which was created as a specialized educational agency of the existing National Institutes of Health. NIOSH conducts occupational safety and health research and develops criteria for new OSHA standards. NIOSH can conduct workplace inspections, issue subpoenas, and question employees, and employers but it does not have the power to issue citations or penalties.

State safety plans

Notwithstanding OSH Act enforcement through the above noted federal agencies, OSHA encourages individual states to take responsibility for OSHA administration and enforcement within their own respective boundaries. Each state can request and be granted the right to adopt state safety and health regulations and enforcement mechanisms. In Section 18(b), the OSH Act provides that any state "which, at any time, desires to assume responsibility for development and the enforcement therein of occupational safety and health standards relating to any... issue with respect to which a federal standard has been promulgated...shall submit a state plan for the development of such standards and their enforcement."[13] For a state plan to be placed into effect, the state must first develop and submit its proposed program to the Secretary of Labor for review and approval. The Secretary must certify that the state plan's standards are "at least as effective" as the

---

[12] 29 U.S.C.A. §654(a)(1).
[13] Id.

federal standards and that the state will devote adequate resources to administering and enforcing standards.[14]

In most state plans, the state agency has developed more stringent safety and health standards than OSHA[15] and has usually developed more stringent enforcement schemes.[16] The Secretary of Labor has no statutory authority to reject a state plan if the proposed standards or enforcement scheme are more strict than the OSHA standards but can reject the state plan if the standards are below the minimum limits set under OSHA standards.[17] These states are known as "state plan" states and territories.[18] As of 1991 there were 21 states and two territories with approved and functional state plan programs.[19] Employers in state plan states and territories must comply with their state's regulations; federal OSHA plays virtually no role in direct enforcement.

OSHA does, however, have an approval and oversight role with regard to state plan programs. OSHA must approve all state plan proposals prior to enactment, and they maintain oversight authority to "pull the ticket" of any/all state plan programs at any time they are not achieving the identified prerequisites. Enforcement of this oversight authority was recently observed following the fire resulting in several workplace fatalities at the Imperial Foods facility in Hamlet, North Carolina. Following this incident, federal OSHA assumed jurisdiction and control over the state plan program in North Carolina and made significant modifications to this program before returning the program to state control.

---

[14] Id. §667(c). After an initial evaluation period of at least three years during which OSHA retains concurrent authority, a state with an approved plan gains exclusive authority over standard setting, inspection procedures, and enforcement of health and safety issues covered under the state plan. *See also Noonan v. Texaco*, 713 P.2d 160 (Wyo. 1986); Plans for the Development and Enforcement of State Standards, 29 C.F.R. §667(f) (1982) and §1902.42(c)(1986). Although the state plan is implemented by the individual state, OSHA continues to monitor the program and may revoke the state authority if the state does not fulfill the conditions and assurances contained within the proposed plan.

[15] Some states incorporate federal OSHA standards into their plans and add only a few of their own standards as a supplement. Other states, such as Michigan and California, have added a substantial number of separate and independently promulgated standards. *See generally* Employee Safety and Health Guide (CCH) §§ 5000-5840 (1987)(compiling all state plans). Some states also add their own penalty structures. For example, under Arizona's plan, employers may be fined up to $150,000 and sentenced to one and one-half years in prison for knowing violations of state standards that cause death to an employee and may also have to pay $25,000 in compensation to the victim's family. If the employer is a corporation, the maximum fine is $1 million. *See* Ariz. Rev. Stat. Ann. §§ 13-701, 13-801, 23-4128, 23-418.01, 13-803 (Supp. 1986).

[16] For example, under Kentucky's state plan regulations for controlling hazardous energy (i.e., lockout/tagout), locks would be required rather than locks or tags being optional as under the federal standard. Lockout/tagout is discussed in more detail in Chapter 2.

[17] 29 U.S.C. §667.

[18] 29 U.S.C.A. §667; 29 C.F.R. §1902.

[19] The states and territories operating their own OSHA programs are Alaska, Arizona, California, Hawaii, Indiana, Iowa, Kentucky, Maryland, Michigan, Minnesota, Nevada, New Mexico, North Carolina partial federal OSHA enforcement, Oregon, Puerto Rico, South Carolina, Tennessee, Utah, Vermont, Virginia, Virgin Islands, Washington, and Wyoming.

OSHA standards and the general duty clause

Promulgation of standards

The OSH Act requires that a covered employer comply with specific occupational safety and health standards and all rules, regulations, and orders issued pursuant to the OSH Act that apply to the workplace.[20] The OSH Act also requires that all standards be based on research, demonstration, experimentation, or other appropriate information.[21] The Secretary of Labor is authorized under the Act to "promulgate, modify, or revoke any occupational safety and health standard,"[22] and the OSH Act describes the procedures that the Secretary must follow when establishing new occupational safety and health standards.[23]

The OSH Act authorizes three ways to promulgate new standards. From 1970 to 1973, the Secretary of Labor was authorized in Section 6(a) of the Act[24] to adopt national consensus standards and establish federal safety and health standards without following lengthy rule-making procedures. Many of the early OSHA standards were adapted mainly from other areas of regulation, such as the National Electric Code and American National Standards Institute (ANSI) guidelines. However, this promulgation method is no longer in effect.

The usual method of issuing, modifying, or revoking new or existing OSHA standards is set out in Section 6(b) of the OSH Act and is known as informal rule making. It requires notice to interested parties, through subscription in the *Federal Register* of the proposed regulation and standard, and provides an opportunity for comment in a nonadversarial administrative hearing.[25] The proposed standard can also be advertised through magazine articles and other publications, thus informing interested parties of the proposed standard and regulation. This method differs from the requirements of most other administrative agencies that follow the Administrative Procedure Act[26] in that the OSH Act provides interested persons an opportunity to request a public hearing with oral testimony. It also requires the Secretary of Labor to publish in the *Federal Register* a notice of the time and place of such hearings.

Although not required under the OSH Act, the Secretary of Labor has directed, by regulation, that OSHA follow a more rigorous procedure for comment and hearing than other administrative agencies.[27] Upon notice and request for a hearing, OSHA must provide a hearing examiner in order to

---

[20] 29 U.S.C. §655(b).
[21] 29 U.S.C.A. §655(b)(5).
[22] 29 U.S.C. § 1910.
[23] 29 C.F.R. §1911.15. (By regulation, the Secretary of Labor has prescribed more detailed procedures than the OSH Act specifies to ensure participation in the process of setting new standards, 29 C.F.R. §1911.15.)
[24] 29 U.S.C. § 1910
[25] 29 U.S.C. §655(b).
[26] U.S.C. §553.
[27] 29 C.F.R. §1911.15.

listen to any oral testimony offered. All oral testimony is preserved in a verbatim transcript. Interested persons are provided an opportunity to cross-examine OSHA representatives or others on critical issues. The Secretary must state the reasons for the action to be taken on the proposed standard, and the statement must be supported by substantial evidence in the record as a whole.

The Secretary of Labor has the authority not to permit oral hearings and to call for written comment only. Within 60 days after the period for written comment or oral hearings has expired, the Secretary must decide whether to adopt, modify, or revoke the standard in question. The Secretary can also decide not to adopt a new standard. The Secretary must then publish a statement of the reasons for any decision in the *Federal Register.* OSHA regulations further mandate that the Secretary provide a supplemental statement of significant issues in the decision. Safety and health professionals should be aware that the standard as adopted and published in the *Federal Register* may be different from the proposed standard. The Secretary is not required to reopen hearings when the adopted standard is a "logical outgrowth" of the proposed standard.[28]

The final method for promulgating new standards, and the one most infrequently used, is the emergency temporary standard permitted under Section 6(c).[29] The Secretary of Labor may establish a standard immediately if it is determined that employees are subject to grave danger from exposure to substances or agents known to be toxic or physically harmful and that an emergency standard would protect the employees from the danger. An emergency temporary standard becomes effective on publication in the *Federal Register* and may remain in effect for six months. During this six-month period, the Secretary must adopt a new permanent standard or abandon the emergency standard.

Only the Secretary of Labor can establish new OSHA standards. Recommendations or requests for an OSHA standard can come from any interested person or organization, including employees, employers, labor unions, environmental groups, and others.[30] When the Secretary receives a petition to adopt a new standard or to modify or revoke an existing standard, he or she usually forwards the request to NIOSH and the National Advisory Committee on Occupational Safety and Health (NACOSH),[31] or the Secretary may use a private organization such as the American National Standards Institute (ANSI) for advice and review.

---

[28] Taylor Diving & Salvage Co. v. Department of Labor, 599 F.2d 622 7 O.S.H. Cases 1507 (5th Cir. 1979).

[29] 29 U.S.C. §655(c).

[30] Id. at §655(b)(1).

[31] Id. at §656(a)(1). NACOSH was created by the OSH Act to "advise, consult with, and make recommendations...on matters relating to the administration of the Act." Normally, for new standards, the Secretary has established continuing committees and *ad hoc* committees to provide advice regarding particular problems or proposed standards.

The general duty clause

As stated above, the OSH Act requires that an employer must maintain a place of employment free from recognized hazards that are causing or are likely to cause death or serious physical harm, even if there is no specific OSHA standard addressing the circumstances. Under Section 5(a)(1), the general duty clause, an employer may be cited for a violation of the OSH Act if the condition causes harm or is likely to cause harm to employees, even if OSHA has not promulgated a standard specifically addressing the particular hazard. The general duty clause is a catch-all standard encompassing all potential hazards that have not been specifically addressed in the OSHA standards. For example, if a company is cited for a workplace violence incident and there is no workplace violence standard to apply, the hazard will be cited under the general duty clause.

OSHA enforcement

Under the OSH Act, Congress provided for civil and criminal penalties for employers who failed to comply with the promulgated standards and regulations. Over the years, the monetary penalties have been modified and the criminal sanctions seldom utilized. Many employers who initially addressed the requirements of the OSH Act and OSHA standards to avoid the OSHA penalties have found that compliance makes sense and is "good business." These employers have moved to a higher level in safety and loss prevention, where simply avoiding penalties by OSHA for noncompliance is no longer the objective but a given. These employers have moved to a higher level where safeguarding their human assets pays dividends not only in personnel but in terms of dollars saved.

Employers who have not heeded the warning of Congress and OSHA have found that failure to comply with the OSHA standards and create a safe and healthful workplace for their employees can be extremely costly. The OSH Act gave OSHA the power to issue monetary penalties, often reaching several million dollars, and, in egregious cases, the ability to pursue criminal sanctions.

Monetary fines and penalties

The OSH Act provides for a wide range of penalties, from a simple notice with no fine to criminal prosecution. The Omnibus Budget Reconciliation Act of 1990 multiplied maximum penalties sevenfold. Violations are categorized and penalties may be assessed as outlined in Table 4.1.

Each alleged violation is categorized and the appropriate fine issued by the OSHA area director. It should be noted that each citation is separate and may carry with it a monetary fine. The gravity of the violation is the primary factor in determining penalties.[32] In assessing the gravity of a violation, the

---

[32] *OSHA Compliance Field Operations Manual (OSHA Manual)* at XI-C3c (Apr. 1977).

Table 4.1   Violation and Penalty Schedule

Penalty	Old Penalty Schedule (in dollars)	New Penalty Schedule (1990) (in dollars)
De minimis notice	0	0
Nonserious	0–1,000	0–7,000
Serious	0–1,000	0–7,000
Repeat	0–10,000	0–70,000
Willful	0–10,000	5,000 minimum
		70,000 maximum
Failure to abate notice	0–1,000 per day	0–7,000 per day
New posting penalty		0–7,000

compliance officer or area director must consider (1) the severity of the injury or illness that *could* result and (2) the probability that an injury or illness *could* occur as a result of the violation.[33] Specific penalty assessment tables assist the area director or compliance officer in determining the appropriate fine for the violation.[34]

After selecting the appropriate penalty table, the area director or compliance officer determines the degree of probability that the injury or illness will occur by considering:

1. The number of employees exposed
2. The frequency and duration of the exposure
3. The proximity of employees to the point of danger
4. Factors such as the speed of the operation that require work under stress
5. Other factors that might significantly affect the degree of probability of an accident[35]

OSHA has defined a serious violation as "an infraction in which there is a substantial probability that death or serious harm could result ... unless the employer did not or could not with the exercise of reasonable diligence, know of the presence of the violation."[36] Section 17(b) of the OSH Act requires that a penalty of up to $7,000 be assessed for every serious violation cited by the compliance officer.[37] In assembly line enterprises and manufacturing facilities with duplicate operations, if one process is cited as possessing a serious violation, it is possible that each of the duplicate processes or machines may be cited for the same violation. Thus, if a serious violation is found in one machine and there are many other identical

---

[33] Id.
[34] Id. at XI-C3c(2).
[35] Id. at (3)(a).
[36] 29 U.S.C. §666.
[37] Id. at §666(b).

machines in the enterprise, a very large monetary fine for a single serious violation is possible.[38]

Currently the greatest monetary liabilities are for "repeat violations," "willful violations," and "failure to abate" cited violations. A *repeat* violation is a second citation for a violation that was cited previously by a compliance officer. OSHA maintains records of all violations and must check for repeat violations after each inspection. A *willful* violation is the employer's purposeful or negligent failure to correct a known deficiency. This type of violation, in addition to carrying a large monetary fine, exposes the employer to a charge of an "egregious" violation and the potential for criminal sanctions under the OSH Act or state criminal statutes if an employee is injured or killed as a direct result of the willful violation. *Failure to abate* a cited violation has the greatest cumulative monetary liability of all. OSHA may assess a penalty of up to $1,000 per day per violation for each day in which a cited violation is not brought into compliance.

In assessing monetary penalties, the area or regional director must consider the good faith of the employer, the gravity of the violation, the employer's history of compliance, and the size of the employer. Mr. Joseph Dear, the Assistant Secretary of Labor, recently stated that OSHA will start using its egregious case policy, which has seldom been invoked in recent years.[39] Under the egregious violation policy, when violations are determined to be conspicuous, penalties are cited for each violation, rather than combining the violations into a single, smaller penalty.

In addition to the potential civil or monetary penalties that could be assessed, OSHA regulations may be used as evidence in negligence, product liability, workers' compensation, and other actions involving employee safety and health issues.[40] OSHA standards and regulations are the baseline requirements for safety and health that must be met, not only to achieve compliance with the OSHA regulations, but also to safeguard an organization against other potential civil actions.

Criminal liability

The OSH Act provides for criminal penalties in four circumstances.[41] In the first, anyone inside or outside of the Department of Labor or OSHA who gives advance notice of an inspection, without authority from the Secretary, may be fined up to $1,000, imprisoned for up to six months, or both. Second, where any employer or person, who intentionally falsifies

---

[38] For example, if a company possesses 25 identical machines, and each of these machines is found to have the identical serious violation, this would theoretically constitute 25 violations rather than one violation on 25 machines, and a possible monetary fine of $175,000 rather than a maximum of $7,000.

[39] *Occupational Safety & Health Reporter,* V. 23, No. 32, Jan. 12, 1994.

[40] See Infra at §1.140

[41] 29 U.S.C. §666(e) — (g). *See also, OSHA Manual,* supra note 62 at VI-B.

statements or OSHA records that must be prepared, maintained, or submitted under the OSH Act, may, if found guilty, be fined up to $10,000,
imprisoned for up to six months, or both. Third, any person responsible
for a violation of an OSHA standard, rule, order, or regulation, causing the
death of an employee may, upon conviction, be fined up to $10,000, impris-
oned for up to six months, or both. If convicted for a second violation,
punishment may be a fine of up to $20,000, imprisonment for up to one
year, or both.[42] Finally, if an individual is convicted of forcibly resisting or
assaulting a compliance officer or other Department of Labor personnel, a
fine of $5,000, three years in prison, or both can be imposed. Any person
convicted of killing a compliance officer or other OSHA or Department of
Labor personnel acting in his or her official capacity may be sentenced to
prison for any term of years or life.

OSHA does not have authority to impose criminal penalties directly;
instead, it refers cases for possible criminal prosecution to the U.S. Department of Justice. Criminal penalties must be based on violation of a specific
OSHA standard; they may not be based on a violation of the general duty
clause. Criminal prosecutions are conducted like any other criminal trial,
with the same rules of evidence, burden of proof, and rights of the accused.
A corporation may be criminally liable for the acts of its agents or employees.[43] The statute of limitations for possible criminal violations of the OSH
Act, as for other federal noncapital crimes, is five years.[44]

Under federal criminal law, criminal charges may range from murder to
manslaughter to conspiracy. Several charges may be brought against an
employer for various separate violations under one federal indictment.

Following a criminal conviction for a federal felony, the sentence to be
imposed is defined under the Federal Sentencing Guidelines.[45] Under these
guidelines, judges have very little leeway in determining the sentence. Each
felony has an offense level. A sentencing table[46] is used to factor in the
criminal history, and a fine table provides a minimum and maximum fine
range. Deduction from or addition to the sentencing structure based on
numerical values is permitted depending on the situation. Departures from
the range provided in the guidelines are rare. Of particular interest to safety
and health professionals facing this type of situation is the fact that the
Federal Sentencing Guidelines provide very little flexibility through which
the court can consider factors involved in a work-related injury or fatality.
In essence, murder is murder, regardless of whether it happened on the street
or in the workplace.[47]

---

[42] A repeat criminal conviction for a willful violation causing an employee death doubles the
possible criminal penalties.
[43] 29 C.F.R. §5.01(6).
[44] *U.S. v. Dye Const. Co.*, 510 F.2d 78, 2 O.S.H. Cases 1510 (10th Cir. 1975).
[45] 18 U.S.C. §3551 *et seq.*
[46] Id. at §3553(a).

Rights and responsibilities under the OSH Act

The OSHA inspection

OSHA performs all enforcement functions under the OSH Act. Under Section 8(a) of the act, OSHA compliance officers have the right to enter any workplace of a covered employer without delay, inspect and investigate a workplace during regular hours and at other reasonable times, and obtain an inspection warrant if access to a facility or operation is denied.[48] Upon arrival at an inspection site (any company facility), the compliance officer must present his or her credentials to the owner or designated representative of the employer before starting the inspection. An employer representative and an employee and/or union representative may accompany the compliance officer on the inspection. Compliance officers can question the employer and employees and inspect required records, such as the OSHA Form 200, which records injuries and illnesses.[49] Most compliance officers cannot issue on-the-spot citations, they only have authority to document potential hazards and report or confer with the OSHA area director before issuing a citation.

A compliance officer or other employee of OSHA may not provide advance notice of the inspection under penalty of law.[50] The OSHA area director is, however, permitted to provide notice under the following circumstances:

1. In cases of apparent imminent danger, to enable the employer to correct the danger as quickly as possible
2. When the inspection can most effectively be conducted after regular business hours or where special preparations are necessary
3. To ensure the presence of employee and employer representatives or appropriate personnel needed to aid in inspections
4. When the area director determines that advance notice would enhance the probability of an effective and thorough inspection[51]

---

[47] For example, if convicted of involuntary manslaughter, the base offense level is 10 if the conduct was criminally negligent, and 14 if the conduct was reckless. If the individual possessed a clean criminal history (Criminal History Category I), the potential sentence for an offense level 10 is between 6 and 12 months, and the potential fine is between of $2,000 and $20,000. For an offense level 14, incarceration time can range from 15 to 21 months, and the fine can range from $4,000 to $40,000. The base level may be reduced through a variety of different activities, such as acknowledgment of activity, cooperation with prosecution, and so on. The Federal Sentencing Guidelines determine the base level, departures or deductions are factored, and the total calculated. The final offense level is applied to the Sentencing Guideline Table, and the sentence and/or fine is determined by the judge within the range provided by the Sentencing Guidelines. (See Appendix H).

[48] *See infra* §§1.10 and 1.12.

[49] 29 C.F.R. §1903.8.

[50] 29 U.S.C. §17(f). The penalty for providing advance notice, upon conviction, is a fine of not more than $1,000, imprisonment for not more than six months, or both.

[51] *Occupational Safety and Health Law*, 208-09 (1988).

Compliance officers can also take environmental samples and take or obtain photographs related to the inspection. Additionally, compliance officers, can use other "reasonable investigative techniques," including personal sampling equipment, dosimeters, air sampling badges, and other equipment.[52] Compliance officers must, however, take reasonable precautions when using photographic or sampling equipment to avoid creating hazardous conditions (i.e., a spark-producing camera flash in a flammable area) or disclosing a trade secret.[53]

An OSHA inspection has four basic components: (1) the opening conference, (2) the walk-through inspection, (3) the closing conference, and (4) the issuance of citations, if necessary. In the opening conference, the compliance officer may explain the purpose and type of inspection to be conducted, request records to be evaluated, question the employer, ask for appropriate representatives to accompany him or her during the walk-through inspection, and ask additional questions or request more information. The compliance officer may, but is not required to, provide the employer with copies of the applicable laws and regulations governing procedures and health and safety standards. The opening conference is usually brief and informal, its primary purpose is to establish the scope and purpose of the walk-through inspection.

After the opening conference and review of appropriate records, the compliance officer, usually accompanied by a representative of the employer and a representative of the employees, conducts a physical inspection of the facility or work site.[54] The general purpose of this walk-through inspection is to determine whether the facility or work site complies with OSHA standards. The compliance officer must identify potential safety and health hazards in the workplace, if any, and document them to support issuance of citations.[55]

The compliance officer uses various forms to document potential safety and health hazards observed during the inspection. The most commonly used form is the OSHA-1 Inspection Report, in which the compliance officer records information gathered during the opening conference and walk-through inspection, including:

- The establishment's name
- Inspection number
- Type of legal entity

---

[52] 29 C.F.R. §1903.7(b) (revised by 47 Fed. Reg. 5548 [1982.])

[53] *See e.g.*, 29 C.F.R. §1903.9. Under §15 of the OSH Act, all information gathered or revealed during an inspection or proceeding that may reveal a trade secret as specified under 18 U.S.C. §1905 must be considered confidential, and breach of that confidentiality is punishable by a fine of not more than $1,000, imprisonment of not more than one year, or both; and removal from office or employment with OSHA.

[54] It is highly recommended by the authors that a company representative accompany the OSHA inspection during the walk-through inspection.

[55] *OSHA Manual, supra* note 62, at III-D8.

- Type of business or plant
- Additional citations
- Names and addresses of all organized employee groups
- The authorized representative of employees
- The employee representative contacted
- Other persons contacted
- Coverage information (state of incorporation, type of goods or services in interstate commerce, etc.)
- Date and time of entry
- Date and time that the walk-through inspection began
- Date and time closing conference began
- Date and time of exit
- Whether a follow-up inspection is recommended
- The compliance officer's signature and date
- The names of other compliance officers
- Evaluation of safety and health programs (checklist)
- Closing conference checklist
- Additional comments

Two additional forms are usually attached to the OSHA Inspection Report. The OSHA-1A form, known as the narrative, is used to record information gathered during the walk-through inspection; names and addresses of employees, management officials, and employee representatives accompanying the compliance officer on the inspection; and other information. A separate worksheet, known as OSHA-1B, is used by the compliance officer to document each condition that he or she believes could be an OSHA violation. One OSHA-1B worksheet is completed for each potential violation noted by the compliance officer.

When the walk-through inspection is completed, the compliance officer conducts an informal meeting with the employer or the employer's representative to "informally advise (the employer) of any apparent safety or health violations disclosed by the inspection."[56] The compliance officer informs the employer of the potential hazards observed and indicates the applicable section of the standards allegedly violated, advises that citations may be issued, and informs the employer or representative of the appeal process and rights.[57] Additionally, the compliance officer advises the employer that the OSH Act prohibits discrimination against employees or others for exercising their rights.[58]

In an unusual situation, the compliance officer may issue a citation(s) on the spot. When this occurs, the compliance officer informs the employer of the abatement period, in addition to the other information provided at the closing conference. In most circumstances, the compliance officer will

---

[56] 29 C.F.R. §1903.7(e).
[57] *OSHA Manual supra* note 62, at III-D9.
[58] 29 U.S.C. §660(c)(1).

leave the workplace and file a report with the area director who has authority, through the Secretary of Labor, to decide whether a citation should be issued, compute any penalties to be assessed, and set the abatement date for each alleged violation. The area director, under authority from the Secretary, must issue the citation with "reasonable promptness."[59] Citations must be issued in writing and must describe with particularity the violation alleged, including the relevant standard and regulation. There is a six-month statute of limitations, and the citation must be issued or vacated within this period. OSHA must serve notice of any citation and proposed penalty by certified mail, unless there is personal service, to an agent or officer of the employer.[60]

After the citation and notice of proposed penalty is issued, but before the notice of contest by the employer is filed, the employer may request an informal conference with the OSHA area director. The general purpose of the informal conference is to clarify the basis for the citation, modify abatement dates or proposed penalties, seek withdrawal of a cited item, or otherwise attempt to settle the case. This conference, as its name implies, is an informal meeting between the employer and OSHA. Employee representatives must have an opportunity to participate if they so request. Safety and health professionals should note that the request for an informal conference does not "stay" (delay) the 15-working-day period to file a notice of contest to challenge the citation.[61]

Under the OSH Act, an employer, employee, or authorized employee representative (including a labor organization) is given 15 working days from when the citation is issued to file a "notice of contest." If a notice of contest is not filed within 15 working days, the citation and proposed penalty become a final order of the Occupational Safety and Health Review Commission (OSHRC), and are not subject to review by any court or agency. If a timely notice of contest is filed in good faith, the abatement requirement is tolled (temporarily suspended or delayed) and a hearing is scheduled. The employer also has the right to file a petition for modification of the abatement period (PMA) if the employer is unable to comply with the abatement period provided in the citation. If OSHA contests the PMA, a hearing is scheduled to determine whether the abatement requirements should be modified.

When the notice of contest by the employer is filed, the Secretary must immediately forward the notice to the OSHRC, which then schedules a hearing before its administrative law judge (ALJ). The Secretary of Labor is labeled the "complainant," and the employer the "respondent." The ALJ may affirm, modify, or vacate the citation, any penalties, or the abatement date. Either party can appeal the ALJ's decision by filing a petition for discretionary review (PDR). Additionally, any member of the OSHRC may

---

[59] Id. §658.
[60] *Fed. R. Civ. P.* 4(d)(3).
[61] 29 U.S.C. §659(a).

"direct review" of any decision by an ALJ, in whole or in part, without a PDR. If a PDR is not filed and no member of the OSHRC directs a review, the decision of the ALJ becomes final in 30 days. Any party may appeal a final order of the OSHRC by filing a petition for review in the U.S. Court of Appeals for the circuit in which the violation is alleged to have occurred or in the U.S. Court of Appeals for the District of Columbia circuit. This petition for review must be filed within 60 days from the date of the OSHRC's final order.

Types of violations

Violations

The OSHA monetary penalty structure is classified according to the type and gravity of the particular violation. Violations of OSHA standards or the general duty clause are categorized as *"de minimis,"*[62] "other" (nonserious),[63] "serious,"[64] "repeat,"[65] and "willful."[66] (see Table 4.1 for penalty schedules.) Monetary penalties assessed by the Secretary vary according to the degree of the violation. Penalties range from no monetary penalty to 10 times the imposed penalty for repeat or willful violations.[67] Additionally, the Secretary may refer willful violations to the U.S. Department of Justice for imposition of criminal sanctions.[68]

De minimis violations

When a violation of an OSHA standard does not immediately or directly relate to safety or health, OSHA either does not issue a citation or issues a *de minimis* citation. Section 9 of the OSH Act provides that "[the] Secretary may prescribe procedures for the issuance of a notice in lieu of a citation with respect to *de minimis* violations which have no direct or immediate relationship to safety or health."[69]

A *de minimis* notice does not constitute a citation and no fine is imposed. Additionally, there usually is no abatement period, and thus there can be no violation for failure to abate.

The *OSHA Compliance Field Operations Manual* (*OSHA Manual*)[70] provides two examples of when de minimis notices are generally appropriate:

---

[62] 29 U.S.C.§§658(a), 666(c).
[63] Id. §666(j).
[64] Id. at §666(c).
[65] Id. at (a).
[66] Id.
[67] Id. at (b).
[68] Id. at (e).
[69] Id. §658(a).
[70] *Supra* note 62.

1. "in situations involving standards containing physical specificity wherein a slight deviation would not have an immediate or direct relationship to safety or health"[71]
2. "where the height of letters on an exit sign is not in strict conformity with the size requirements of the standard"[72]

OSHA has found *de minimis* violations in cases where employees, as well as the safety records, are persuasive in exemplifying that no injuries or lost time have been incurred.[73] Additionally, in order for OSHA to conserve valuable resources to produce a greater impact on safety and health in the workplace, it is highly likely that the Secretary will encourage use of the *de minimis* notice in marginal cases and even in other situations where the possibility of injury is remote and potential injuries would be minor.

Other or nonserious violations

"Other" or nonserious violations are issued where a violation could lead to an accident or occupational illness, but the probability that it would cause death or serious physical harm is minimal. Such a violation, however, does have a direct or immediate relationship to the safety and health of workers.[74] Potential penalties for this type of violation range from no fine up to $7,000 per violation.[75]

In distinguishing between a serious and a nonserious violation, the OSHRC has stated that "a nonserious violation is one in which there is a direct and immediate relationship between the violative condition and occupational safety and health but no such relationship that a resultant injury or illness **is death or serious physical harm.**"[76]

The *OSHA Manual* provides guidance and examples for issuing nonserious violations. It states that:

> "an example of nonserious violation is the lack of a guardrail at a height from which a fall would more probably result in only a mild sprain or cut or abrasion; i.e., something less than serious harm."[77]
>
> "A citation for serious violation may be issued or a group of individual violations (which) taken by themselves would be nonserious, but together would be

---

[71] Id. at VII-B3a.

[72] Id.

[73] *Hood Sailmakers*, 6 O.S.H. Cases 1207 (1977).

[74] *OSHA Manual supra* note 62, at VIII-B2a. The proper nomenclature for this type of violation is "other" or "other than serious." Many safety and health professionals classify this type of violation as nonserious for explanation and clarification purposes.

[75] A nonserious penalty is usually less than $100 per violation.

[76] *Crescent Wharf & Warehouse Co.*, 1 O.S.H. Cases 1219, 1222 (1973).

[77] *OSHA Manual supra* note 62, at VIII-B2a.

> serious in the sense that in combination they present
> a substantial probability of injury resulting in death or
> serious physical harm to employees."[78]
>
> "A number of nonserious violations (which) are
> present in the same piece of equipment which, consid-
> ered in relation to each other, affect the overall gravity
> of possible injury resulting from an accident involving
> the combined violations...may be grouped in a manner
> similar to that indicated in the preceding paragraph,
> although the resulting citation will be for a nonserious
> violation."[79]

The difference between a serious and a nonserious violation hinges on subjectively determining the probability of injury or illness that might result from the violation. Administrative decisions have usually turned on the particular facts of the situation. The OSHRC has reduced serious citations to nonserious violations when the employer was able to show that the probability of an accident, and the probability of a serious injury or death, was minimal.[80]

Serious violations

Section 17(k) of the OSH Act defines a serious violation as one in which:

> "there is a substantial probability that death or serious
> physical harm could result from a condition which
> exists, or from one or more practices, means, methods,
> operations, or processes which have been adopted or
> are in use, in such place of employment unless the
> employer did not, and could not with exercise of rea-
> sonable diligence, know of the presence of the viola-
> tion."[81]

To prove that a violation is within the serious category, OSHA must only show a substantial probability that a foreseeable accident would result in serious physical harm or death. Thus, contrary to common belief, OSHA does not need to show that a violation would create a high probability that an accident would result. Because substantial physical harm is the distin-guishing factor between a serious and a nonserious violation, OSHA has

---

[78] Id. at B2b(1).
[79] Id. at (2).
[80] *See Secretary v. Diamond In.*, 4 O.S.H. Cases 1821 (1976); *Secretary v. Northwest Paving*, 2 O.S.H. Cases 3241 (1974); *Secretary v. Sky-Hy Erectors & Equip.*, 4 O.S.H. Cases 1442 (1976). *But see Shaw Constr. v. OSHRC*, 534 F.2d 1183, 4 O.S.H. Cases 1427 (5th Cir. 1976) (holding that serious citation was proper whenever accident was merely possible.)
[81] 29 U.S.C. §666(j).

defined "serious physical harm" as "permanent, prolonged, or temporary impairment of the body in which part of the body is made functionally useless or is substantially reduced in efficiency on or off the job." Additionally, an occupational illness is defined as "illness that could shorten life or significantly reduce physical or mental efficiency by inhibiting the normal function of a part of the body."[82]

After determining that a hazardous condition exists and that employees are exposed or potentially exposed to the hazard, the *OSHA Manual* instructs compliance officers to use a four-step approach to determine whether the violation is serious:

1. Determine the type of accident or health hazard exposure that the violated standard is designed to prevent in relation to the hazardous condition identified
2. Determine the type of injury or illness which it is reasonably predictable could result from the type of accident or health hazard exposure identified in step 1
3. Determine that the type of injury or illness identified in step 2 includes death or a form of serious physical harm
4. Determine that the employer knew or with the exercise of reasonable diligence could have known of the presence of the hazardous condition.[83]

The *OSHA Manual* provides examples of serious injuries, including amputations, fractures, deep cuts involving extensive suturing, disabling burns, and concussions. Examples of serious illnesses include cancer, silicosis, asbestosis, poisoning, and hearing and visual impairment.[84]

Prudent professionals should be aware that OSHA is not required to show that the employer actually knew that the cited condition violated safety or health standards. The employer can be charged with constructive knowledge of the OSHA standards. OSHA also does not have to show that the employer could reasonably foresee that an accident would happen, although it does have the burden of proving that the possibility of an accident was not totally unforeseeable. OSHA does need to prove, however, that the employer knew or should have known of the hazardous condition and that it knew there was a substantial likelihood that serious harm or death would result from an accident.[85] If the Secretary cannot prove that the cited violation

[82] *OSHA Manual supra* note 62, at IV-B-1(b)(3)(a), (c).
[83] Id. at VIII-B1b(2)(c). In determining whether a violation constitutes a serious violation, the compliance officer is functionally describing the *prima facie* case that the Secretary would be required to prove, i.e., (1) the causal link between the violation of the safety or health standard and the hazard, (2) reasonably predictable injury or illness that could result, (3) potential of serious physical harm or death, and (4) the employer's ability to foresee such harm by using reasonable diligence.
[84] Id. at VIII-B1c(3)a.
[85] Id. at (4). *See also, Cam Indus.*, 1 O.S.H. Cases 1564 (1974); *Secretary v. Sun Outdoor Advertising*, 5 O.S.H. Cases 1159 (1977).

meets the criteria for a serious violation, the violation may be cited in one of the lesser categories.

Willful violations

The most severe monetary penalties under the OSHA penalty structure are for willful violations. A willful violation can result in penalties of up to $70,000 per violation, with a minimum required penalty of $5,000. Although the term "willful" is not defined in OSHA regulations, courts generally have defined a willful violation as "an act voluntarily with either an intentional disregard of, or plain indifference to, the Act's requirements."[86] Further, the OSHRC defines a willful violation as "action taken knowledgeably by one subject to the statutory provisions of the OSH Act in disregard of the action's legality. No showing of malicious intent is necessary. A conscious, intentional, deliberate, voluntary decision is properly described as willful."[87]

There is little distinction between civil and criminal willful violations other than the due process requirements for a criminal violation and the fact that a violation of the general duty clause cannot be used as the basis for a criminal willful violation. The distinction is usually based on the factual circumstances and the fact that a criminal willful violation results from a willful violation which caused an employee death.

According to the *OSHA Manual*, the compliance officer "can assume that an employer has knowledge of any OSHA violation condition of which its supervisor has knowledge; he can also presume that, if the compliance officer was able to discover a violative condition, the employer could have discovered the same condition through the exercise of reasonable diligence."[88]

Courts and the OSHRC have agreed on three basic elements of proof that OSHA must show for a willful violation. OSHA must show that the employer (1) *knew or should have known that a violation existed*, (2) *voluntarily chose not to comply with the OSH Act to remove the violative condition, and* (3) *made the choice not to comply with intentional disregard of the OSH Act's requirements or plain indifference to them properly characterized as reckless.*

Although these elements of proof appear fairly straightforward and clear, several unresolved issues continue to be litigated: the supervisor's role in identifying and correcting the hazardous condition, what the employer actually knew regarding the hazardous condition, and the good faith of the employer.

Regarding the role of a first-line supervisor or other member of the management team, an employer may not be responsible for its supervisors' actions if they are contrary to consistently and adequately enforced work

---

[86] *Cedar Constr. Co. v. OSHRC*, 587 F.2d 1303, 6 O.S.H. Cases 2010, 2011 (D.C. Cir. 1971). Moral turpitude or malicious intent are not necessary elements for a willful violation. *U.S. v. Dye Constr.*, 522 F.2d 777, 3 O.S.H. Cases 1337 (4th Cir. 1975); *Empire-Detroit Steel v. OSHRC*, 579 F.2d 378, 6 O.S.H. Cases 1693 (6th Cir. 1978).

[87] *P.A.F. Equip. Co.*, 7 O.S.H. Cases 1209 (1979).

[88] *OSHA Manual supra* note 62, at VIII-B1c(4).

regulations or rules.[89] Conversely, many courts have upheld willful violations based on the supervisor's knowledge of the hazardous condition and
his or her inaction.[90] Additionally, hazards within plain view of the supervisor have been found to be within the "knew or should have known"
category and are potentially willful violations.[91]

Inaction can constitute a willful violation as well as an overt disregard
for OSHA standards. In *Georgia Electric Co. v. Marshall*,[92] the Fifth Circuit
held that "it is precisely because the Company made no effort whatsoever
to make anyone with supervisory authority at the job site aware of the OSHA
regulations that the Company can be said to have acted with plain indifference and thereby acted willfully."[93] Additionally, in *Donovan v. Williams
Enterprises*,[94] the court upheld a willful violation in finding that "employee
safety was never discussed with the company president or any of its supervisory personnel until OSHA inspection of the project began."[95]

Imputed knowledge to the employer has been the basis for willful violation findings by courts and the OSHRC. In *Bergin Corp.*, the OSHRC found
a willful violation because of the employer's poor judgment in hiring a
supervisor. The employer hired a person experienced in excavation work
and instructed him to provide safety instruction to his employees. When a
trench caved in on an employee who was being trained, the OSHRC found
that the employer's reliance on the experienced supervisor and the instruction to provide safety training were not adequate to remove the willful
violation. Courts and the OSHRC have affirmed findings of willful violations
in many other circumstances, ranging from deliberate disregard of known
safety requirements[96] through fall protection equipment not being provided.[97] Other examples of willful violations include cases where safety
equipment was ordered but employees were permitted to continue work
until the equipment arrived,[98] inexperienced and untrained employees were
permitted to perform a hazardous job,[99] and where an employer failed to
correct a situation that had been previously cited as a violation.

Repeat and failure to abate violations

"Repeat" and "failure to abate" violations are often quite similar and confusing to professionals. When, upon reinspection by OSHA, a violation of a
previously cited standard is found but the violation does not involve the

[89] *See e.g., Central Soya De Puerto Rico v. Secretary*, 653 F.2d 38 (1st Cir. 1981).
[90] Id.
[91] *Central Soya*, 653 F.2d 38.
[92] 595 F.2d 309, 320, 7 O.S.H. Case 1343, 1350 (5th Cir. 1979).
[93] Id. at 320, 7 O.S.H. Cases at 1350.
[94] 744 F.2d 170, 11 O.S.H. Cases 2241 (D.C. Cir. 1984).
[95] Id.
[96] *Universal Auto Radiator Mfg. Co. v. Marshall*, 631 F.2d 20, 8 O.S.H. Cases 2026 (3d Cir. 1980).
[97] *Haven Steel Co. v. OSHRC*, 738 F.2d 397, 11 O.S.H. Cases 2057 (10th Cir. 1984).
[98] *Donovan v. Capital City Excavating Co.*, 712 F.2d 1008, 11 O.S.H. Cases 1581 (6th Cir. 1983).
[99] *Ensign-Bickford Co. v. OSHRC*, 717 F.2d 1419, 11 O.S.H. Cases 1657 (D.C. Cir. 1983).

same machinery, equipment, process, or location, this would constitute a repeat violation. If, upon reinspection by OSHA, a violation of a previously cited standard is found but evidence indicates that the violation continued uncorrected since the original inspection, this would constitute a failure to abate violation.[100]

The most costly civil penalty under the OSH Act is for repeat violations. The OSH Act authorizes a penalty of up to $70,000 per violation but permits a maximum penalty of 10 times the maximum authorized for the first instance of the violation. Repeat violations can also be grouped within the willful category (i.e., a willful repeat violation) to acquire maximum civil penalties.

In certain cases where an employer has more than one fixed establishment and citations have been issued, the *OSHA Manual* states,

> "the purpose for considering whether a violation is repeated, citations issued to employers having fixed establishments (e.g., factories, terminals, stores) will be limited to the cited establishment. For employers engaged in businesses having no fixed establishments, repeated violations will be alleged based upon prior violations occurring anywhere within the same Area Office Jurisdiction."[101]

When a previous citation has been contested but a final OSHRC order has not yet been received, a second violation is usually cited as a repeat violation. The *OSHA Manual* instructs the compliance officer to notify the assistant regional director and to indicate on the citation that the violation is contested.[102] If the first citation never becomes a final OSHRC order (i.e., the citation is vacated or otherwise dismissed), the second citation for the repeat violation will be removed automatically.[103]

As noted previously, a failure to abate violation occurs when, upon reinspection, the compliance officer finds that the employer has failed to take necessary corrective action and thus the violation continues uncorrected. The penalty for a failure to abate violation can be up to $7,000 per day to a maximum of $70,000. Prudent professionals should also be aware that citations for repeat violations, failure to abate violations, or willful repeat violations can be issued for violations of the general duty clause. The *OSHA Manual* instructs compliance officers that citations under the general duty clause are restricted to serious violations or to willful or repeat violations that are of a serious nature.[104]

---

[100] *OSHA Manual supra* note 62, at VIII-B5c.
[101] Id. at IV-B5(c)(1).
[102] Id. at VIII-B5d.
[103] Id.
[104] Id. at XI-C5c.

Failure to post violation notices

A new penalty category, the failure to post violation notices, carries a penalty of up to $7,000 for each violation. A failure to post violation notices occurs when an employer fails to post notices required by the OSHA standards, including the OSHA poster, a copy of the year-end summary of the OSHA 200 form, a copy of OSHA citations when received, and copies of other pleadings and notices. OSHA has recently initiated a program whereby the compliance officer will provide a copy of the required poster to the employer and if the employer immediately posts the poster no citation will be issued.

Criminal penalties

In addition to civil penalties, the OSH Act provides for criminal penalties of up to $10,000 and/or imprisonment for up to six months. A repeated willful violation causing an employee death can double the criminal sanction to a maximum of $20,000 and/or one year of imprisonment. Given the increased use of criminal sanctions by OSHA in recent years (however, there have been no criminal sanctions used in workplace violence incidents to date), prudent professionals should advise their employers of the potential for these sanctions being used when the safety and health of employees is disregarded or put on the back burner.

OSHA, as an agency, does not prosecute employers for criminal violations of the OSH Act. In fact, the Secretary of Labor does not even have authority under the OSH Act to impose criminal sanctions. Instead, the Secretary of Labor refers all cases meeting criminal sanction criteria to the U.S. Department of Justice for prosecution. The Justice Department prosecutes an OSH Act case as it would any other criminal action. The same rules of evidence and rights of the accused, although a substantially different burden of proof than in an OSHRC hearing, apply. The statute of limitations for the Justice Department to file criminal charges against an employer for OSHA violations is the same as any other noncapital federal crime, five years. Criminal sanctions cannot be brought by OSHA for violations of the general duty clause. For OSHA to refer a case to the Justice Department for possible criminal prosecution, the employer must be alleged to have willfully violated a *specific* OSHA standard, and the willful act must fall within the criminal categorization and be serious in nature.

OSHA can refer violations to the Justice Department for criminal prosecution under the following circumstances:

1. When anyone inside or outside the Department of Labor or OSHA gives advance notice of an inspection, without authority from the Secretary, a fine of up to $1,000, imprisonment for up to six months, or both may be imposed upon conviction.[105]

---

[105] 29 U.S.C. §666(f).

2. When any employer or other person intentionally falsifies statements or OSHA records that must be prepared, maintained, or submitted under the OSH Act, a fine of up to $10,000, imprisonment for up to six months, or both may be imposed upon conviction.[106]
3. When any person willfully violates an OSHA standard, rule, order, or regulation and the violation causes the death of an employee, a fine of up to $10,000, imprisonment for up to six months, or both may be imposed upon conviction. If convicted for a second violation, a fine of up to $20,000, imprisonment for up to one year, or both may be imposed.[107]
4. When an individual forcibly resists or assaults a compliance officer or other Department of Labor personnel, upon conviction a penalty of $5,000 and/or three years in prison can be imposed.
5. When any person kills a compliance officer or other OSHA or Department of Labor personnel acting in their official capacity, upon conviction a prison sentence for any term of years or life may be imposed.

Criminal liability for a willful OSHA violation can attach to an individual or a corporation. In addition, corporations may be held criminally liable for the actions of their agents or officials.[108] Supervisors, managers, and other corporate officials may also be subject to criminal liability under a theory of aiding and abetting the criminal violation in their official capacity with the corporation.[109]

Prudent professionals should also be aware that an employer could face two prosecutions for the same OSHA violation without the protection of double jeopardy. The OSHRC can bring an action for a civil willful violation using the monetary penalty structure described previously and the case then may be referred to the Justice Department for criminal prosecution of the same violation.[110]

Prosecution of willful criminal violations by the Justice Department has been rare in comparison to the number of inspections performed and violations cited by OSHA on a yearly basis (as can be seen in Figure 4.1). However, the use of criminal sanctions has increased substantially in the last few years. With adverse publicity being generated as a result of workplace accidents and deaths[111] and Congress emphasizing reform, a decrease in criminal prosecutions is unlikely.

The law regarding criminal prosecution of willful OSH Act violations is still emerging. Although few cases have actually gone to trial, in most situations the mere threat of criminal prosecution has encouraged employers to

---

[106] Id. at (g).
[107] Id. at (e).
[108] U.S. v. Crosby & Overton, No. CR-74-1832-F (S.D. Cal. Feb. 24, 1975.)
[109] 18 U.S.C. §2.
[110] These are uncharted waters. Employers may argue due process and double jeopardy, but OSHA may argue that it has authority to impose penalties in both contexts. There are currently no cases on this issue.
[111] Jefferson, *Dying for Work*, A.B.A. J. 46 (Jan. 1993).

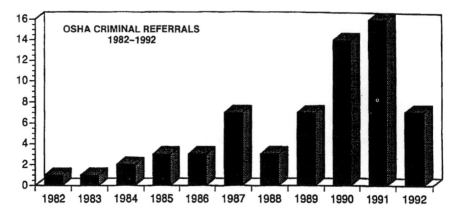

Figure 4.1 OSHA criminal referrals. (From The Bureau of National Affairs, 0095-3237/93, 1993.)

settle cases with assurances that criminal prosecution would be dismissed. Many state plan states are using criminal sanctions permitted under their state OSH regulations more frequently.[112] State prosecutors have also allowed use of state criminal codes for workplace deaths.[113]

Prudent professionals should exercise caution when faced with an on-the-job fatality (**Note:** Criminal sanctions cannot be utilized if cited under the general duty clause.) The potential for criminal sanctions and criminal prosecution is substantial if a willful violation of a specific OSHA standard is directly involved in the death. The OSHA investigation may be conducted from a criminal perspective in order to gather and secure the appropriate evidence to later pursue criminal sanctions.[114] A prudent professional facing a workplace fatality investigation should address the OSHA investigation with legal counsel present and reserve all rights guaranteed under the U.S. Constitution.[115] Obviously, under no circumstances should a health and safety professional condone or attempt to conceal facts or evidence which consists of a cover-up.

Employer's rights

All enforcement functions under the OSH Act rest with OSHA which is under the direction of the Department of Labor. All OSHA compliance officers can, under Section 8 of the Act, inspect any public or private-sector workplace covered by the Act.[116] The compliance officer must present his or

[112] *See*, Levin, *Crimes Against Employees: Substantive Criminal Sanctions Under the Occupational Safety and Health Act*, 14 Am. Crim. L. Rev., 98 (1977).
[113] See Chapter 8.
[114] *See L.A. Law: Prosecuting Workplace Killers*, A.B.A. #J. 48, (Los Angeles prosecutor's "roll out" program could serve as model for OSHA.)
[115] *See infra* §1.1.1
[116] 29 U.S.C. §657.

her credentials to the owner, operator, or agent in charge before proceeding with the inspection tour.[117] The employer and a union or employee representative have the right to accompany the compliance officer during the inspection.[118] After the inspection, a closing conference is held in which the compliance officer and the employer discuss safety and health conditions and possible violations. Most compliance officers cannot issue on-the-spot citations but must first confer with the regional or area director.

When a compliance officer observes a violation in an employer's work place and files this observation on his or her report, the area director then decides whether to issue a citation. The area director normally computes any penalties and sets abatement dates for each violation. The citation is mailed via U.S. Postal Service to the employer as soon as possible after the inspection, but in no event can it be sent more than six months after the alleged violation occurred. All citations must be in writing and must describe with particularity the violation alleged, including the relevant standard and regulation.

The OSH Act enforcement scheme includes both civil and criminal penalties for violations. Violators of specific standards or of the general duty clause may face civil penalties according to the range provided in Table 4.1. Penalties may be assessed only within the range set forth under the Act. The Act currently provides for imprisonment of up to six months for willful violations that cause the death of an employee.[119] OSHA normally reserves use of criminal sanctions for egregious circumstances and normally relies on the monetary fines provided under the civil penalties. The good faith of the employer, the gravity of the violation, the employer's history of compliance, and the size of the employer are all considered in assessing the penalty. The area director can compromise, reduce, or remove a violation.

Once a citation is issued, the employer, any employee, or any authorized union representative has 15 working days to file a notice of contest. If the employer does not contest the violation, abatement date, or proposed penalty, the citation becomes final and is not subject to review by any court or agency. If a timely notice of contest is filed in good faith, the abatement requirement is tolled and a hearing is scheduled. An employer may also file

---

[117] Id. §657(a). *But see Marshall v. Barlow's,* 436 U.S. 307, 6 O.S.H. Cases 1571 (1978). (Fourth Amendment protection requires OSHA compliance officers to obtain a search warrant if entry onto an employer's premises is forbidden by the owner, operator, or agent of the company.)

[118] 29 U.S.C. §657(a). The compliance officer can make notes, diagrams, and take photographs during the inspection tour as long as the equipment or procedure cited is not a trade secret of the company; the compliance officer may also talk with workers while they are on the job if the conversation does not disrupt production or may acquire the employee's names, addresses, and telephone numbers from the employer for later contact outside the workplace; the employer may make any notes, diagrams, or take any photographs necessary during the inspection.

[119] 29 U.S.C. §666(e). These criminal sanctions apply only to violations of specific standards, not to violations of the general duty clause. *See also* Id. §666(f)–(g) (denoting criminal penalties for giving advance notice of inspections and for making false statements or certifications in OSHA safety reports.)

a petition of modification of the abatement period (PMA) if it cannot comply with any abatement that has become a final order. If the Secretary of Labor or an employer contests the petition of modification of abatement, a hearing is held to determine whether any abatement requirement, even if part of an uncontested citation, should be modified.[120]

The Secretary of Labor must immediately forward any notice of contest to the OSHRC. In cases before the OSHRC, the Secretary of Labor is usually referred to as the complainant and has the burden of proving the violation. Conversely, the employer is usually called the respondent. The hearing is presided over by an administrative law judge (ALJ), who renders a decision either affirming, modifying, or vacating the citation, penalty, or abatement date. The ALJ's decision then automatically goes before the OSHRC. The aggrieved party may file a petition for discretionary review of the ALJ's decision, but even without this discretionary review any OSHRC member may call for review of any part or all of the ALJ's decision. If, however, no member of the OSHRC calls for a review within 30 days, the ALJ's decision is final. Through either review route, the OSHRC may reconsider the evidence and issue a new decision. At this point, any person adversely affected by the OSHRC's final order may file, within 60 days of the decision, a petition for review in the U.S. Court of Appeals for the circuit in which the alleged violation had occurred or in the U.S. Court of Appeals for the District of Columbia Circuit.[121]

The inspection, violation and appeal procedures under virtually all state programs are virtually identical to those of OSHA. After exhausting the state's administrative route, an adversely affected employer may file, usually within 60 days of the decision, a petition for review in the state supreme court or the U.S. Court of Appeals for the circuit in which the employer is located. It should also be noted that the employer's discrimination against the employee's union representative was in effect discrimination against the represented employees.[122]

What activities are protected?

The OSH Act prohibits discharging or otherwise discriminating against an employee who has filed a complaint, instituted or testified in any proceeding, or otherwise exercises any right afforded by the Act.[123] The Act also specifically gives employees the right to contact OSHA and request an inspection without retaliation from the employer if the employee believes a violation of a health or safety standard threatens physical harm or creates an imminent

---

[120] *Occupational Safety and Health Law*, (2d. 1983), summarized and reprinted in *Employment Law*, Rothstein, Knapp, and Liebman, Foundation Press (1987) at 509.

[121] Id. at 512.

[122] Id.

[123] 29 C.F.R. §1977.5.

danger.[124] Employees exercising the right to contact OSHA with a complaint can also remain anonymous to the employer and public under the Act.[125]

Employees are also protected against discrimination under the Act when testifying in proceedings under or related to the act, including inspections,[126] employee-contested abatement dates,[127] employee-initiated proceedings for promulgating new standards,[128] employee applications for modifying or revoking variances,[129] employee-based judicial challenges to OSHA standards,[130] or employee appeals from decisions by the OSHRC.[131] An employee "need not himself directly institute the proceedings" but may merely set "into motion activities of others which result in proceedings under or related to the Act."[132]

When testifying in any proceeding related to the Act, employees are protected against discrimination by employers. The protection is extended to proceedings instituted or caused to be instituted by the employee, as well as "any statement given in the course of judicial, quasijudicial, and administrative proceedings, including inspections, investigations, and administrative rule making or adjudicative functions."[133]

The Act also provides protection against discrimination for employees who petition for hearings on variance requests,[134] request inspections,[135] challenge abatement dates,[136] accompany the OSHA inspector during the inspection,[137] participate in and challenge OSHRC decisions[138] and citation contests,[139] and bring actions for injunctive relief against the Secretary of Labor for imminent danger situations.[140]

There are few reported cases of discrimination and, although employee rights appear to be straightforward under the act, determining when the protection attaches to the employee and the situation remains an unresolved area.

In *Dunlop v. Hanover Shoe Farms*,[141] the employer argued that the employee was terminated for just cause before a complaint was filed with OSHA. The court, in rejecting the employer's argument, found that the employee's complaint of unsafe and unhealthful working conditions five days before his termination was the first step in exercising his rights under

---

[124] OSH Act §8(f)(1); 29 U.S.C. §657(f)(1).
[125] *See* generally 29 C.F.R. Part 1977.
[126] OSH Act §8; 29 U.S.C. §657.
[127] OSH Act §10(c); 29 U.S.C. §659(c).
[128] OSH Act §6(b); 29 U.S.C. §655(b).
[129] OSH Act §6(b); 29 U.S.C. §655(d).
[130] OSH Act §6(d); 29 U.S.C. §655(f).
[131] OSH Act §11(a); 29 U.S.C. §660(a).
[132] 29 C.F.R. §1977.10(b).
[133] 29 C.F.R. §1977.11.
[134] OSH Act §6(f); 29 U.S.C. §655(f).
[135] OSH Act §6(f); 29 U.S.C. §657(f). *See also*, 29 C.F.R. §1903.10 and 1903.11.
[136] OSH Act §10(a); 29 U.S.C. §569(a).
[137] OSH Act §8(e); 29 U.S.C. §657(e).
[138] OSH Act §11(a); 29 U.S.C. §660(a).
[139] OSH Act §10(c); 29 U.S.C. §659(c).
[140] OSH Act §13(d); 29 U.S.C. §662(d).
[141] 441 F. Supp. 385 (M.D. Pa. 1976).

the act, and thus the employer discriminated against the employee when it discharged him for exercising those rights.[142]

Waiver of rights

In *Marshall v. N.L. Industries*,[143] the court addressed the issue of when an employee waives discriminatory rights provided under the OSH Act. In this case, the Seventh Circuit Court of Appeals held that an employee's acceptance of an arbitration award did not preclude the Secretary of Labor from bringing an action against the employer based upon the same facts.[144] Specifically, the employee refused to load metal scraps into a melting kettle because the payloader did not have a windshield or enclosed cab to protect him from the molten metal. The employer discharged the employee, and the employee filed a complaint with OSHA and a grievance with his union. An arbitrator awarded the employee reinstatement without back pay, and the employee accepted the award. The lower court found that acceptance of the award constituted a voluntary waiver of the right to statutory relief under the act.[145] The Seventh Circuit reversed, finding that "the OSHA legislation was intended to create a separate and general right of broad social importance existing beyond the parameters of an individual labor agreement and susceptible of full vindication only in a judicial forum."[146]

Filing a complaint against an employer

Specific administrative rules govern the nondiscrimination provisions of the OSH Act. An employee who believes he or she has been discriminated against may file a complaint with the Secretary within 30 days of the alleged violation which the Secretary will then investigate.[147] The purpose of the 30-day limitation is "to allow the Secretary to decline to entertain complaints that have become stale."[148] This relatively short period can be tolled under special circumstances[149] and has no effect on other causes of action. When an employee has filed a complaint, the Secretary must notify him or her as to whether an action will be filed on his or her behalf in federal court. At least one court has ruled that OSHA may bring discrimination action against corporate officers as individuals[150] and against the corporation itself and the officers in their official capacities.[151]

[142] Id.
[143] 618 F.2d 1220 (7th Cir. 1980).
[144] Id.
[145] Id.
[146] Id.
[147] In *Taylor v. Brighton Corp.*, 616 F.2d 256 (6th Cir. 1980).
[148] 29 C.F.R. §1977.15(d)(2).
[149] 29 C.F.R. §1977.15(d)(3).
[150] *Donovan v. RCR Communications*, 12 O.S.H. Cases 1427 (M.D. Fla. 1985)
[151] *Moore v. OSHRC*, 591 F.2d 991 (4th Cir. 1979).

Regarding an employee's right to refuse unsafe or unhealthy work, the Supreme Court, in *Whirlpool Corp. v. Marshall*[152] stated: "circumstances may exist in which the employee justifiably believes that the express statutory arrangement does not sufficiently protect him from death or serious injury. Such circumstances will probably not often occur, but such a circumstance may arise when (1) the employee is ordered by the employer to work under conditions that the employee reasonably believes pose an imminent risk of death or serious bodily injury, and (2) the employee has reason to believe that there is not sufficient time or opportunity either to seek effective redress from the employer or to apprise OSHA of the danger."[153]

In this case, two employees refused to perform routine maintenance tasks that required them to stand on a wire mesh guard approximately 20 feet above the work surface. The mesh screen was designed to catch appliance components that might fall from an overhead conveyor. While performing this activity in the past, several employees had punctured the screen, and one employee died after falling through the mesh guard. The employees refused to perform the task, and the employer reprimanded them. The district court denied relief but the Sixth Circuit reversed the decision.[154] The Supreme Court, in affirming the Sixth Circuit, found the act's provisions were "designed to give employees full protection in most situations from the risk of injury or death resulting from an imminently dangerous condition at the worksite."[155]

Private litigation under the OSH Act

Although there is no common law basis for actions under the OSH Act, OSHA regulations are used in many tort actions, such as negligence and product liability suits, as evidence of the standard of care and conduct to which the party must comply. Additionally, documents generated in the course of business that are required under the OSH Act are usually discoverable under the Freedom of Information Act (FOIA) and can be used as evidence of a deviation from the required standard of care.

According to Section 653(b)(4) of the OSH Act:

> "Nothing in this Act shall be construed to supersede or in any manner affect any workmen's compensation law or to enlarge or diminish or affect in any other manner the common law or statutory rights, duties, or liabilities of employers and employees under any law with respect to injuries, diseases, or death of employees arising out of, or in the course of, employment."[156]

---

[152] 445 U.S. 1, 100 S. Ct. 883, 63 L.Ed.2d 154 (1980).
[153] Id.
[154] Id.
[155] Id.
[156] 29 U.S.C. §653 (b) (4).

This language prevents injured employees or families of employees killed in work-related accidents from directly using the OSH Act or OSHA standards as an independent basis for a cause of action (i.e., wrongful death actions).[157] However, many federal and state courts have found that Section 653(b)(4) does not bar application of the OSH Act or OSHA standards in workers' compensation litigation or application of the doctrine of negligence or negligence *per se* to an OSHA violation.[158] These decisions do distinguish between use of an OSHA standard as the basis for a standard of care in a state or federal common law action and the OSH Act or OSHA standards creating a separate and independent cause of action.

Negligence actions

OSHA standards are most widely used in negligence actions. The plaintiff in a negligence action must prove the four elements: duty, breach of duty, causation, and damages. *Black's Law Dictionary* defines negligence *per se* as:

> "conduct, whether of action or omission, that may without any argument or proof as to the particular surrounding circumstances, either because it is in violation of a statute or valid municipal ordinance, or because it is so palpably opposed to the dictates of common prudence that it can be said without hesitation or doubt that no careful person would have been guilty of it."[159]

In simpler terms, if a plaintiff can show that an OSHA standard applied to the circumstances and the employer violated the OSHA standard, the court can eliminate the plaintiff's burden of proving the negligence elements of duty and breach through a finding of negligence *per se*.

The majority of courts have found that relevant OSHA standards and regulations are admissible as evidence of the standard of care,[160] and thus

---

[157] *Byrd v. Fieldcrest Mills*, 496 F.2d 1323, 1 O.S.H. Cases 1743 (4th Cir. 1974).

[158] *Pratico v. Portland Terminal Co.*, 783 F.2d 255, 12 O.S.H. Cases 1567 (1st. Cir. 1985). ("Our review of the legislative history of OSHA suggests that it is highly unlikely that Congress considered the interaction of OSHA regulations with other common law and statutory schemes other than workers' compensation. The provision is satisfactorily explained as intended to protect workers' compensation acts from competition by a new private right of action and to keep OSHA regulations from having any affect on the operation of the workers' compensation scheme itself."); Frohlick *Crane Serv. v. OSHR Cases*, 521 F.2d 628 (10th Cir. 1975); *Dixon v. International Harvester Co.*, 754 F.2d 573 (5th Cir. 1985); *Radon v. Automatic Fasteners*, 672 F.2d 1231 (5th Cir. 1982); *Melerine v. Avondale Shipyards*, 659 F.2d 706, 10 O.S.H. Cases 1075 (5th Cir. 1981).

[159] *Black's Law Dictionary*, West Publishing Co., Fifth Edition (1983).

[160] Id., *See also, Teal v. E.I. DuPont de Nemours & Co.*, 728 F.2d 799, 11 O.S.H. Cases 1857 (6th Cir. 1984); *Johnson v. Niagara Machine & Works*, 666 F.2d 1223 (8th Cir. 1981); *Knight v. Burns, Kirkley & Williams Construction Co.*, 331 So.2d 651, 4 O.S.H. Cases 1271 (Ala. 1976).

violation of OSHA standards can be used as evidence of an employer's negligence or negligence *per se*. It should be noted, however, that courts have prohibited use of OSHA standards and regulations, and evidence of their violation, if the proposed purpose of the OSHA standards use conflicts with the purposes of the OSH Act,[161] unfairly prejudices a party,[162] or is meant to enlarge a civil cause of action.[163] The Fifth Circuit, reflecting the general application, approves the admissibility of OSHA standards as evidence of negligence but permits the court to accept or reject the evidence as it sees fit.[164]

In using OSHA standards to prove negligence *per se*, professionals should be aware that numerous courts have recognized the OSHA standards as the reasonable standard of conduct in the workplace. With this recognition, a violation by the employer would constitute negligence *per se* to the employee.[165] A few other courts have held, however, that violations of OSHA standards can never constitute negligence *per se* because of Section 653(b)(4) of the Act.[166]

In *Walton v. Potlatch Corp.*,[167] the court set forth four criteria to determine whether OSHA standards and regulations could be used to establish negligence *per se*:

1. The statute or regulation must clearly define the required standard of conduct
2. The standard or regulation must have been intended to prevent the type of harm the defendant's act or omission caused
3. The plaintiff must be a member of the class of persons the statute or regulation was designed to protect
4. The violation must have been the proximate cause of the injury[168]

If the court provides an instruction on negligence *per se* rather than an instruction on simple negligence, the effect is that the jury cannot consider the reasonableness of the employer's conduct. In essence the court has already established a violation that constituted unreasonable conduct on the part of the employer and that the conduct was prohibited or required under a specific OSHA standard. Thus, as a matter of law, the jury will not be permitted to address the reasonableness of the employer's actions.

---

[161] *Cochran v. Intern. Harvester Co.*, 408 F. Supp. 598, 4 O.S.H. Cases 1385 (W.D. Ky. 1975)(OSHA standards not applicable where plaintiff worker was independent contractor); *Trowell v. Brunswick Pulp & Paper Co.*, 522 F. Supp. 782, 10 O.S.H. Cases 1028 (D.S.C. 1981) (Motion in Limine prevented use of OSHA regulations as evidence).

[162] *Spankle v. Bower Ammonia & Chem. Co.*, 824 F.2d 409, 13 O.S.H. Cases 1382 (5th Cir. 1987)(Trial judge did not err in prohibiting OSHA regulations to be admitted which he thought were unfairly prejudicial under Fed. R. Evid. 403.)

[163] *Supra* n. 240.

[164] *Melerine v. Avondale Shipyards, supra* n. 237.

[165] *Supra* n. 237.

[166] *Wendland v. Ridgefield Construction Service*, 184 Conn. 173, 439 A.2d 954 (1981); *Hebel v. Conrail*, 273 N.E.2d 652 (Ind. 1985); *Cowan v. Laughridge Construction Co.*, 57 N.C. App. 321, 291 S.E.2d 287 (1982).

[167] *Walton v. Potlatch Corp.*, 781 P.2d 229, 14 O.S.H. Cases 1189 (Idaho 1989).

[168] 741 P.2d at 232.

OSHA standards as defense

Under appropriate circumstances, an employer may be able to use OSHA standards and regulations as a defense. Simple compliance with required OSHA standards is not in itself a defense, and the use of OSHA standards as a defense has received mixed treatment by the courts. However, at least one court has held that violation of a state OSHA plan by an employee could be considered in determining the employee's comparative negligence in a liability case.[169] Use of OSHA standards and regulations to demonstrate an appropriate standard of care in third-party product liability actions, workers' compensation litigation, and other actions may be permitted and should be explored by prudent professionals in incidents of workplace violence or other appropriate circumstances.

The use of OSHA citations and penalties in tort actions has also received mixed treatment by the courts. In *Industrial Tile, v. Stewart*,[170] the Alabama Supreme Court stated:

> "We hold that it was not error to admit the regulation if the regulations are admissible as going to show a standard of care, then it seems only reasonable that the evidence of violation of the standards would also be admissible as evidence that the defendant failed to meet the standards that it should have followed. Clearly, the fact that Industrial Tile had been cited by OSHA for violating the standards, and the fact that Industrial Tile paid the fine, are relevant to the conduct of whether it violated the standards of care applicable to its conduct. It was evidenced from a number of witnesses that the crane violated the 10-foot standards. It seems to us that evidence that Industrial Tile paid the fine without objection was properly admitted into evidence as a declaration against interest."[171]

Other courts have found that OSHA citations and fines are inadmissible under the hearsay rule of the Federal Rules of Evidence.[172] However, this can normally be overcome easily by offering a certified copy of the citations and penalties to the court, under the investigatory report exception to the Federal Rules of Evidence.[173]

Besides direct litigation with OSHA and negligence actions, OSHA standards used as evidence of the standard of care, and citations used to show

[169] *Zalut v. Andersen & Ass.*, 463 N.W.2d 236 (Mich. Ct. App., 1990).
[170] 388 So. 2d 171 (Ala. 1980).
[171] Id. at *Lowe v. General Motors*, 624 F.2d 1373 (5th Cir. 1980) (applied to National Traffic & Motor Vehicle Safety Act standards.)
[172] Fed. R. Evid., 28 U.S.C.A. §803.
[173] Id.

a breach of the duty of care have also been used in product liability cases,[174] construction site injury actions against general contractors,[175] and toxic tort actions.[176] Other actions where OSHA standards and citations have been found admissible include[177] Federal Tort Claim Act actions,[178] against OSHA in the area of inspections, and actions under the Federal Employers' Liability Act.

In summation, prudent professionals can expect an inspection by OSHA and related federal, state, and local agencies following a workplace violence incident. There is no substitution for a thorough knowledge of the rights, responsibilities, and perimeters of each governmental agency and the potential penalties and use in litigation. OSHA and correlating state plan programs can cite under the general duty clause now for workplace violence, and this author fully expects a standard addressing this growing area in the near future.

---

[174] *Spangler v. Kranco*, 481 F.2d 373 (4th Cir. 1973); *Bunn v. Caterpillar Tractor Co.*, 415 F. Supp. 286 (W.D. Pa. 1976); *Scott v. Dreis & Krump Mfg. Co.*, 26 Ill. App. 3d 971, 326 N.E.2d 74 (1975); *Bell v. Buddies Super-Market*, 516 S.W.2d 447 (Tex. Civ. App. 1974); *Brogley v. Chambersburg Engineering Co.*, 452 A.2d 743 (Pa. Super. Ct. 1982) (**Note:** OSHA standards are usually used as evidence of acceptable standards of machine design, industrial standard of care, or of reasonable conduct by employer or industry.)

[175] *Secretary v. Grossman Steel & Aluminum Corp.* "The general contractor normally has responsibility to assure that the other contractors fulfill their obligations with respect to employee safety which affects the entire site. The general contractor is well situated to obtain abatement of hazards, either through its own resources or through its supervisory role with respect to other contractors. It is therefore reasonable to expect the general contractor to assure compliance with the standards insofar as all employees on the site are affected. Thus, we will hold the general contractor responsible for violations it could reasonably have been expected to prevent or abate by reason of its broad supervisory capacity." *Secretary v. Grossman Steel & Aluminum Corp.*, 4 O.S.H. Cases 1185 (1976).

[176] *See, e.g., Hebel v. Conrail*, 475 N.E.2d 652 (Ind. 1985); *Sprankle v. Bower Ammonia & Chemical Co.*, 824 F.2d 409, 13 O.S.H. Cases 1382 (5th Cir. 1987). (**Note:** Toxic tort cases can utilize various theories ranging from failure to warn under a strict liability or negligence theory to wanton misconduct.)

[177] *See, e.g., Blessing v. U.S.*, 447 F. Supp. 1160 (E.D. Pa. 1978) (Allegations of negligent OSHA inspections state a viable Federal Tort Claim Act claim under Pennsylvania law.); *Mandel v. U.S.*, 793 F.2d 964 (8th Cir. 1986).

[178] 20 U.S.C. §2671 *et seq. See also, Blessing.*

chapter five

Workers' compensation liability

> *"When you come right down to it, almost any problem
> eventually becomes a financial problem."*

Frederic G. Donner

> *"Beware of little expenses; a small leak will sink a great
> ship."*

Benjamin Franklin

When an employee is injured as a result of a workplace violence incident, the primary area of recompense for the employee and a major area of legal protection for the employer is the individual state's workers' compensation laws. In most states, employees will receive a structured amount for the physical and mental injuries incurred in the workplace violence incident which the employer would be required to pay through its workers' compensation insurance plan. The employer, however, would be protected against additional legal actions. Workers' compensation costs in these circumstances can be very high in terms of dollars. However, these fixed costs are usually low in comparison to a potential jury verdict. It should be noted that in some states workers' compensation is provided for injured employees and a second level of recovery through the courts may also be permitted.

Overview

Most state workers' compensation systems are fundamentally a no-fault mechanism through which the employee who incurs work-related injuries and illnesses is compensated with monetary and medical benefits. Either party's potential negligence is not an issue so long as an employer/employee relationship exists. In essence, workers' compensation is a compromise in that employees are guaranteed a percentage of their wages (usually 2/3 of their wages with a minimum and maximum amount) and full payment for their medical costs when injured on the job. Employers are guaranteed a reduced monetary cost for these injuries or illnesses and are usually provided protection from additional or future legal action by the employee based on the injury.

The typical workers' compensation system has the following features:

1. Every state in the United States has a workers' compensation system. There may be variations in the amounts of benefits, the rules, administration, etc. from state to state. In most states, workers' compensation is the exclusive remedy for on-the-job injuries and illnesses.
2. Coverage for workers' compensation is limited to *employees* who are injured *on the job*. The specific locations as to what constitutes the work premises and "on the job" may vary from state to state.
3. Negligence or fault by either party is largely inconsequential. No matter whether the employer is at fault or the employee is negligent, the injured employee generally receives workers' compensation coverage for any injury or illness incurred on the job.
4. Workers' compensation coverage is automatic, i.e., employees are not required to sign up for worker's compensation coverage. By law, employers are required to obtain and carry workers' compensation insurance or be self insured.
5. Employee injuries or illnesses that "arise out of 'and/or' are in the course of employment" are considered compensable. These definition phrases have expanded this beyond the four corners of the workplace to include work-related injuries and illnesses incurred on the highways, at various in- and out-of-town locations, and other such remote locales. These two concepts — "arising out of" the employment and "in the course of" the employment — are the basic burdens of proof for the injured employee. Most states require both. Prudent professionals are strongly advised to review the case law in his/her state to see the expansive scope of these areas. That is, the injury or illnesses must "arise out of," i.e., there must be a causal connection between the work and the injury, or illness must be "in the course of" the employment; this relates to the time, place, and circumstances of the accident in relation to the employment. The key issue is a *work connection* between the employment and the injury/illness.
6. Most workers' compensation systems include wage-loss benefits (sometimes known as time-loss benefits), which are usually between 50 and 75% of the employee's average weekly wage. These benefits are normally tax free and are commonly called temporary total disability (TTD) benefits.
7. Most workers' compensation systems require payment of all medical expenses, including such expenses as hospital expenses, rehabilitation expenses, and prostheses expenses.
8. In situations where an employee is killed, such as workplace violence situations, workers' compensation benefits for burial expenses and future wage-loss benefits are usually paid to the dependents.

9. When an employee incurs an injury or illness that is considered permanent in nature, most workers' compensation systems provide a dollar value for the percentage of loss to the injured employee. This is normally known as permanent partial disability (PPD) or permanent total disability (PTD).

10. In accepting workers' compensation benefits, the injured employee is normally required to waive any common law action to sue the employer for damages from the injury or illness.

11. If the employee is injured by a third party, the employer usually is required to provide workers' compensation coverage but can be reimbursed for these costs from any settlement that the injured employee receives through legal action or other methods.

12. Administration of the workers' compensation system in each state is normally assigned to a commission or board. The commission/board generally oversees an administrative agency located within state government which manages the workers' compensation program within the state.

13. The Workers' Compensation Act in each state is a statutory enactment which can be amended by the state legislatures. Budgetary requirements are normally authorized and approved by the legislatures in each state.

14. The workers' compensation commission/board in each state normally develops administrative rules and regulations (i.e., rules of procedure, evidence, etc.) for the administration of workers' compensation claims in the state.

15. In most states, employers with one or more employees are normally required to carry workers' compensation coverage. Employers are generally allowed several avenues through which to acquire this coverage. Employers can elect to acquire workers' compensation coverage from private insurance companies, from state-funded insurance programs, or become "self-insured" (i.e., after posting bond, the employer pays all costs directly from its coffers).

16. Most state workers' compensation provides a relatively long statute of limitations. For *injury* claims, most states grant between 1 and 10 years in which to file the claim for benefits. For work-related *illnesses*, the statute of limitations may be as high as 20 to 30 years from the time the employee first noticed the illness or the illness was diagnosed. An employee who incurred a work-related injury or illness is normally not required to be employed with the employer when the claim for benefits is filed.

17. Workers' compensation benefits are generally separate from the employment status of the injured employee. Injured employees may continue to maintain workers' compensation benefits even if the employment relationship is terminated, the employee is laid off, or other significant changes are made in the employment status.

18. Most state workers' compensation systems have some type of administrative hearing procedures. Most workers' compensation acts have designed a system of administrative "judges" (normally known as administrative law judges or ALJs) to hear any disputes involving workers' compensation issues. Appeals from the decision of the administrative law judges is normally to the workers' compensation commission/board. Some states permit appeals to the state court system after all administrative appeals have been exhausted.

Prudent professionals should be aware that the workers' compensation system in every state is administrative in nature. Thus there is a substantial amount of paperwork that must be completed in order for benefits to be paid in a timely manner. In most states, specific forms have been developed.

The most important form to initiate workers' compensation coverage in most states is the first report of injury/illness form. This form may be called a "First Report" form, an application for adjustment of claim, or it may have some other name or acronym like the SF-1 or Form 100. This form, often divided into three parts in order that information can be provided by the employer, employee, and attending physician, is often the catalyst that starts the workers' compensation system reaction. If this form is absent or misplaced, there is no reaction in the system and no benefits are provided to the injured employee.

Under most workers' compensation systems, there are many forms that need to be completed in an accurate and timely manner. Normally, specific forms must be completed if an employee is to be off work or is returning to work. These include: forms for the transfer from one physician to another, forms for independent medical examinations, forms for the payment of medical benefits, and forms for the payment of permanent partial or permanent total disability benefits. Prudent professionals responsible for workers' compensation are advised to acquire a working knowledge of the appropriate legal forms used in their state's workers' compensation program.

In most states, information regarding the rules, regulations, and forms can be acquired directly from the state workers' compensation commission/board. Other sources for this information include your insurance carrier, self-insured administrator, or state-fund administrator.

Prudent professionals should be aware that workers' compensation claims have a "long tail," i.e., they stretch over a long period of time, especially in incidents of workplace violence. Under the OSHA recordkeeping system, every year injuries and illnesses are totaled on the OSHA Form 200 log and a new year begins. This is not the case with workers' compensation. Once an employee sustains a work-related injury or illness, the employer is responsible for the management and costs until such time as the injury or illness reaches maximum medical recovery or the time limitations are exhausted. When an injury reaches maximum medical recovery, the employer may be responsible for payment of permanent partial or permanent total disability benefits prior to closure of the claim. Additionally, in some states, the medical benefits can remain open indefinitely and cannot be settled or closed with the claim. In many circumstances, the workers' compensation claim for a work-related injury or illness may remain open for several years and thus require continued management and administration for the duration of the claim process.

Some states allow the employer to take the deposition of the employee claiming benefits, while others strictly prohibit it. Some states have a schedule of benefits and have permanent disability awards strictly on a percentage of disability from that schedule. Others require that a medical provider outline the percentage of *functional* impairment due to the injury/illness (based on the Americans Medical Association [AMA] Guidelines), then using this information, as well as the employee's age, education, and work history, the ALT determines the amount of *occupational* impairment upon which permanent disability benefits are awarded. Still other states have variations on those systems.

In summary, professionals who are responsible for the management of a workers' compensation program should become knowledgeable in the rules, regulations, and procedures under their individual state's workers' compensation system. Professionals who have facilities or operations in several states should be aware that, although the general concepts may be the same, each state's workers' compensation program has specific rules, regulations, schedules, and procedures that may vary greatly between states. There is no substitute for knowing the rules and regulations under your state's workers' compensation system, especially when you are "under the gun" after a workplace violence incident.

Administrative hearing process

Within the framework of most workers' compensation systems, arbitration systems of varying types usually have been established to decide disputes in an informal and cost-effective manner. In most systems, the initial level of adjudication is a hearing before an administrative law judge, followed by an appeal stage before an appellate panel. Appeals from the appellate panel are normally to the commission/board. In some states, the final appeal stage lies with the commission or board, while in other states appeals to the state court system are allowed.

Following a workplace violence incident, representatives of the employer are normally involved during the initial hearing phase before the administrative law judge. In some organizations, the safety, workers' compensation or human resource professional may be responsible for the presentation of evidence at the hearing, while in other organizations legal counsel presents the case.

Workers' compensation hearings before an administrative law judge are often informal in comparison to a court of law. These hearings are often held in conference rooms in government buildings or even in hotel conference rooms. Most administrative law judges are granted wide discretion as to courtroom procedure, rules of evidence, and other procedural aspects of the hearing. Professionals should be prepared for the administrative law judge to be actively involved in the hearing and to ask questions of the parties and witnesses.

Written documentation, diagrams, photographs, videotapes, and other evidence gathered following the workplace violence incident are normally presented to the administrative law judge for review and acceptance into evidence. Thus, it is vitally important that this evidence is properly gathered following the workplace violence incident.

In most states, the administrative law judge will not render an immediate decision in the case. The administrative law judge will conclude the hearing at the end of closing statements and provide a written decision to the parties via mail. Appeals from the written decision normally must be filed within a relatively short period from the receipt of the written decision (commonly 30 days).

Preparation is the key to success in a workers' compensation administrative hearing. Presentations should be concise and to the point, information and evidence should always be at your fingertips for immediate location, and, above all, always be professional during the hearing.

Under most workers' compensation systems, negligence by the employee is inconsequential. Thus, employers will want to assume liability for workplace injuries or fatalities following a workplace violence incident under workers' compensation. Although the costs can be high under workers' compensation, these costs are certain and can be budgeted and the potential of a multimillion dollar verdict against the company is minimized.

Conversely, injured employees or the estate of individuals killed in a workplace violence incident may attempt to avoid workers' compensation coverage. The primary theories to avoid compensability in most states are as follows:

1. The injury or illness was not work-related, i.e., the injury or illness did not arise out of or in the course of employment.
2. The employee incurring the injury or illness is excluded from coverage under the workers' compensation act. This exclusion may be voluntary, i.e., opted out of coverage at an earlier date, or involuntary through the provisions of the specific act.
3. The injured individual is not an employee within the definition of the act. The individual may be an independent contractor or subcontractor.
4. The employee may have been injured while being lent to another employer.
5. The employee may be a dual employee working for two or more employers.
6. The employer did not have workers' compensation coverage.

If the injured employee or estate of a deceased employee can avoid workers' compensation coverage, then the injured employee or estate is free to sue in civil court. If successful, the damages to the employee or estate can be substantially higher than those benefits provided under most state workers' compensation systems.

However, in most states, the rules of evidence are relaxed in workers' compensation adjudications. Virtually any information related to the accident, the injury or illness, the employment status, or other related information may be heard by the administrative law judge. Hearsay evidence is admissible and can often be used to support a position.[1] This is usually within the ALT's discretion.

Although the cost of workers' compensation is a concern for most employers, this cost can be low in comparison with the potential of a multimillion dollar jury award. Additionally, following the initial trauma of a workplace violence incident, the injured employees or the families of killed employees will require financial support. Most workers' compensation systems provide for death benefits as well as burial costs. These costs are fixed and paid promptly. Thus, it is important that the necessary paperwork be initiated as soon as feasible following a workplace violence incident to ensure prompt and proper payment of workers' compensation benefits to these employees or their families.

For the employer, prompt initiation of workers' compensation benefits can prevent other types of litigation. As noted above, workers' compensation is usually the sole remedy for injured or killed employees, and thus prompt initiation can prevent the generation of other types of direct litigation.

---

[1]   *Greenfarb v. Arre*, 62 N.J. Super. 420, 163 A.2d 173 (1960). Hearsay evidence was admissible and was capable alone of supporting an award of compensation for a 60-year-old employee with a heart condition who died while lifting a 300-pound weight. *But see Carroll v. Knickerbocker Ice Co.*, 218 N.Y. 435, 113 N.E. 507 (1916). Hearsay evidence was admissible but alone could not support an award. A residuum of legal evidence was required.

(Modified for the purposes of this text)

PIZZA HUT OF AMERICA, INC.;
Orson Thomas; and Ronald
Pulda, Petitioners,
v.
Ray KEEFE and Paula
Keefe, Respondents.
No. 93SC251.

Supreme Court of Colorado,
En Banc.

June 30, 1995.
Rehearing Denied Aug. 21, 1995.

Parents brought action for wrongful death of their child from prenatal injuries, allegedly caused by mother's employment, against mother's employer and two co-workers. The District Court, Denver County, Larry J. Naves, J., entered summary judgment for employer and co-workers, and parents appealed. The Court of Appeals, 868 P.2d 1092, reversed and remanded. Certiorari was granted. The Supreme Court, Scott, J., held that exclusivity provision of Workers' Compensation Act did not bar action.

Court of Appeals affirmed.

Rovira, C.J., dissented and filed opinion in which Erickson and Vollack, JJ., joined.

Fortune Law Firm, P.C., Lowell Fortune. Denver, for petitioners.

James D. King & Associates, P.C., James F. Scherer, Denver, for respondents.

Turner and Meiklejohn, P.C., Scott A. Meiklejohn, Denver, for amicus curiae Colorado Trial Lawyers Ass'n.

Justice SCOTT delivered the Opinion of the Court.

The respondents Ray and Paula Keefe brought this action against Pizza Hut of America, Inc., Orson Thomas and Ronald Pulda (collectively "petitioners" or "Pizza Hut"), seeking damages for the wrongful death of their child, Shanae Keefe. The trial court ruled that the damage claims were barred by the exclusive remedy provision of the Colorado Workers' Compensation Act, section 8-41-102, 3B C.R.S. (1994 Supp.) (the "Act"). The court of appeals held that the exclusive remedy provision did not apply to the wrongful death claim because the wrongful death of the child was not "for and on account of" the personal injury of the employee and the judgment was reversed and cause remanded with directions.

[1] We granted certiorari to determine whether the exclusive remedy provision of the Act bars a tort claim against an employer for prenatal injury occurring in the workplace.[1] Because we find that a non-employee child who suffered prenatal injuries as the result of the negligence of the mother's employer is not limited to remedies available under our workers' compensation law, we affirm the judgment of the court of appeals.

I.

Paula Keefe was employed by Pizza Hut from August 1990 to March 1991 as an assistant manager. Orson Thomas and Ronald Pulda were Paula Keefe's supervisors. Paula Keefe learned she was pregnant in August 1990 and in October 1990 she began suffering medical complications related to her pregnancy. In response to these compli-cations, her treating physician imposed a series of work restrictions, limiting her hours of work and the types of tasks she was allowed to perform. Despite those and other medical precautions, her daughter Shanae Keefe, was born three months prematurely and died ten days later of medical complications arising from her premature birth. The Keefe's claim that Pizza Hut coerced Paula to work hours and perform tasks in violation of her medical work restrictions, resulting in Shanae's prema-ture birth and subsequent death.

On April 18, 1991, respondents brought a wrongful death action against Pizza Hut, Orson Thomas and Ronald

---

[1] Our order granting certiorari set forth the following question for review: "Does the exclusive remedy provision of the Colorado Workers' Compensation Act bar a tort claim against an employer for prenatal injury occurring in the workplace?"

Pulda in the District Court for the City and County of Denver under Section 13-21-202, 6A C.R.S. (1987).[2] In their complaint respondents alleged that Pizza Hut coerced Paula to perform her normal work despite knowing about her medical restrictions. Respondents claimed damages for the premature birth and subsequent death of their child. In addition, respondents sought personal damages for emotional distress under an outrageous conduct theory. The complaint alleged that the mother sustained "bodily injury of a severe and permanent nature" because of the employer's wrongful conduct.

Pizza Hut filed a motion for summary judgment claiming that the Keefes' wrongful death action was barred by the exclusivity provisions of the Act. The trial court awarded summary judgment to Pizza Hut on all counts, concluding that Pizza Hut was immune from liability under the Act.

The court of appeals reversed the judgment of the trial court in *Keefe v. Pizza Hut of America, Inc.* 868 P.2d 1092 (Colo.App.1993) (not selected for publication), and remanded the case to the trial court with directions. The court of appeals found that the workers' compensation statute by its terms did not operate to bar a tort claim against an employer for the wrongful death of an employee's child. The court of appeals reasoned that "Section 8-41-102 would not bar a claim against an employer by an employee's child for injuries sustained while visiting the employee at the workplace because there would be no injury to the employee." *Id.* 868 P.2d at 1094. For similar reasons, the court of appeals found that the exclusive remedy provision would not bar an employee's claim for the death of a child visiting the workplace, reasoning that a wrongful death claim is not "for and on account of" the personal injury or death of the employee, but rather is for and on account of the child's death. *Id.* at 1094. Finding that the wrongful death claim derived from the injuries and death of a

---

[2] Section 13-21-202 provides in relevant part: Action notwithstanding death. When the death of a person is caused by a wrongful act, neglect, or default of another, and the act, neglect, or default is such as would, if death had not ensued, have entitled the party injured to maintain an action and recover damages in respect thereof, then, and in every such case, the person who or the corporation which would have been liable, if death had not ensued, shall be liable in an action for damages notwithstanding the death of the party injured.

non-employee, the Keefes' child, and not an injury to Pamela Keefe, the employee the court of appeals concluded that the claim was not barred by the exclusive remedy provision. *Id.*

Because the child died after birth, leaving the parents with a wrongful death claim separate and distinct from any claim a parent may have for personal injuries, we affirm the judgment of the court of appeals.

II.

It is well-settled in Colorado that an injured worker's exclusive remedy for injuries that arise out of or in the course of employment and are proximately caused by the employment is recovery under the workers' compensation statute, Section 8-41-102, 3B C.R.S. (1994 Supp.), which reads as follows:

> Liability of employer complying. An employer who has complied with the provisions of [the Colorado Worker's Compensation Act], including the provisions relating to insurance, shall not be subject to the provisions of Section 8-41-101 [abrogating defenses of assumption of the risk and negligence of employee of fellow servant]; nor shall such employer or the insurance carrier, if any, insuring the employer's liability under said articles be subject to any other liability for the death of or personal injury to any employee, except as provided in said articles; and all causes of action, actions at law, suits in equity, proceedings, and statutory and common law rights and remedies *for and on account of such death or personal injury to any such employee* and accruing to any person are abolished except as provided in said articles.

(Emphasis added.) The exclusivity provision effectively abolishes all claims accruing to any person on account of injury to an employee. *See Triad Painting Co. v Blair,* 812 P.2d 638 (Colo.1991); *Continental Sales Corp. v.*

Stookesberry, 170 Colo. 16, 459 P.2d 566 (1969). Where the statutory bar applies, it constitutes a complete defense to civil tort liability of an employer.

[2,3] Under the exclusive remedy provision of the Act, certain injuries or damages sustained by non-employees are barred if they "derive from" the injury to the employee. This principle, known as the derivative-injury doctrine, is based upon the language in the statute barring claims "for and on account of" death of or personal injury to an employee and "accruing to any person." Under the derivative-injury doctrine, a non-employee's claims may be barred even though the workers' compensation law provides no substitute remedy to the injured non-employee, as it does to the injured employee. *See Bell v. Macy's California,* 212 Cal.App.3d 1442, 261 Cal.Rptr. 447, 455 n. 7 (1989).

[4-6] We have previously applied the derivative-injury doctrine to bar recovery by certain non-employees. For example, a wrongful death action brought against an employer by an employee's heirs, based upon the death of an employee which occurred in the course and scope of the employee's employment, is barred by the statute, since such an action is for and on account of the death of an employee. *See Ryan v. Centennial Race Track, Inc.,* 196 Colo. 30, 35, 580 P.2d 794, 797 (1978). Similarly, the exclusive remedy provision of the Act bars contribution and indemnity claims against an employer by third parties who are liable to an injured employee, as these claims also arise out of the death of or personal injury to an employee. *See Williams v. White Mountain Constr. Co. Inc.,* 749 P.2d 423, 428 (Colo. 1988); *Hilzer v. MacDonald,* 169 Colo. 230, 237, 454 P.2d 928, 931-32 (1969). The derivative-injury doctrine also precludes civil actions by an employee's spouse against an employer for loss of consortium arising out of personal injuries suffered by the employee in the course of his or her employment. The rationale of such preclusion is that the spouse's rights in such cases are strictly derivative of, and arise out of, the personal injury suffered by the employee. *See Alexander v. Morrison-Knudsen Co.,* 166 Colo. 118, 124, 444 P.2d 397, 400 (1968), *cert. denied,* 393 U.S. 1063, 80 S.Ct. 715, 21 L.Ed.2d 706 (1969); *Rodriquez v. Nurseries, Inc.,* 815 P.2d 1006 (Colo.App.1991).

[7] The exclusive remedy statute by its terms does not apply, however, to the civil tort liability of employers for negligence or wrongful acts resulting in injury or death to persons not employed when the non-employee's claims do not derive from the injury to the employee. Thus, we must determine whether the injuries and the subsequent death of Shanae Keefe derived from an injury to her mother.

[8,9] Colorado, like other jurisdictions, recognizes a child's right to bring a cause of action for prenatal injuries. *See Empire Casualty Co. v. St. Paul Fire and Marine Ins. Co.,* 764 P.2d 1191, 1196 (Colo.1988); *see generally* Restatement (Second) of Torts § 869 (1979); William L. Prosser, *Torts* § 55 (5th ed. 1984), Roland F. Chase, Annotation, *Liability for Prenatal Injuries,* 40 A.L.R.3d 1222 (1971). If a child dies after birth as a result of prenatal injuries, a surviving parent may bring a wrongful death claim derived from the child's injuries. *See Callahan v. Slavsky,* 153 Colo. 291, 384 P.2d 674 (1963); Prosser § 55. Thus, for purposes of our analysis, it makes no difference that Shanae Keefe's injuries were sustained before her birth.[3]

[10] We also find that there is no difference, for the purpose of determining the applicability of the exclusive remedy statute, between an action for prenatal injury to an employee's child, and a wrongful death action for prenatal injury to an employee's child which results in the child's death. Therefore, the issue under consideration here is whether the Keefes' daughter would have had a right of action against Pizza Hut for her injuries had she survived those injuries.

[11] The petitioners contend that since the mother also claimed an injury in this case, it follows that immunity is grounded upon the employee mother's injury in the workplace. We conclude, however, that regardless of whether the mother was injured, the injury to the child was separate and distinct and subjects the employer

---

[3] Pizza Hut contends that a fetus *in utero* is inseparable from its mother and any injury to the child therefore can only occur as the result of some injury to the mother. The facts of this case do not require us, however, to answer today the difficult question of whether a fetus is a separate and distinct person from the mother, since in this case, the baby was in fact born and hence was at the time of her death a separate person.

to separate liability. In this case, the child's right of action arises out of and on account of her own personal injuries, and not any personal injury suffered by the mother. The mother and child happened to be injured at the same time — the fact that the mother may have been injured, however, is not a bar to tort recovery for the child or a basis for limiting the child's recovery to the workers' compensation law. The exclusivity provisions do not constitute a bar to a claim asserted by a third-party victim, even though both the employee and the victim were injured together as a result of the same negligent act in a single transaction.

[12, 13] Legally, the child, when born, stands in the same position as any other nonemployee member of the public. Civil actions for recovery of damages for personal injuries to non-employees whose injuries are not derivative of an employee's injuries, including non-employee children of employees, are not affected by the language of Section 8-41-102. For example, Section 8-41-102 would not bar a claim against an employer by an employee's child or any other non-employee for injuries sustained while visiting the employee at the workplace, because there would be no injury to the employee. *See Thompson v. Pizza Hut of America, Inc.*, 767 F.Supp. 916, 918 (N.D.Ill.1991); *Bell*, 261 Cal.Rptr. at 453 (1989); *Cushing v. Time Saver Stores, Inc.* 552 So.2d 730, 732 (La. App. 1989), *Cert. denied*, 556 So.2d 1281 (La.1990). It follows that Section 8-41-102 does not bar a claim by a non-employee child who sustained prenatal injuries at the workplace. While we agree with the petitioners that the underlying policy of the exclusivity provisions of the Act is to provide a no-fault system of compensation which limits the employer's overall liability, we note that the Act does not alter the employer's liability to non-employees so injured as the result of the employer's negligence. Accordingly, we conclude that the exclusionary language of Section 8-41-102 does not apply to Shanae Keefe.[4]

III.

[14] The Keefes' wrongful death claim was brought pursuant to the provisions of Section 13-21-202, 6A C.R.S. (1987). Section 13-21-201 transfers the cause of action created by Section 13-21-202 to the decedent's heirs, who

in this case are the decedent's parents. The cause of action created by this statute arises out of torturous acts which injured the decedent and resulted in the decedent's death; the survivors' right of action is derivative of and dependent upon the right of action which the decedent would have had, had she survived her injuries. The derivative nature of the wrongful death action here places it outside the parameters of the exclusionary language contained in Section 8-41-201, because the claim against the employer is not derivative of any personal injury to the employee mother herself, but rather is derived from, and based upon, an injury to the employee's child. Thus, the Keefes' are not barred from bringing suit against Pizza Hut on their daughter's behalf. Accordingly, we affirm the judgment of the court of appeals.

ROVIRA, C.J., dissents, and ERICKSON and VOLLACK, J.J., join in the dissent.

Chief Justice ROVIRA, dissenting:

[4] Our holding is consistent with a majority of other jurisdictions which have addressed this issue. *See, e.g., Thompson*, 767 F.Supp. at 918-19 ("to have status to bring the cause of action it makes no difference whether or not the fetus is viable at the time the injury occurs"); *Namislo v. Akzo Chemicals, Inc.*, 620 So.2d 573, 575 (Ala.1993) (exclusivity provisions of Workers' Compensation Act did not bar personal injury action of employee's daughter against employer for injuries daughter claimed to have sustained *in utero* as a result of employer's negligence); *Cushing*, 552 So.2d at 732 (a child who sustains injuries while *in utero* "is entitled to assert a cause of action in tort against his mother's employer in the same way that a child already born, who was injured on the mother's job site, could assert such a claim"); *Womack v. Buchhorn*, 384 Mich. 718, 187 N.W.2d 218, 222 (1971) (common law negligence action can be brought on behalf of a surviving child negligently injured during the fourth month of pregnancy); *Witty v. American General Capital Distributors, Inc.*, 697 S.W.2d 636, 641 (Tex.App.1985) (Texas Worker's Compensation Act does not bar a claim asserted by a third-party victim, including an unborn fetus, even though both the employee and the victim were injured together as a result of the same negligent act of the employer), *rev'd on other grounds*, 727 S.W.2d 503 (Tex.1987). *But see Bell*, 261 Cal.Rptr. at 454 (fetus *in utero* is inseparable from its mother, and any injury to it can only occur as a result of some condition affecting its mother). Although the workers' compensation exclusive remedy statutes from other jurisdictions are not always worded in precisely the same way as the Colorado statute, the analysis applied in those cases typically does not turn on the exact language of the statute, but rather depends on the scope of the derivative-injury doctrine. The appellate decisions of other states are therefore helpful.

The majority holds that the injuries and subsequent death of Shanae Keefe did not derive from an injury to her mother. Its holding is based on premise that "regardless of whether the mother was injured, the injury to the child was separate and distinct from the mother's injuries and subjects the employer to separate liability." Maj. op. at 101. Because I believe the majority improperly dissociates Shanae's injuries from their source, I respectfully dissent.

While I agree that a child has a cause of action for *in utero* injuries, I disagree with the majority's conclusion that Shanae Keefes' injuries sustained while *in utero* can be divorced from the injuries sustained by her mother which provide the entire basis for this action. Though the majority recognizes the well settled derivative-injury principles in the context of workers' compensation, it fails to properly incorporate them into its analysis.

As pertinent here the Workers' Compensation Act of Colorado (Act) provides coverage for:

> *[A]ll* causes of action, actions at law, suits in equity, proceedings, and statutory and common law rights and remedies *for and on account of* such death or personal injury to any such employee and accruing to *any person....* § 8-41-102, 3B C.R.S. (1994 Supp.)

(Emphasis added.)

The plain language of the statute indicates its exclusivity for all causes of action that accrue on account of an injury to an employee. *E.g.; Williams v. White Mountain Const. Co.,* 749 P.2d 423, 425 (Colo.1988) (the Act is meant to be the exclusive remedy for all work-related injuries).

Under the derivative-injury doctrine, no cause of action exists for any party who sustains injury as a direct result of an employee's work-related injury. *See, e.g., Alexander v. Morrison-Knudsen,* 166 Colo. 118, 444 P.2d 397 (1968) (when the claim of an employee's dependent derives from the injury to the employee, the injuries are covered by the Act), *cert. denied,* 393 U.S. 1063,

80 S.Ct. 715, 21 L.Ed.2d 706 (1969). For instance, there is no cause for action for surviving heirs of an employee killed during the course and scope of employment, even though the heirs sustain their own distinct injuries. *Ryan v. Centennial Race Track, Inc.*, 196 Colo. 30, 35, 580 P.2d 794, 796-97 (1978). Similarly, no cause of action exists for loss of consortium based on a work-related injury, even though loss of consortium is an injury recognized apart from the injured spouse. *Alexander*, 166 Colo. at 124, 444 P.2d at 400. These injuries, though "separate and distinct" from the injury to the worker, are nevertheless covered by the Act because they derive from or are "on account of such death or personal injury to any such employee." § 8-41-102, 3B C.R.S. (1994 Supp.). Thus, the question in a derivative-injury action is not whether the third party sustained a "separate and distinct" injury, but rather, whether the third party's injury derived from a work-related source.

Throughout these proceedings Paula Keefe has claimed that she was coerced into performing work in excess of that permitted by her doctor, and that as a result she sustained bodily injury of a severe and permanent nature. She and her husband also allege that as a result of the defendant's conduct the child was born prematurely and died. The majority concludes that "it makes no [analytical] difference that Shanae Keefe's injuries were sustained before her birth." Maj. op. at 101. Though I agree that viability is not determinative of whether a cause of action exists, the fetal status is relevant to whether the injuries are derivative.[5] Injuries sustained *in utero* are by definition injuries sustained while the fetus is in the womb. *Stedman's Medical Dictionary*, at 798 (25th ed. 1990). While *in utero* the fetus is inseparable from its mother. *See Bell v. Macy's California*, 212 Cal.App.3d 1442, 261 Cal. Rptr. 447 (1989). Indeed, I can think of no situation where third-party injuries derive more directly from the employee's injury than injuries sustained by a fetus while *in utero*. Because Shanae Keefe was injured "on account of" her

---

[5] The majority dismissed the fact that the injuries were sustained before the baby's birth by relying on the fact that in this particular case the baby was born alive. Maj. op. at 101 n. 3. I believe this conclusion ignores the Act's coverage of injuries that *accrue* in the workplace.

mother's work-related injuries, her injuries are covered by the Act.

Examination of the complaint reveals the derivative nature of Shanae's injuries. It contains no allegation of direct injury to Shanae, but rather claims her injuries occurred as a result of her mother's treatment. Paragraph 12 alleges that "[d]espite the medical work restrictions imposed on Mrs. Keefe by her treating physicians, the Defendants *compelled Mrs. Keefe to work* on a full-time schedule, and to otherwise violate the medical work restrictions which had been imposed." (Emphasis supplied.) Paragraph 13 continues "that the death of Shanae Keefe was a proximate result of the negligence and wrongful acts of the Defendants, and their employees and agents." Paragraph 14 then states that such wrongful conduct included, but was not limited to, "Defendant's failure to conform Mrs. Keefe's work schedule and job duties to the medical requirements which were communicated to them by Mrs. Keefes' treating physicians, and their failure to formulate and implement an adequate medical disability policy." According to the complaint, Mrs. Keefe's work-related injuries caused premature delivery and ultimately resulted in Shanae's death.

I also disagree with the majority's conclusion that Shanae's injuries are analogous to injuries that may be sustained by a child when visiting a parent at the workplace. Any such injury does not necessarily occur as a result of an injury to the employee. Indeed, the parent may not be present when the child is injured. Here, however, Shanae's injuries occurred solely as a result of her mother's work-related injuries. There was no negligent treatment of Shanae independent of the harm inflicted upon her mother.

Further, an employer's relationship to the unborn children of its employees differs from its relationship to typical non-employees. While employers have the option to reduce their risk of liability to third persons by restricting access to the workplace, when a pregnant employee is involved such restrictions are generally unavailable. *See International Union, UAW v. Johnson Controls, Inc.*, 499 U.S. 187, 111 S.Ct. 1196, 113 L.Ed.2d 158 (1991) (striking a fetal protection policy in a battery manufacturing plant).[6] Under the majority's broad holding an injured fetus has a cause of action for any work-related injury regardless of whether the injury pertained to the pregnancy.[3] Thus, an employer is effectively subject to two standards of liability in the workplace over which the employer has no control, at a time when injury to the employee carries with it the high likelihood that injury to the fetus will also occur. *See Bell*, 261 Cal.Rptr. at 455 (considering the policy implications of allowing independent tort actions for *in utero* injuries sustained by a fetus while in the workplace, including the possibility of "financially driven discrimination by liability conscious employers"). In my view the court's holding today only increases the tension between workers' compensation law, tort law and employment discrimination law. *See* Susan S. Grover, *The Employer's Fetal Injury Quandary after Johnson Controls*, 81 Ky. L.J. 639 (1992-93).

Because I conclude Shanae Keefe's injuries derived from her mother's work-related injuries and are covered by the Act, I dissent.

I am authorized to say that Justice ERICKSON and Justice VOLLACK join in this dissent.

---

[6] In *Johnson Controls* the Court struck a fetal protection policy based on its discrimination against women. There, the Court explained that "[i]t is no more appropriate for the courts than it is for individual employers to decide whether a woman's reproductive role is more important to herself and her family than her economic role." *Id.* 499 U.S. at 194-95, 111 S. Ct. at 1201. While I agree that the decision to work while pregnant is personal to the woman, I believe that along with this autonomy must go an acknowledgment.

(Modified for the purposes of this text)

Perry ZEIGER, Plaintiff and Appellant,
v.
STATE of California et al., Defendants
and Respondents.
No. 3 Civ. C019828.
Court of Appeal, Third District.
Sept. 25, 1997.

Subcontractor's employee sued state and general contractor for renovation project on state-owned building
for negligence based on injuries sustained in fall from
scaffolding. The Superior Court, Sacramento County, No.CV
535306, Joe S. Gray, J., granted summary judgment for
defendants. Plaintiff appealed. The Court of Appeal,
Raye, J., held that: (1) *Privette* decision, abrogating
the peculiar risk doctrine insofar as it allowed independent contractor's employee to recover for injuries
from contractor's non-negligent employer, does not preclude a independent contractor's employee who has received
workers' compensation benefits from recovering against an
owner or general contractor for owner's or general contractor's own negligence, and (2) fact issues existed as
to whether state and general contractor breached their
duties to plaintiff, precluding summary judgment.

Reversed.

Scotland, J., filed a dissenting opinion.

———————————

Law Offices of Winchell & Truett, Harold J. Truett,
III, San Francisco, for Plaintiff and Appellant.

Law Office of Lea & Arruti, Robert Lea and Daniel
S. Glass, Sacramento, for Defendants and Respondents.

RAYE, Associate Justice.

The State of California and its general contractor,
John Otto, Inc., were granted summary judgments in the
negligence action filed by a subcontractor's employee,
Perry Zeiger. It is undisputed that plaintiff complained
repeatedly to the project manager for the state, the
superintendent for the general contractor, the

subcontractor's foremen, and even the safety hygienist for the job about the dangerous condition of the scaffolding set up by the subcontractor to perform asbestos abatement. Plaintiff sustained personal injuries when a plank on the scaffold gave away.

On appeal, defendants maintain the plaintiff's action is barred by *Privette v. Superior Court* (1993) 5 Cal.4th 689, 21 Cal.Rptr.2d 72, 854 P.2d 721 wherein the Supreme Court abrogated the peculiar risk doctrine insofar as it imposed liability on a non-negligent landowner who retained no control over the work of an independent contractor and was without fault for injuries sustained by the contractor's employee. In the trial court, plaintiff conceded that *Privette* doomed the first cause of action based on peculiar risk.

The issue presented by this appeal is whether the rationale of *Privette* and its progeny precludes any claims of negligence by a subcontractor's employee against an owner or general contractor if the employee has received workers' compensation benefits. We conclude it does not and shall reverse the summary judgment herein.

FACTUAL BACKGROUND

We extract the pertinent facts from the moving and opposing papers and the documents and deposition transcripts incorporated into those papers.

The State of California contracted with John Otto, Inc. to renovate the Veteran's Affairs Building in Sacramento. Otto assumed sole responsibility for supervising "all construction means, methods, techniques, sequences and procedures." Otto also agreed to initiate, maintain, and supervise all safety precautions and programs and to provide reasonable protection to prevent injury to all employees on the job and all other persons who may be affected thereby. Otto was obligated to comply with all safety laws, ordinances, rules, and regulations. It had an affirmative contractual obligation to attempt to prevent accidents.

During the renovation, asbestos was discovered. Otto subcontracted with Jerry Eaton, Inc. to remove the asbestos on the first, second, fourth, and fifth floors.

Eaton agreed to comply with all health and safety laws and regulations as well as the accident prevention and safety programs of the State and Otto. It assumed sole responsibility for providing a safe place to work for its employees.

Dennis Hazelton, the superintendent for Otto, described Otto's broad responsibility for safety on the job site. Although he personally bore the ultimate responsibility for safety, his foremen were also trained to provide a safe working environment and his assistant, Eric Alward, conducted weekly safety meetings. He believed he was responsible for the safety of the subcontractors and their employees on the job site and if he saw a subcontractor create a risk of harm he had the authority to demand alteration or stop work. Either he or Alward inspected the premises on a daily basis. Both were certified to enter the asbestos containment areas.

Eaton assembled scaffolding within the containment areas. Otto did not supply the scaffolding nor assist in the assembly. Orlando Martinez, the hygienist on the project, monitored the safety of the asbestos work. He expressed to Dan Maffuccio, the project manager for the state, his concerns about the dangerous condition of the scaffolding. According to Hazelton, Maffuccio chastised Martinez, contending the safety in the containment areas was none of his business. Maffuccio insisted that Otto and Eaton shared responsibility for the safety of the workers inside the containment areas.

Martinez also reported his concerns about the rolling scaffolding and the absence of hand rails to Eaton's supervisor on September 24 and 30, on October 6 and 8, and to Dennis Hazelton about one week before the accident. Hazelton informed Martinez the rails were not compelled because the scaffolding was under seven feet, below the OSHA minimum standards for hand rails. Martinez believed there was an exception to the rule cited by Hazelton compelling rails when scaffolding was placed on an unlevel surface.

Plaintiff, at 6-feet 4-inches tall, claims the scaffolding was about 8 feet. During the two weeks he worked in containment before the fall, he complained repeatedly to Eaton foremen, Otto foremen and the job superintendent, the hygienist, and the project manager from the state. Plaintiff observed an Eaton foreman take tie wire hanging out of the ceiling, wrap it around the edge of the scaffolding, and hang the plank. The result, according to plaintiff, was a rickety and unsafe scaffolding.

His many complaints were fruitless. In response to his warnings about the danger of the scaffolding an Eaton foreman "told me to get the F off the scaffolding and go home if I didn't like it. And I told him, 'I don't have a job anywhere else but here.'" The foreman told him to shut his mouth and get to work. Other foremen issued similar threats.

Plaintiff then complained to Otto's foreman, Eric Alward, about one week before the accident but nothing was done to remedy the danger. Reviewing pictures plaintiff took of the scaffolding the day following the fall, Hazelton testified at his deposition the config-uration was unsafe because it was not set up properly. Had he seen the scaffolding, he would have immediately stopped all work on the scaffold and he would have ordered it dismantled and erected properly. Hazelton recounted it was a hazard because of the potential for collapse.

Plaintiff, recognizing the same dangers, attempted to stabilize the scaffolding. Every time he changed the configuration, Eaton personnel changed it back. A super-visor finally admonished him that "if [he] kept screwing around with adjusting the scaffold...he was just going to send [him] down the road."

On the morning of October 2, 1992, plaintiff was standing on a tie-wired plank attempting to take down false ceilings above him. The plank fell out from under him and he fell to the floor.

DISCUSSION

[1] A motion for summary judgment is properly granted only if the papers submitted show there is no triable issue of material fact and the moving party is entitled to judgment as a matter of law. (Code Civ. Proc., § 437c, subd. (c).) Since the summary judgment motion raises only questions of law regarding the construction and

effect of the supporting and opposing papers, we independently review such papers to determine whether the
moving party's papers establish facts to justify a
judgment on the issues framed by the pleadings and
whether the opposition demonstrates the existence of a
triable, material factual issue. *(AARTS Productions, Inc.
v. Crocker National Bank* (1986) 179 Cal.App.3d 1061,
1064, 225 Cal.Rptr.203.)

Since plaintiff concedes his first cause of action
based on peculiar risk no longer remains viable, the
second cause of action for negligence frames the issues.
Plaintiff alleges in pertinent part: "That at such time
and place and prior thereto Defendants, and each of
them, so negligently and carelessly owned, possessed,
operated, constructed, inspected, maintained, contracted,
subcontracted, supervised, controlled, engineered,
designed, performed, and planned said remodeling and
demolition work and supplied men and materials for the
remodeling and construction so as to cause plaintiff to
fall and to sustain the injuries and damages complained
of."

[2] A construction worker injured on the job site
is entitled to compensation under the workers' compensation system. By statute, receipt of these benefits
comprises the victim's exclusive remedy against his
employer, the subcontractor. Does the availability of
workers' compensation benefits also immunize the landowner and general contractor for their personal negligence? The general rule is that when work is turned over
to an independent contractor, neither the owner nor
general contractor is liable to the contractor's employees for the independent contractor's negligence. *(Srithong v. Total Investment Co.* (1994) 23 Cal.App.4th 721,
725, 28 Cal.Rptr.2d 672; *Caswell v. Lynch* (1972) 23
Cal.App.3d 87, 90, 99 Cal.Rptr. 880.) The general rule,
however, has been eroded by exceptions. Numerous cases,
as well as the Restatement of Torts 2d, raise multiple
theories of liability against landowners and general
contractors for injuries to employees of independent
contractors. *(Felmlee v. Falcon Cable TV* (1995) 36
Cal.App.4th 1032, 1035, 1040, 43 Cal.Rptr.2d 158; *Whitford v. Swinerton & Walberg Co.* (1995) 34 Cal.App.4th
1054, 1055-1058, 40 Cal.Rptr.2d 688; *Smith v. ACandS,
Inc.* (1994) 31 Cal.App.4th 77, 94, 37 Cal.Rptr.2d 457;

Owens v. Giannetta-Heinrich Constr. Co. (1994) 23
Cal.App.4th 1662, 1666, 29 Cal.Rptr.2d 11; *Srithong v.
Total Investment Co., supra,* 23 Cal.App.4th at p. 725,
28 Cal.Rptr.2d 672; *Johnson v. Tosco Corp.* (1991) 1
Cal.App.4th 123, 139, 1 Cal.Rptr.2d 747; *Osborn v. Mission
Ready Mix* (1990) 224 Cal.App.3d 104, 273 Cal.Rptr. 457;
West v. Guy F. Atkinson Constr. Co. (1967) 241 Cal.App.2d
296, 297-300, 59 Cal.Rptr.286.)

A landowner has been held liable to a subcontractor's
employee on theories of premises liability, peculiar
risk, negligent hiring, and control over the operative
details of a construction project. (*Austin v. Riverside
Portland Cement Co.* (1955) 44 Cal.2d 225, 232-233, 282
P.2d 69.) The landowner also has been held vicariously
liable based on a nondelegable duty. (*Srithong v. Total
Investment Co., supra,* 23 Cal.App.4th at p. 726, 28
Cal.Rptr.2d 672.) Similarly, a general contractor has
been liable for its personal negligence in failing to
take adequate precautions when peculiar risks are presented, for negligent hiring, and for the failure to
exercise due care supervising a project over which it
retained some degree of control. (*Caswell v. Lynch,
supra,* 23 Cal.App.3d at p.91, 99 Cal.Rptr. 880.) The
general contractor's liability may also be vicarious.
(*West v. Guy F. Atkinson Constr. Co., supra,* 251
Cal.App.2d at p. 300, 59 Cal.Rptr.286.)

A. Privette v. Superior Court and the Peculiar Risk Doctrine

"Under the peculiar risk doctrine, a person who hires
an independent contractor to perform work that is
inherently dangerous can be held liable for tort damages
when the contractor's negligent performance of the work
causes injuries to others. By imposing such liability
without fault on the person who hires the independent
contractor, the doctrine seeks to ensure that injuries
caused by inherently dangerous work will be compensated,
that the person for whose benefit the contracted work
is done bears responsibility for any risks of injury to
others, and that adequate safeguards are taken to prevent
such injuries." (*Privette v. Superior Court, supra,* 5
Cal.4th at p. 691, 21 Cal.Rptr.2d 72, 854 P.2d 721.)

Franklin Privette, a schoolteacher, owned a duplex. He hired a roofer to install a new roof on the duplex, only after verifying the roofer was licensed, was reliable, and carried workers' compensation. One of the roofer's employees carried hot tar up a ladder at his foreman's request and was injured when he fell off the ladder. Privette was not present when the injury occurred; nor did he participate in the foreman's decision to send the victim up a ladder with a five-gallon bucket of hot tar. (*Id.* at p. 692, 21 Cal.Rptr.2d 72, 854 P.2d 721.)

In a narrowly crafted holding, the Court restricted the application of the peculiar risk doctrine in California. The Court wrote: "When an employee of the independent contractor hired to do dangerous work suffers a work-related injury, the employee is entitled to recovery under the state's workers' compensation system. That statutory scheme, which affords compensation regardless of fault, advances the same policies that underlie the doctrine of peculiar risk. Thus, when the contractor's failure to provide safe working conditions results in injury to the contractor's employee, additional recovery from the person who hired the contractor — a *non-negligent party* — advances no societal interest that is not already served by the workers' compensation system. Accordingly, we join the majority of jurisdictions in precluding such recovery under the doctrine of peculiar risk." (*Id.* at p. 692, 21 Cal.Rptr.2d 72, 854 P.2d 721, emphasis added.)

The scope of *Privette* is best determined by understanding the nature of the societal interests the Court was seeking to promote. When liability is commensurate with fault, a tidy symmetry is achieved: a victim is compensated and the tortfeasor pays. But the peculiar risk doctrine as explained in *Privette* involves two innocent parties, "the person who contracted for the work and the hapless victim of the contractor's negligence." (*Id.* at p. 694, 21 Cal.Rptr.2d 72, 854 P.2d 721.) Hence, as tort law strains and stretches to achieve its primary objective to provide compensation to innocent victims, a tension arises between the competing societal interests to compensate and to fasten liability on one at fault and, if not at fault, to impose liability upon one who benefits from or controls the injury producing activity.

By enacting a comprehensive no fault system of recovery for work related injuries, the Legislature has assured that the hapless and innocent victim will not be left without recourse. Consequently, workers' compensation benefits satisfy the fundamental social policy to provide compensation to victims.

It is the equally important societal interest in promoting workplace safety as well as the availability of equitable indemnity, according to *Privette*, which makes the shifting of liability to a non-negligent party tolerable. (*Id.* at p. 695, 21 Cal.Rptr.2d 72, 854 P.2d 721.) The court believed that by spreading the risk of loss to those who benefit from the hired work there was a greater incentive to provide a safe work place. (*Ibid.*) Moreover, the non-negligent defendant, saddled with a tortfeasor's liability, is entitled to recover the damages it was obligated to pay on the wrongdoer's behalf through the vehicle of equitable indemnity. (*Id.* at p. 696, 21 Cal.Rptr.2d 72, 854 P.2d 721.)

Workers' compensation benefits, however, upset the symmetry of compensation. A third party, forced to pay an injured worker as a consequence of either its relationship to the tortfeasor or of its own negligence, cannot pursue an equitable indemnity claim against the negligent employer. The Supreme Court lamented that inequitable shifting of liability produces the "anomalous result that a non-negligent person's liability for an injury is greater than that of the person whose negligence actually caused the injury." (*Id.* at p. 698, 21 Cal.Rptr.2d 72, 854 P.2d 721.) Although an injured worker is compensated under the workers' compensation system, application of the peculiar risk doctrine before *Privette* allowed the victim to receive a windfall from the innocent party who hired the tortfeasor but who could not recover from the wrongdoer through equitable indemnity. (*Id.* at p. 698, 21 Cal.Rptr.2d 72, 854 P.2d 721.) The Court concluded the workers' compensation system of recovery "achieves the identical purposes that underlie recovery under the doctrine of peculiar risk: It ensures compensation for injury by providing swift and sure compensation to employees for any workplace injury; it spreads the risk created by the performance of dangerous work to those who contract for and thus benefit from such work,

by including the cost of workers' compensation insurance
in the price for the contracted work; and it encourages
industrial safety." (*Id.* at p. 701, 21 Cal.Rptr.2d 72,
854 P.2d 721.)

In sum, the Supreme Court in *Privette*, abolished the
arcane concept of peculiar risk to avoid doubly compen-
sating a victim at the expense of a non-negligent
defendant. Since the fundamental social policy of providing compensation is achieved by the workers' compensation system, the Court found no justification for
imposing vicarious liability on one who had hired an
independent contractor. *Privette* is predicated on an
equation which balances the equities of two innocent,
non-negligent parties. It also spoke to the importance
of workplace safety, but in the context of two non-
negligent parties. In that circumstance it deferred to
the workers' compensation system to protect the health
and safety of workers. The Court did not speak, however,
to the issue of workplace safety when an owner or general
contractor was at fault in allowing an unsafe working
condition.[7]

B. Liability Predicated on Fault: The Importance of Control

Although a well established body of law involving
the liability of owners and general contractors has
evolved along with peculiar risk cases, the theories are
often confused or intermingled and, therefore, mask the

---

[7] Since plaintiff abandoned his cause of action predicated on the
peculiar risk doctrine, we need not explore the thorny nuances of
"direct" liability as described in Section 413 and "vicarious"
liability as described in Section 416. Suffice it to say a footnote
in *Privette* has spawned a contentious debate. (*Yanez v. U.S.* (9th
Cir.1995) 63 F.3d 870; *Owens v. Giannetta-Heinrich Construction Co.,*
supra, 23 Cal.App.4th 1662, 29 Cal.Rptr.2d 11.) The Supreme Court
has granted hearing in *Toland v. Sunland Housing Group, Inc.* (1995)
49 Cal.App.4th 212, 47 Cal.Rptr.2d 373 "limited to whether, after
Privette.... a subcontractor's employee has a cause of action for direct
liability against a general contractor or developer." (Hrg. granted 51
Cal.Rptr.2d 84, 912 P.2d 535 (1996).) To the extent "direct" liability
refers to the Section 413 variation of peculiar risk, it is irrelevant
to our disposition of this appeal. If, however, "direct" implicates other
theories of liability predicated on either a landowner or general
contractor's personal negligence or fault, then the outcome of *Toland*
might encompass this case as well.

determinative factor in the analysis. Whatever the theory, the cases turn on an assessment of fault and fault frequently derives from the retention of control.

Hard v. Hollywood Turf Club (1952) 112 Cal.App.2d 263, 246 P.2d 716 and *Austin v. Riverside Portland Cement Co.* (1955) 44 Cal.2d 225, 282 P.2d 69 reach opposite results based on an analysis of control. In *Hard*, the court of appeal reversed a judgment against a general contractor because there was no evidence the contractor had any control over the painter of the scaffolding upon which he worked. The court rejected the notion a general contractor has a duty to act as an insurer of every subcontractor's activities. (*Id.* at p. 271, 246 P.2d 716.) The court wrote: "The lawmakers must have reflected that to impose the duty upon the general contractor directly to oversee all labor being performed and to inspect every device imported for use by the workmen on a particular construction and to impose nondelegable duties upon him to enforce all statutory safety provisions would be to place an extremely onerous burden upon him who has the general control over the ultimate result of the labor done yet *without control or management of the means utilized to achieve the purpose planned.*" (*Ibid.*, emphasis added.)

The Supreme Court distinguished an action against a general contractor or owner for its own negligence from an action based on a subcontractor's negligence in *Austin v. Riverside Portland Cement Co., supra* 44 Cal.2d at p. 229, 282 P.2d 69. The defendant cement company owned and operated a manufacturing plant. Defendant hired an independent contractor to perform extensive construction work on the plant. Defendant retained control of its premises and plant, although it had no control over the contractor's employees. Nevertheless, defendant had an employee on the job. Its supervisory personnel knew the contractor used large cranes near power lines and often worked at night and it had the authority to have the lines deenergized. Plaintiff was severely burned when a crane came into contact with an overhead conduit of electricity. (*Id.* at pp. 229-230, 282 P.2d 69.)

The Supreme Court explained: "It is true that California rather than defendant was maintaining the power line, yet defendant had control to the extent that it

:ould have the line deenergized and as it controlled
.ts premises it could provide adequate lighting for night
ork and proper warnings. Hence, the care required of
.t must be measured by the hazard inherent in highly
:harged wires together with the duty it owed to [the
:ubcontractor's] employees who were as to defendant,
.nsofar as it had control of the premises, invitees or
:usiness visitors rather than licensees or trespassers;
.hat their status was invitees is clear." (*Id.* at p. 232,
82 P.2d 69.) The Court went on to quote Section 414 of
.he Restatement 2d and affirmed the judgment against the
.efendant based, not on the negligence of its contractor,
:ut for its own breach of duty. (*Id.* at pp. 232-233,
82 P.2d 69.)

Cases decided since *Hard* and *Austin* also focus on
:ontrol (See, e.g., *Sabin v. Union Oil Co.* (1957) 150
:al.App.2d 606, 310 P.2d 685; *Kingery v. Southern Cal.
:dison Co.* (1961) 190 Cal.App.2d 625, 12 Cal.Rptr. 173;
est v. Guy F. Atkinson Constr. Co., supra, 251 Cal.App.2d
:96, 59 Cal. Rptr. 286; *Johnson v. Cal-West Constr.
:o.* (1962) 204 Cal.App.2d 610, 22 Cal.Rptr. 492.) "Ordi-
.arily where an owner or general contractor does nothing
:ore with respect to the portion of the construction
ob done by an independent subcontractor than exercise
:eneral supervision to bring about the satisfactory
:ompletion of the job the general contractor owes no
:uty to assure the safety of the subcontractor's employees
.nd is not liable, therefore, for the subcontractor's
:orts... This does not mean that where the general
:ontractor himself is negligent he is not liable. The
:ontrary is true. [Citation.] But a necessary element
:f legal negligence is a duty of care. Therefore, the
:eneral contractor is liable only when he has assumed
.nd has violated such a duty." (*West v. Guy F. Atkinson
:onstr. Co.,* supra, 251 Cal.App.2d at p. 299, 59 Cal.Rptr.
:86.)

In *Caswell v. Lynch,* supra, 23 Cal.App.3d 87, 99
:al.Rptr. 880, the court ascribed a duty of care to
:mployers of contractors commensurate with their knowl-
.dge and control. "Certain provisions of the Labor Code
:citations] treat employees of subcontractors as 'employ-
:es' of contractors for purposes of imposing liability
:o provide a safe place of employment, but they do not
:mpose an affirmative duty to supervise safety, and they

do not impute to the general contractor knowledge of a negligent subcontractor. To fasten on a general contractor the duty of supervising and inspecting labor and equipment used is thus to require the general contractor to become, in effect, the safety inspector on the job. (*Hard v. Hollywood Turf Club,* [*supra,*] 112 Cal.App.2d [at p.] 266 [246 P.2d 716][].) In the language of *West v. Guy F. Atkinson Constr. Co.,* [*supra.*] 251 Cal.App.2d 296, 59 Cal.Rptr. 286[], 'there is no vicarious lability.' [¶] We do not read the foregoing authorities as immunizing, however, the general contractor from liability in those cases where the general contractor has himself knowledge, actual or constructive, of a condition hazardous to those who come upon the premises, and such degree of control as permits reasonable preventive or protective measures to be taken by him." (*Id.* at p. 91, 99 Cal.Rptr.880.)

Courts have failed to clearly articulate the theories of liability and the overlapping roles of the target tortfeasors. Premises liability has been blurred with peculiar risk, negligent hiring or supervision may be blurred with a nondelegable duty. But the pivotal factor, whether labeled under a section of the Restatement or as a duty of care under the common law, is control. *Privette* is consistent with a long line of authority because the landowner, the school teacher, retained no control over the subcontractor he hired to install a new roof on his duplex.

C. Liability Predicated on Fault After Privette

Although *Privette* has been cited in a variety of contexts not relevant here (*Doney v. TRW, Inc.* (1995) 33 Cal.App.4th 245, 39 Cal.Rptr.2d 292; *Owens v. Giannetta-Heinrich Construction Co., supra,* 23 Cal.App.4th-1662, 29 Cal.Rptr.2d 11; *Srithong v Total Investment Co.* Supra, 23 Cal. App. 4th 721, 28 Cal.Rptr.2d 672), the courts have continued to sustain the viability of subcontractors' employees' negligence actions against owners and general contractors unrelated to theories of peculiar risk. Moreover, in many of the cases, the courts discuss at length the intricacies and peculiarities of peculiar risk and briefly report the well accepted proposition that all other theories of liability remain intact.

In *Smith v. ACandS, Inc., supra,* 31 Cal.App.4th 77, 37 Cal.Rptr.2d 457, the court reversed a jury verdict in favor of an injured employee of a subcontractor having concluded that a utility company was improperly held vicariously liable under the peculiar risk doctrine. The court remanded, however, for a retrial on the utility company's general negligence. "We have reviewed the entire record, and are satisfied that there was sufficient evidence to find PG & E negligent. While there was no direct evidence that PG & E knew asbestos's dangers, the jury could conclude that PG & E should have foreseen the dangers given publication of various medical and industrial studies of asbestos diseases and promulgation of industrial safety standards and that PG & E breached its duty to exercise ordinary care in the management of its premises by failing to hire careful and competent insulators or to take other reasonable precautions." (*Id.* at p. 97, 37 Cal.Rptr.2d 457.) *Smith* thereby sanctions an action by a subcontractor's employee against the employer of the contractor for general negligence in the aftermath of *Privette.*

In *Felmlee v. Falcon Cable TV, supra,* 36 Cal.App.4th 1032, 43 Cal.Rptr.2d 158, a general contractor retained a subcontractor to maintain and repair its cable television lines. An employee of the subcontractor was injured while working on the lines. The case went to the jury on general negligence principles and a verdict was returned in favor of the general contractor. The court acknowledged, "The jury was free to consider whether [the general contractor] was directly negligent in failing to correct any foreseeable, dangerous condition of the cables which may have contributed to the cause of the [worker's] injuries." (*Id.* at p. 1040, 43 Cal.Rptr.2d 158.)

Similarly, in *Fire Ins. Exchange v. American States Ins. Co.* (1995) 39 Cal.App.4th 653, 46 Cal.Rptr.2d 135, a general contractor's active negligence was at issue as was the landowner's liability under a theory of peculiar risk. The case involved a fight between the owner's insurer and the general contractor's insurer.

The court wrote: "Here, the causes of action for negligence and peculiar risk overlapped. Gebhardt's complaint alleged that the defendants, including Alam, 'were, and are, the owners, employers, contractors, and/or developers of the project site at which plaintiff, Russell Gebhardt, was injured.' Alam functioned as a general contractor and personally took charge of the general plan and method of construction. Gebhardt claimed that Alam refused to provide metal scaffolding and ordered Gebhardt's employer to construct the scaffolding from wood. Alam supervised the work and visited the job site twice a day. Attorney Howard Cho, who defended Azam, Alam and Yelvington, opined that Alam had potential liability under the peculiar risk doctrine because he functioned as the general contractor. As indicated, both the owner of land and a general contractor who hires an independent contractor may have liability pursuant to the peculiar risk doctrine. [Citation.] [¶] The court found that Alam Sher, Azam Sher, and Cynthia Yelvington were all at risk of liability in the Gebhardt suit. Azam and Cynthia had greater exposure than Alam under the doctrine of peculiar risk before *Privette*. Alam had greater exposure than they based on active negligence. All had significant risk of exposure and there was no utility for settlement to allocate their individual contribution to the overall risk.' The court made an implied finding that the liability of Alam on one hand, and Azam and Yelvington on the other hand, was approximately the same for settlement purposes. In the exercise of its equitable discretion, the trial court concluded that the umbrella policies should be equally prorated because the policy limits were the same. This was not an abuse of discretion." (*Id.* at pp. 660-661, 46 Cal.Rptr.2d 135.)

On facts very similar to the facts of the case before us, the court in *Whitford v. Swinerton & Walbert Co.*, *supra*, 34 Cal.App.4th 1054, 40 Cal.Rptr.2d 688 reached the same result. The general contractor "bore the responsibility of formulating the work plan and the safety program on the project and ensuring that the workers had a safe workplace." (*Id.* at p. 1055, 40 Cal.Rptr.2d 688.) It hired a safety officer to maintain safety including compliance with all federal safety regulations. By contract, the general contractor was obligated to follow all codes and regulations.

The general contractor was aware of one of the subcontractors' practice of lowering the top cable of a safety railing, even though the practice was a safety violation. The general contractor did not direct any of its employees to investigate the problem or to solve it. (*Id.* at p. 1056, 40 Cal.Rptr.2d 688.) The plaintiff was injured while working near an open and unprotected elevator shaft.

As in the instant case, the injured employee asserted two causes of action, one for peculiar risk and the other for negligence. A jury found the general contractor was negligent and its negligence was a cause of the victim's injuries. The jury also found the general contractor was vicariously liable for the subcontractor's negligence under a peculiar risk theory.

The court concluded *Privette* should be applied retroactively but the decision did not impact the outcome of the case. Because the trial court had issued a remittitur reducing the damages as compelled by Section 1431.2 of the Civil Code (Proposition 51), the judgment was based on the general contractor's negligence, not on any negligence imputed to the general contractor under the peculiar risk doctrine. The court acknowledged the impact of *Privette* on peculiar risk cases, but did not expand the rationale to immunize a general contractor for its own negligence.

The Ninth District Court of Appeals is in accord. (*Yanez v. U.S.* (9th Cir.1995) 63 F.3d 870.) Isabel Yanez lost her arm and suffered third degree burns from an explosion at a munitions factory. After reaching a number of settlements and receiving workers' compensation benefits, Yanez brought suit in federal court alleging the government's negligent inspections and lax enforcement of safety regulations caused her injuries. The district court granted the government's motion for judgment on the pleadings, but the court of appeals, applying California law, reversed. (*Id.* at p. 875.)

The second part of the opinion addresses the issue of control and generates an impassioned dissent on whether *Privette* applies to Section 414 cases. "Yanez argues next that because the evidence supports a finding that the United States exercised control over the [munitions contractor's] activities, the district court erred by granting summary judgment to the defendant on her negligent control claim." (*Id.* at p. 874.) This is precisely the issue now before us.

Yanez presented evidence the government knew of widespread safety violations at the factory but failed to act. The court concluded the evidence raised the issue whether the government's knowledge gave rise to a duty to order correction, to take corrective action, or to exercise its right to halt the work. The duty emanated from Section 414, a section having nothing to do with the peculiar risk doctrine.

The majority reversed the summary judgment on the control claim based on Section 414 without commenting on the applicability of *Privette*. The court held: "In *Holman v. State*[(1975), 53 Cal.App.3d 317, 124 Cal.Rptr. 773], a California Court of Appeal held that Section 414 liability applies where a principal has actual knowledge of a dangerous condition and the authority to correct the dangerous condition. Here, evidence adduced by Yanez indicates that government inspectors may have had actual knowledge of safety violations and the dangers posed by the violations, yet failed to exercise their right to order the contractor to correct them, and if it did not, to order the work halted. In California, these facts, if true, are sufficient to establish liability under Section 414." (63 F.3d at p. 875.)

According to Justice Noonan in his dissent, the half-life benevolently bestowed by the majority on the plaintiff's claim for negligent control conflicts with the reasoning of *Privette*. Justice Noonan emphasizes the exclusivity of workers' compensation and the inability to obtain equitable indemnity from a negligent employer. He bemoans the "unwarranted windfall" employees of independent contractors reap by allowing a second recovery after workers' compensation benefits are paid. Finally, he writes that the person or entity who hires an independent contractor indirectly pays for the cost of workers' compensation through the contract price. (63 F.3d at p. 876.)

Though Justice Noonan's dissent offers compelling policy reasons for expanding the scope of *Privette*, it finds little support in the *Privette* holding. His views were embraced by the *Toland* majority and by the Fourth

District Court of Appeal in *Voigts v. Brutoco Engineering and Construction Co.* (1996) 49 Cal.App.4th 354, 57 Cal.Rptr.2d 87.[8] While these cases await hearing in the Supreme Court, we must resolve this negligence case, which does not involve liability predicated on peculiar risk in particular, or vicarious liability in general. We agree with the majority in *Yanez* that a subcontractor's employee's negligence claim against a general contractor or owner survives *Privette*.

Consequently, with the exception of the two cases pending before the Supreme Court, the remaining cases decided since *Privette* continue to sustain the viability of negligence cases against general contractors and landowners by injured workers of subcontractors. The Supreme Court in *Privette* acknowledged the vicarious nature of peculiar risk liability whether or not it emanated from the owner's or contractor's failure to direct its subcontractor to take special precautions as described in Section 413 of the Restatement or its nondelegable duty as described in Section 416. Some confusion has resulted from the use of the word "direct" within the specialized context of peculiar risk. Perhaps the division of the peculiar risk doctrine in the Restatement between sections 413 and 416 created some of the confusion. Undoubtedly, by granting hearing in *Toland* to determine whether a "subcontractor's employee has a cause of action for direct liability against a general contractor or developer," the Supreme Court will clarify the semantic confusion evolving around "direct liability."

D. Conclusion

[3] For our purposes, in resolving a case which does not involve peculiar risk, we hold that neither the language nor the rationale of *Privette* or its progeny obliterates any of the long established theories of liability of an owner or general contractor for his independent negligence. The law both before and after *Privette* clearly sanctions a cause of action predicated on premises liability, negligent hiring or supervision,

---

[8] The Fourth District has adopted reasoning similar to the majority opinion in *Toland*. (*Voigts v. Brutoco Engineering & Construction Co., supra*, 49 Cal.App.4th 354, 57 Cal.Rptr.2d 87 review granted 59 Cal.Rptr.2d 669, 927 P.2d 1173 (1996).)

or retention of control. We need only apply that law to the facts before us.

[4, 5]   Here the State is the owner of the premises and its project manager, Dan Mattuccio, was aware of the dangerous condition of the scaffolding. According to the declarations submitted in opposition to the motion for summary judgment, Mattuccio, chastised the site hygienist for raising safety concerns. Consequently, the State's duties arise as owners of the premises with notice of a dangerous condition. "The general rule in that regard is that an owner or occupier of premises, who, by invitation express or implied, whether the invitation is pursuant to a written contract or otherwise, induces, or knowingly permits, a workman to enter the premises for the performance of duties mutually benefi-cial to both parties, is required to use reasonable care to protect the workman by supplying him with a reasonably safe place in which to work and to furnish and maintain appliances in connection therewith which are reasonably safe for the purposes embraced therein." (*Kingery v. Southern Cal. Edison Co.* (1961) 190 Cal.App.2d 625, 632, 12 Cal.Rptr. 173.)

[6]   Plaintiff raises a triable issue of material fact concerning defendant Otto's liability as well. Otto's superintendent, Dennis Hazelton, testified in his deposition he had responsibility for safety on the jobsite and the authority to demand alterations or stop work. He, as well as his assistant, were aware of the dangerous condition of the scaffolding. He testified that if he had seen the scaffolding, he would have immediately stopped all work. This constitutes evidence Otto retained sufficient control as described in Section 414 of the Restatement to give rise to a duty of due care. Moreover, the notice of the danger triggered a similar duty. "Acknowledging the origin of the language in precedents dealing with a landowner's duties toward invitees, the court referred jointly to the duties of an owner or general contractor: 'An owner or general contractor exercising supervision over a project owes a common law duty to the employees of independent contractors to exercise ordinary care, to furnish them with a reasonably safe place in which to work.' [Citation.] [¶] The language is sanctioned by ample judicial precedent applying to a contractor's duty toward employees of a subcontractor,

but it has uncertain application, demanding further clarification, with respect to the duty of the employer of an independent contractor toward the contractor's employees. In a particular case, the language may be consistent with the employer's duties toward the contractor's employees arising from retained control over the work, or, on the assumption that the employer possesses the land on which the work is done, with the duties of a landowner toward invitees." (*Johnson v. Tosco Corp., supra,* 1 Cal.App.4th 123, 139, 1 Cal.Rptr.2d 747.)

Privette embodies a balanced social policy given the interests of two non-negligent parties and the availability of a comprehensive workers' compensation system. The introduction of fault, however, jeopardizes and impedes workplace safety. This case illustrates why *Privette* should not be expanded beyond the vicarious liability umbrella of the peculiar risk doctrine.

The deposition transcripts and declarations submitted in opposition to the motion for summary judgment depict a disregard for the safety of the employees on the construction site. If a jury believes the evidence presented at the hearing on the summary judgment, the subcontractor repeatedly threatened to fire plaintiff for raising safety concerns, the general contractor, although aware that the scaffolding could collapse from improper installation remained indifferent to the imminent danger, and the state, as owner of the premises, knew of the dangers and actively discouraged corrective action. Unlike the school teacher in *Privette*, who had no knowledge of the manner in which the job was performed and no control over the contractor's performance, Otto and the state, under the facts adduced by plaintiff, had the authority and the duty to direct the subcontractor to remedy known dangers.

Workers' compensation in a case such as this is an ineffective prophylactic. Plaintiff sought prevention, not treatment, by lodging complaints with all levels of the construction pyramid.

In this context, neither the inability to collect equitable indemnity nor the inclusion of insurance costs in the contract price outweighs the importance of providing a safe workplace. In *Privette*, the landowner's inability to seek indemnity from the subcontractor was inequitable because the contractor was a tortfeasor, whereas the landowner was not; the landowner's liability was purely vicarious. In this case, plaintiff seeks only the opportunity to prove the general contractor and owners were themselves negligent. Hence, plaintiff seeks only to hold defendants liable for their own wrongdoing.

Under defendant's reading of *Privette*, the general contractor and owner of the premises, though knowingly perpetuating an unsafe workplace, are immunized from liability. Hence, workers' compensation becomes a disincentive to maintaining safe working conditions. *Privette* simply does not compel that result.

At trial a jury will determine whether as a matter of fact either the state or Otto breached their duties to the plaintiff. In reversing the summary judgments, we hold as a matter of law, that *Privette* does not immunize either the state as owner or Otto as the general contractor from their personal negligence. The judgments are reversed. Appellant shall recover his costs on appeal.

SCOTLAND, Associate Justice, dissenting.

For reasons stated in *Toland v. Sunland Housing Group, Inc.* (1995) 49 Cal.App.4th 212, 221-226, 47 Cal.Rptr.2d 373 (hg.granted, 51 Cal.Rptr.2d 84, 912 P.2d 535 (1996)), I must dissent in this case.

I concurred in the *Toland* opinion, which held that *Privette v. Superior Court* (1993) 5 Cal.4th 689, 21 Cal.Rptr.2d 72, 854 P.2d 721 abrogates liability under Section 414 of the Restatement Second of Torts for work-related injuries to a contractor's employee even though the landowner retains control over the work.

The rationale of *Toland* applies equally to bar an action by the injured employee of a subcontractor against the general contractor who retains and exercises control over the work.

The workers' compensation system imposes statutory liability for injuries caused to an injured worker regardless of fault. As part of this compensation bargain, the employer, who must provide workers' compensation coverage irrespective of fault, cannot be held personally liable for the injury. In my view, the rationale for this public policy extends to a general contractor, who in effect pays the workers' compensation indirectly via the cost of its contract with the subcontractor. Because the general contractor pays for workers' compensation coverage as part of the subcontract price, thus insuring that the subcontractor's employee is compensated for work-related injuries regardless of fault, the general contractor should get the benefit of the workers' compensation bargain. That benefit bars a cause of action against the general contractor for damages based upon its alleged negligence.

Accordingly, I would affirm the trial court's judgment.

chapter six

Negligent hiring

"The employer generally gets the employees he deserves."

Sir Walter Bilbey

"I tell you, sir, the only safeguard of order and discipline in the modern world is a standardized worker with interchangeable parts. That would solve the entire problem of management."

Jean Giraudoux

One of the newest tort theories being utilized in the area of workplace violence is that of negligent hiring. This theory has its foundations as an exception to the fellow servant rule and operates to find liability against the employer where an employee is improperly hired where the employer knew or should have known the applicant was not fit for the job and ultimately causes injury to another employee. The general rule under the fellow servant doctrine is that the employer would be exempt from liability because of the negligence, carelessness, or intentional misconduct of a fellow employee. However, the courts in 28 states and the District of Columbia have recognized exceptions to this general rule under the theory of negligent hiring.[1]

The foundation for the theory of negligent hiring can be traced back to the case of *Western Stone Company v. Whalen*.[2] In this case, the Illinois Supreme Court found that an employer had a duty to exercise reasonable and ordinary care in the employment and selection of careful and skillful co-employees.[3] In recognition of this exception, the tort of negligent hiring has been expanded significantly by the courts to find that an employer may be liable for the injurious acts of an employee if these acts were within the scope of the employment.[4] This theory was expanded even further when courts began finding the employer liable even when the employee's acts were outside the

---

[1] Peterson, Donald J. and Douglas Massengill, The negligent hiring doctrine — a growing dilemma for employers, 15 *Employee Relations Law Journal* at 410 Note 1 (1989-1990).
[2] 151 Ill. 472, at 484, 38 N.E. 241 (1894).
[3] Id.

scope of the workplace or the employment setting.[5] In the early cases, the theory of negligent hiring developed into what we would today call negligent security. Many of the cases dealt with maintenance personnel or rental property managers with access to individuals' dwellings through master keys and other means.[6] In these cases, the court generally found that where the owner or employer knew that the duties of the job required these individuals to go into the personal residences of the individuals, the employer had a duty to use reasonable care in selecting an employee reasonably fit to perform these duties.[7] In more recent cases, the doctrine of negligent hiring has been significantly expanded to cover a wide variety of areas. For example, employers have been found liable in cases where they have employed truck drivers with known felony backgrounds who ultimately assaulted individuals, cases involving sexual harassment charges, and situations where off-duty management personnel assaulted others. A similar basis for liability in a number of these cases involved the employer's failure to properly screen and evaluate the individual before offering employment.[8]

In the area of workplace violence, the case of *Yunker v. Honeywell, Inc.*[9] appears to be one of the first to address this issue. In this case, the Minnesota Court of Appeals reversed the lower court's finding that Honeywell, Inc., as a matter of law, did not breach its duty in hiring in supervision an employee who shot and killed a co-worker off the employer's premises. In reversing the summary judgment ruling for the employer, the court not only applied a negligence theory but also made a distinction between the negligent hiring theory and negligent retention theory.

In this case, an individual worked at Honeywell from 1977 to his conviction and imprisonment for the strangulation death of a co-worker in 1979. On his release from prison, the employee reapplied and was rehired as a custodian by Honeywell in 1984. The individual befriended a female co-worker assigned to his maintenance crew and dated the female employee. The female employee severed the relationship and requested a transfer from the particular Honeywell facility.

The individual began to harass and threaten the female employee both at work and at her home. On July 1, 1988, the female employee found a death threat scratched on her locker door at work. The individual did not report to work after that date and Honeywell accepted his formal resignation on July 11, 1988. On July 19, 1988, the individual killed the female co-worker

---

[4] *See e.g., Ballard's Administratrix v. Louisville and Nashville Railroad Co.*, 128 Ky. 826, 110 S.W. 296 (1908).

[5] *See Missouri, Kansas, & Texas Railway Company v. Texas and Day*, 104 Tex. 237, 136 S.W. 435 (1911).

[6] *See Mallory v. O'Neil*, 69 So.2d 313 (Fla. 1954).

[7] *See also, Argonne Apartment House Company v. Garrison*, 42 F.2d 605 (D.C. Cir. 1930); *La Lone v. Smith*, 39 Wash. 2d 167, 234 Pa.2d 893 (1951).

[8] *See i.g., Geise v. Phoenix Company of Chicago, Inc.*, 246 Ill. App. 3d 441, 615 N.E. 2d 1179 (2nd District 1993), reversed on other grounds, 159 Ill. 2d 507, 639 N.E. 2d 1273 (1994).

[9] 496 N.W. 2d 419 (Minn. 1993).

in her driveway at close range with a shotgun. The individual was convicted of first degree murder and sentenced to life imprisonment.

The estate of the female employee brought a wrongful death action against Honeywell based on the theories of negligent hiring, negligent retention, and negligent supervision of a dangerous employee. The district court dismissed the negligent supervision theory because it derives from the *respondeat superior* doctrine, which the court recognized relied on the connection to the employer's premises or chattels.[10] The court additionally defined negligent hiring as predicated upon the negligence of the employer in hiring a person with known propensities, or propensities that should have been discovered by reasonable investigation, which the employer should have conducted prior to hiring the individual.[11] The court went further in distinguishing the doctrine based on the scope of the employer's responsibility associated with the particular job. In this case, the individual was a custodian, which did not expose him to the general public and required only limited interaction with fellow employees.

On appeal, the court in upholding the summary judgment for Honeywell, stated that:

> "To reverse the district court's determination on duty as it relates to hiring would extend Ponticas and essentially hold that ex-felons are inherently dangerous and that any harmful act they commit against persons encountered through employment would automatically be considered foreseeable. Such a rule would deter employers from hiring workers with criminal records and offend our civilized concept that society must make reasonable effort to rehabilitate those who have erred so that they can be assimilated into the community."[12]

Additionally, the court made the distinction between negligent hiring and negligent retention as theories of recovery. The court noted that negligent hiring focuses on the adequacy of the employer's pre-employment investigation of the employee's background, and the court found that there was a record of evidence of a number of episodes in which the individual's post-imprisonment employment at Honeywell demonstrated propensity for abuse and violence toward fellow employees, including sexual harassment of females and threatening to kill a co-worker during an angry confrontation after a minor car accident. The *Yunker* case exemplifies the general trend in the U.S. courts to permit theories of recovery for victims of workplace violence incidents. Employers should be cautious and take the appropriate steps in the hiring and screening phases to possibly avoid this potential area of

---

[10] Id., 496 N.W. 2d 422, citing *Semrad v. Edina Realty Inc.*, 493 N.W. 2d 528 at 534 (Minn. 1992).
[11] Id., 331 N.W. 2d 911.
[12] 496 N.W. 2d 423, quoting *Ponticas v. K.M.S., Inc.*, 331 N.W. 2d 907 (Minn. 1983).

legal liability. The trend to permit recovery under the theory of negligent hiring appears to be expanding in the courts, and employers can no longer rely upon the doctrine of the fellow servant rule to protect them in this area, especially following an incident of workplace violence.

Under the negligent hiring theory, most courts look to the case of *Ponticas v. K.M.S. Investments* for support for this cause of action. Under this decision, the employer has a "duty to exercise reasonable care in viewing all of the circumstances in hiring individuals who, because of their employment, may pose a threat of injury to members of the public."[13] Additionally, applying the "knew or should have known" standard to the duty and breach elements of the negligence standard, the court would examine whether the failure to check the background is "directly related to the severity of the risk."[14]

In the vast majority of negligent hiring cases, the failure to check backgrounds and other hiring criteria ultimately resulted in violent conduct on the part of the employee. However, several jurisdictions have applied this theory to a number of different factual situations. For example, in *Rosen-House v. Sloan's Supermarket*,[15] the court found the supermarket did not exercise due care when it hired an employee to deliver groceries, and the employee raped and robbed a customer on his fourth day on the job. And in *DiCosala v. Boy Scouts of America*,[16] in which the nephew of the Boy Scout camp counselor was accidentally shot in the neck while visiting the counselor's living quarters, the court found that the Boy Scouts knew the counselor kept guns on the premises and that the premises were open to everyone on the campgrounds. The New Jersey Supreme Court, in reversing and remanding the case for trial, recognized the negligent hiring theory for "incompetent, unfit or dangerous employee(s)."[17]

The primary defense, in addition to the deteriorating fellow servant rule, in cases of negligent hiring is the employer's foreseeability, or lack thereof, of the injury. In *Edwards v. Robinson-Humphrey Company*[18] the court found that it was not foreseeable by the company that a bond salesman would threaten a customer. The court found, despite the fact that the saleman lied about his credentials and qualifications, that even if the company knew about the lies, the company would not have known of the saleman's violent nature.

Another defense which may be available in some circumstances is the argument that the employer did not conduct a reasonable investigation into the employee's background given the totality of the circumstances.

---

[13] 331 N.W. 2d 906 (Minn. 1983) *Note:* This case involved an apartment caretaker who was hired by an apartment management company. The caretaker, Dennis Graffice, had past violent behaviors and felony convictions, and the apartment management company did not check his background or references. The apartment management company provided Mr. Graffice with a passkey, and he ultimately entered an apartment and sexually assaulted the tenant.

[14] Id.

[15] 540 N.Y.S. 2d 120 (Sup. Ct. 1988).

[16] 450 A.2d 508 (N.J. 1982).

[17] Id.

[18] 298 So. 2d 600 (Fla. App. 1983).

For example, there may be no need to conduct a criminal background check for the position of grave digger. Under this defense, the courts look to the totality of the circumstances, which can include the type of job, employee's history, nature of the employee's past crimes, frequency of contact with the public, frequency of contact with other employees, vulnerability of the public or other employees, and related factors.

Given the protections being provided to ex-felons under federal and state laws, employers may utilize these statutory laws as a possible public policy type defense. For example, in Kentucky, Kentucky Revised Statute 335B.2(1) provides the following:

> "...no person shall be disqualified from public employ-
> ment, nor shall a person be disqualified from pursuing,
> practicing, or engaging in any occupation, solely be-
> cause of a prior conviction of a crime, unless the crime
> is one limited to conviction for felonies, high misde-
> meanors, or crimes of moral turpitude."

Additionally, the Equal Employment Opportunity Commission, interpreting the Americans with Disabilities Act, has taken the position that checking an employee's background for arrest records, rather than conviction records, as well as checking into various physical or psychological areas at the pre-employment stage is discriminatory.

Employers may be able to argue these statutory laws as a defense on the basis of public policy given the restrictions placed upon the employer by these federal and state statutory laws. For example, a city government is interviewing for a receptionist. A young man is hired for the position, and his background is checked for prior convictions and the report reflects no convictions. The employee failed to inform the city of his prior arrests for statutory rape and assault which were vacated on a technicality. Additionally, the employee failed to inform the employer of his stay in a rehabilitation center for a mental illness. The employer would possess no knowledge of these events and would be prohibited under statutory law from inquiring into these areas.

Paralleling this defense, employers may argue the potential invasion of the individual's privacy rights. There is an inherent conflict between the employer's right to know and the individual's right to privacy which potentially could be utilized as a defense. Depending on the facts, a defense could be present whereby the individual's privacy rights would be violated if the employer inquired about an unrelated issue such as a domestic situation or juvenile conviction.[19]

---

[19] *But see Bryant v. Livigni*, 619 N.E. 2d 550 (Ill. App. 1993) where the manager of the store spotted a 10-year-old boy urinating on the exterior of the store. The manager yelled at the boy, used racial language, and followed the boy to the car where he pulled a 4-year-old from the car, threw him in the air and caused serious injuries. The manager possessed two prior incidents of throwing milk crates at an employee and has thrown his 13-year-old son, breaking his collarbone. The court found the supermarket liable for both compensatory and punitive damages.

Given the statutory laws and the case law in this area, employers are being placed on a narrow pathway whereby, if appropriate background checks are not made, there can be liability, and if the backgrounds of individuals are checked too closely, liability may be incurred. Prudent professionals should analyze this area carefully and ensure that the perimeters on both sides of the potential liability spectrum are avoided.

(Modified for the purposes of this text)

Emma CARTER, as Administrator of
the Estate of Emma L. Hopkins,
Deceased, Plaintiff-Appellee,
v.
SKOKIE VALLEY DETECTIVE AGENCY, LTD.,
an Illinois corporation,
Defendant-Appellant.
No. 1-91-3732.
Appellate Court of Illinois,
First District, First Division.
Dec. 6, 1993.

Rape and murder victim's mother sued detective agency that employed perpetrator as security guard, alleging that perpetrator was negligently hired. The Circuit Court, Cook County, Walter B. Bieschke, J., entered judgment on jury's verdict for victim's mother, and denied agency's motions for judgment n.o.v., and agency appealed. The Appellate Court, Buckley, J., held that agency's hiring of perpetrator was not proximate cause of victim's death.

Reversed and remanded with instructions.

———————————

Lord, Bissell & Brook (Hugh C. Griffin, Sandra K. Macauley, of counsel), and Purcell & Wardrope, Chtd. (Paul V. Kaulas, of counsel), Chicago, for defendant-appellant.

Levin & Perconti, P.C., Chicago (Steven M. Levin, Joel Ostrow, of counsel), for plaintiff-appellee.

Justice BUCKLEY delivered the opinion of the court:

Following a jury trial, Terry Harris was convicted of the murder, aggravated criminal sexual assault, and aggravated kidnapping of Emma L. Hopkins. (*People v. Harris* (1989), 132 Ill.2d 366, 138 Ill.Dec. 620, 547 N.E.2d 1241.) This case involves the civil action for money damages brought by Hopkins' mother, Emma Carter, as administrator of her daughter's estate, against Harris' employer, the Skokie Valley Detective Agency, Ltd. ("Skokie Valley"). Count I of plaintiff's complaint was a wrongful death claim and count II was a survival claim. Plaintiff alleged that Skokie Valley was negligent in hiring Harris as a security guard and that this negligent hiring was a proximate cause of the kidnap, rape, and murder of her daughter. Skokie Valley moved for a directed verdict at the end of plaintiff's case and then again after the presentation of all the evidence. The trial judge denied both motions. The jury ruled for plaintiff and against Skokie Valley under count I (wrongful death) in the sum of $10,000 and under count II (survival action) in the sum of $500,000. The trial judge denied Skokie Valley's motions for judgment not-withstanding the verdict ("judgment *n.o.v.*") or, in the alternative, a new trial and entered judgment on the jury's verdict. On appeal, Skokie Valley argues: (1) that it was entitled to judgment *n.o.v.* because, as a matter of law, Skokie Valley's alleged negligence was not a proximate cause of plaintiff's decedent's injuries and death; and (2) that the trial judge's inconsistent and prejudicial evidentiary rulings entitled Skokie Valley to a new trial on the issue of liability.

During 1983, Harris worked for the security guard firm of L.J. Stamps. His supervisor was Thomas Davis. Davis stated that Harris resigned in September 1983, because he had found a better paying job. At the end of 1983, the L.J. Stamps agency was taken over by Skokie Valley and, on January 1, 1984, Davis officially began working for Skokie Valley. On October 20, 1984, Harris applied for a job with Skokie Valley. Davis remembered that Harris had left L.J. Stamps the previous year in "good standing" and, therefore, he considered Harris to be essentially a "re-hire." Consequently, he neglected to conduct the detailed investigation and background check which would ordinarily be conducted and which was required by the Private Detective, Private Alarm, and Private Security Act of 1983. (Ill.Rev.Stat. 1991, ch. 111, par. 2651 *et seq.* (now 225 ILCS 445/1 (West 1992)).) Thus, he never discovered that Harris had been fired by North Central Security, another security guard firm, within the last year or that he had several misdemeanor convictions and an arrest warrant outstanding.

Apparently, Harris had been arrested on July 11, 1982, on a misdemeanor charge of unlawful use of weapons because he was carrying a gun while off duty. He pled guilty and was sentenced to a year of conditional discharge. On May 6, 1984, he was again arrested for carrying a gun while off duty. He was again convicted of unlawful use of weapons and this time sentenced to 30 days incarceration and one year probation. On May 29, 1984, Harris pulled a gun on two police officers who were walking up a flight of stairs in his building in connection with a narcotics investigation of another tenant. The officers wrested the gun from Harris' control and, after being arrested and taken to the station, he was found to be in possession of marijuana. He was subsequently convicted of aggravated assault and possession of marijuana. Finally, on September 8, 1984, defendant again was arrested for carrying a gun while off duty. He never appeared on the scheduled court date and, at the time he was arrested for plaintiff's decedent's murder, an arrest warrant was outstanding.

Davis admitted that, if he had known that Harris had several criminal convictions, or was on probation, or had a warrant outstanding for his arrest, he never would have hired him. Neil Meccia, Skokie Valley's majority shareholder, also stated that he would never hire any person who had been convicted of aggravated assault or unlawful use of weapons.

In effect, Davis "rehired" Harris "on the spot" and gave him a Skokie Valley Detective Agency I.D. card and a patch to sew on his uniform. Davis told Harris to report for duty the next day at the Amoco station located at Pulaski and Division streets in Chicago. On October 21, 22, and 23, Harris worked the 2 p.m. to 10 p.m. shift at the Amoco station. The plaintiff's decedent, Emma Hopkins, was employed as a cashier at the Amoco station.

At approximately 1:45 p.m. on October 28, 1984, the date that Harris murdered Emma, Harris arrived at the Amoco station wearing his uniform and carrying his gun. Raphael Davidson, a co-worker of Emma's, stated that Harris told them he was not scheduled to work at the station that day. He testified that Harris explained that he was assigned to work somewhere else and asked Emma to give him a ride if she was going in the same direction. Emma agreed to give him a ride and they left together in her car.

Her body was discovered the next morning at approximately 11 a.m. stuffed under a storage tank at a vegetative oil processing plant located at 1301 East 99th Street in Chicago. Skokie Valley was unable to locate its roster sheets for October 28, 1984, which would have illustrated where Harris was supposed to be working that day. Davis recalled, however, that Harris had requested that weekend off in order to move. Ultimately, the judge gave the jury the "missing evidence" instruction which allows the finder of fact to construe an absent document against the party who fails to produce it at trial. *Saunders v. Department of Public Aid* (1990), 198 Ill.App.3d 1076, 1082, 145 Ill.Dec. 118, 122, 556 N.E.2d 736, 740.

On appeal, Skokie Valley does not contest that it was negligent in hiring Terry Harris as a security guard in view of his convictions for unlawful use of weapons and aggravated assault. Skokie Valley argues, however, that the negligent hiring of Harris was not a proximate cause of plaintiff's decedent's injuries and death as a matter of law. Skokie Valley also contends that, as a matter of public policy, a security guard company's liability should not be extended to the off duty, off premises crimes alleged in this case. Plaintiff maintains, on the other hand, that there was ample evidence from which the jury could have reasonably concluded that Skokie Valley proximately caused the plaintiff's decedent's injuries and death. Plaintiff asserts that, if it were not for Skokie Valley's negligence, "this demonstrably unfit person" would not have been in Emma Hopkins' life.

[1, 2] It is established in Illinois that a cause of action exists against an employer for negligently hiring someone the employer knew or reasonably should have known was unfit for the job in the sense that the employment would place the employee in a position where his unfitness would create a foreseeable danger to others. (*Bates v. Doria* (1986), 150 Ill.App.3d 1025, 1030, 104 Ill.Dec. 191, 195, 502 N.E.2d 454, 458.) In addition, this action for negligent hiring can be successful "even though the employee commits the criminal or intentional act outside the scope of employment," (*Gregor v. Kleiser* (1982), 111 Ill.App.3d 333, 338, 67 Ill.Dec. 38, 42, 443 N.E.2d 1162, 1166.) However, proximate cause is an

essential element of any negligence action and, although
an employer may have negligently hired someone who caused
harm to another, liability will only attach if the
injuries were brought about by reason of the employment
of the unfit employee. (*Bates*, 150 Ill.App.3d at 1031,
104 Ill.Dec. at 195, 502 N.E.2d at 458.) Therefore, "in
ascertaining whether there is a causal connection between
the employer's negligence and plaintiff's injuries, it
is necessary to inquire whether the injury occurred by
virtue of the servant's employment" (*i.e.*: "because of
the employment."). (*Bates*, 150 Ill.App.3d at 1032, 104
Ill.Dec. at 196, 502 N.E.2d at 459.) In other words,
the employment itself must create the situation where
the employee's violent propensities harm the third
person. *Coath v. Jones* (1980), 277 Pa.Super. 479, 482,
419 A.2d 1249, 1250.

> "A proximate cause is one which produces
> the injury through a natural and con-
> tinuous sequence of events unbroken by
> any effective intervening cause. [Cita-
> tion.] If the negligence charged does
> nothing more than furnish a condition
> which made the injury possible and that
> condition causes an injury by the sub-
> sequent independent act of a third party,
> the creation of that condition is not
> the proximate cause of the injury. [Ci-
> tation.] The subsequent independent act
> becomes the effective intervening cause
> which breaks the causal connection, and
> itself becomes the proximate cause."
> (*Escobar v. Madsen Construction Co.*
> (1992), 226 Ill.App.3d 92, 94, 168
> Ill.Dec. 238, 239, 589 N.E.2d 638, 639,
> citing *Kemp v. Sisters of the Third Order
> of St. Francis* (1986), 143 Ill.App.3d
> 360, 361, 97 Ill.Dec. 709, 710, 493
> N.E.2d 372, 373.)

Generally, the existence of proximate cause is a question
of fact for the jury. (*Malorney v. B & L Motor Freight,
Inc.* (1986), 146 Ill.App.3d 265, 269, 100 Ill.Dec. 21,
24, 496 N.E.2d 1086, 1089.) If the evidence is insufficient to establish that the employer's negligence
proximately caused plaintiff's injuries, however, the

employer is entitled to judgment as a matter of law. *Escobar*, 226 Ill.App.3d at 94, 168 Ill.Dec. at 239, 589 N.E.2d at 639.

[3] In this case, as a matter of law, Skokie Valley's negligence in hiring Harris was not a proximate cause of plaintiff's decedent's injuries and death and Skokie Valley was entitled to judgment *n.o.v.* In *Easley v. Apollo Detective Agency, Inc.* (1979), 69 Ill.App.3d 920, 923, 26 Ill.Dec. 313, 315, 387 N.E.2d 1241, 1243, a security guard for the defendant security company who was assigned to the Chicago Beach Tower Apartments used his passkey to enter plaintiff's apartment where he then assaulted her. The evidence showed that the guard had several prior criminal convictions. The *Easley* court reasoned that these facts could support a claim for willful and wanton hiring because, a connection with his employment as an armed security guard, he was in possession of a passkey to the apartments he was employed to protect.

In *Malorney*, 146 Ill.App.3d at 265, 100 Ill.Dec. at 21, 496 N.E.2d at 1086, the defendant trucking company hired an over-the-road driver without checking whether he had criminal convictions. In fact, the driver had a history of convictions for sex-related crimes and, the previous year, had been arrested for aggravated sodomy of two teenage hitchhikers while he was driving an over-the-road truck for another company. Subsequently, while employed for the defendant company, he picked up the plaintiff and repeatedly raped and sexually assaulted her in the sleeping compartment of the over-the-road truck he was provided by defendant. The *Malorney* court held that these facts were sufficient to submit the case to the jury on the issue of negligent hiring and entrustment because the employer supplied defendant with the over-the-road truck which he used to facilitate the assault and it was foreseeable that an over-the-road trucker would pick up hitchhikers.

In *Gregor*, 111 Ill.App.3d at 333, 67 Ill.Dec. at 38, 443 N.E.2d at 1162, the defendant held a party at his parents' house and, according to the allegations of the complaint, hired a friend who he knew had a reputation for violence to act as the "bouncer." During the party, the "bouncer" allegedly attacked the plaintiff and caused severe and permanent injuries. The *Gregor* court held that these allegations were sufficient to state a cause of action under both the theory of *respondeat superior* and the theory of negligent hiring.

In *Bates*, 150 Ill.App.3d at 1025, 104 Ill.Dec. at 191, 502 N.E.2d at 454, however, the court held that summary judgment for defendant was proper where an off-duty sheriff's deputy who the department knew or should have known was psychologically unstable raped and assaulted plaintiff. The court held that plaintiff's injuries were not in any way the result of the sheriff's employment where he was not on duty, he was out of uniform and where the plaintiff did not even know he was a sheriff's deputy until he told her after the assault. In *Escobar*, 226 Ill.App.3d at 92, 168 Ill.Dec. at 238, 589 N.E.2d at 638, the court also held that summary judgment for defendant on a claim for negligent hiring and retention was proper where plaintiff was shot by a co-worker before work with this own weapon and two blocks from the jobsite even though defendant knew that the co-worker was dangerous and had threatened the victim previously. The *Escobar* court reasoned that the employment provided a condition for the attack, but was not what proximately caused the attack.

We believe that the existence of proximate cause in the instant case is not as clear as in *Easley, Malorney,* and *Gregor*, and, on the surface, not as absent as in *Escobar* and *Bates*. However, after analyzing the reasoning of the above cited cases in view of the facts of this case, we believe that, as a matter of law, the fact of Harris' negligent employment was not the cause of plaintiff's decedent's injuries and death. Unlike in *Easley, Malorney,* and *Gregor,* Harris' employment merely furnished a condition which made the rape and murder possible, but it was not the result of "a natural and continuous sequence of events" set in motion by Skokie Valley's negligence and "unbroken by any effective intervening cause." (*Escobar*, 226 Ill.App.3d at 94, 168 Ill.Dec. at 239, 589 N.E.2d at 639, citing *Kemp*, 143 Ill.App.3d at 361, 98 Ill.Dec. at 710, 493 N.E.2d at 373.) It was not the fact Harris was a security guard that got him into Emma's car and proximately caused her injuries and death; it was the fact that she trusted him because she knew him from work where he happened to be employed as a security guard. In order to find liability in this case, we would have to reason that if Skokie Valley had not negligently hired Harris he would not have met Emma and she would not have trusted him enough to let him in her car. If we followed such reasoning then we would have had to find Skokie Valley liable if, after leaving the Amoco station alone, she had seen him on the street out of uniform and offered him a ride or if they ran into each other on a weekend at the supermarket. We do not believe that the concept of proximate cause should be extended this far. As illustrated by *Easley*, *Malorney*, and *Gregor*, there must be a tangible connection between the employee's violent tendencies, the particular job he is hired to do, and the harm to the plaintiff. No such connection is present here. If we were not to hold as we do, "an employer would essentially be an insurer of the safety of every person who happens to come into contact with his employee simply because of his status as an employee." (*Bates*, 150 Ill.App.3d at 1032, 104 Ill.Dec. at 196, 502 N.E.2d at 459.) Such a rule would make any person with a conviction for any violent crime virtually unemployable if the employer knew, or the employment was such that the employer should have known, of the employee's criminal record. No employer would hire such person for even the most menial of jobs if he would be liable for any criminal or intentional acts that the employee might inflict upon another who happened to have come into contact with him as a result of his employment. Consequently, we must reverse the judgment of the circuit court and remand with instructions to enter a judgment n.o.v. for Skokie Valley. In light of our disposition of this issue, we need not address Skokie Valley's second argument that the trial judge's inconsistent and prejudicial evidentiary rulings entitled it to a new trial.

For the foregoing reasons, we reverse the judgment of the circuit court of Cook County and remand with instructions to enter an order consistent with this opinion.

Reversed and remanded with instructions.

CAMPBELL and O'CONNOR, JJ., concur.

chapter seven

Negligent retention

"Life is tons of discipline."

Robert Frost

"People are talking about the new 'civilized' way to fire executives. You kick'em upstairs. They're given a little, a liberal tithe, nothing to do, and a secretary to do it with. What a way to go!"

Malcolm Forbes

Closely allied with the tort theory of negligent hiring is that of negligent retention. In general terms, the theory of negligent retention involves an employer knowing that an employee has a propensity toward violence in the workplace but permitting the employee to retain his/her employment status despite this knowledge by the employer. In *Yunker v. Honeywell, Inc.,*[1] the Minnesota Court of Appeals defined negligent retention as focused "on when the employer was on notice that an employee posed a threat and failed to take steps to ensure the safety of third parties."[2] In most circumstances, the theories of negligent hiring and negligent retention are closely allied. For example, an employer hires an individual with a criminal record but fails to properly screen and identify this individual as having the propensity toward workplace violence. Once the individual is employed, when the employer acquires knowledge of the particular background or propensity and fails to act to safeguard employees in the situation, the theory of negligent retention would be appropriate. In this example, both the theory of negligent hiring as well as negligent retention appear to be applicable.

Looking at the general theory of negligence, four basic elements are required to establish a *prima facie* case, i.e., duty, breach, causation, and damage. Under the negligent retention theory, the duty would be created when the employer had or acquired the knowledge that an employee with

[1] Id.
[2] *Yunker v. Honeywell, Inc.,* 496 N.W. 2d 423.

a propensity toward workplace violence was employed; the breach would apply when the employer failed to act or react to this knowledge; the causation would attach when the individual with a propensity actually assaulted or otherwise harmed fellow employees; and the damages would stem from the injury or death. The pivotal issue in most negligent retention cases involves whether the employer knew of the individual's propensity. In actuality, this is a Catch 22 for many employers. If the employer did not properly screen the individual prior to hiring and the individual performed a workplace violence incident, the negligent hiring theory would apply. If, in the event the employer did not acquire the knowledge during the hiring phase and permitted the employee to continue to work and later acquired information regarding the propensity and failed to react, the theory of negligent retention would apply.

The defenses available under this theory of negligence would parallel those discussed in Chapter 6. Additionally, prudent professionals should also address the basic elements of a negligence action, namely duty, breach, causation, and damages, to ascertain the ability of the plaintiff to prove each of these elements. Following is a basic format addressing the elements of negligent retention.

Duty
1. Did your company have a duty to check the individual's background?
2. Did the company have direct knowledge of the individual's violent tendencies?
3. Did the job require a criminal background investigation?
4. Did or should the company have known of the individual's violent behavior?
5. Did any agent of the company know of the individuals' violent behavior?

Breach of duty
1. What did the company do when knowledge of violent behavior became available?
2. Could the company have easily found out about the violent behavior?
3. Could the company have located the information regarding the violent behavior without violating any laws or invading the individual's privacy?

Causation
1. Was the individual's violent behavior foreseeable?
2. Was the past behavior of the individual indicative of the violent behavior?
3. Is there a causal link between the employer's knowledge, or lack thereof, and the incident?

Damages
1. What were the actual physical or psychological damages?
2. Was compensation provided through workers' compensation or other sources?
3. Are punitive damages permitted under this cause of action?
4. Were all alleged damages examined closely?

Prudent employers should take the appropriate steps to properly screen and evaluate employees during the pre-employment phase of the operation and continually throughout the individual's employment in order to avoid the possibility of liability in the area of negligent retention. Once an individual is employed, the employer appears to possess an affirmative duty to take appropriate steps to safeguard employees in the workplace once the employer knows the employee's propensity toward workplace violence. In essence, the employer must react once the knowledge is acquired in order to safeguard other employees in the workplace from the particular individual's propensity toward violence. However, this duty to act must be weighed in balance with the individual's rights on the job and individual privacy rights.

(Modified for the purposes of this text)

Farris BRYANT, a minor, by his next friend Diana Bryant,
and Diana Bryant, individually, Plaintiffs-Appellees,
v.
Mark LIVIGNI, Defendant and National Super Markets, Inc.,
Defendant-Appellant.
No. 5-92-0141.
Appellate Court of Illinois,
Fifth District.
Sept. 1, 1993.
Rehearing Denied Sept. 29, 1993.

Injured child and child's mother filed action against
store manager on theories of battery and invasion of
privacy and against store on theories of *respondeat
superior*, negligent retention of manager, and willful
and wanton retention of manager as employee. The Circuit
Court, St. Clair County, William B. Starnes and Milton
S. Wharton, JJ., found for plaintiffs and awarded
compensatory and punitive damages against store and store
manager. The store appealed. The Appellate Court Maag,
J., held that: (1) evidence of prior violent incidents
involving store manager was chargeable to his employer;
(2) evidence supported finding that store's retention
of manager as management employee constituted willful
and wanton misconduct supporting award of punitive
damages; and (3) outrageousness of manager's conduct did
not preclude judgment against *respondeat superior*
grounds.

Affirmed.

Welch, J., filed opinion concurring in part and
dissenting in part.

———————

Kortenhof & Ely, John D. Warner, Jr., St. Louis, MO,
for defendant-appellant.

John Long and Greg Weber, Troy, for plaintiffs-
appellees.

Justice MAAG delivered the opinion to the court:

FACTS

In 1970, Mark Livigni was a 16-year-old high school student. He was hired by National Super Markets, Inc., as a bagger and continued to work for National after graduating from high school. Livigni worked himself up to a management position, and in 1985 he was made manager of the National store in Cahokia. Livigni was evaluated on a regular basis by National with respect to his ability to relate to customers and employees. The record reflects that he always received good evaluations while employed at National.

On March 18, 1987, while off duty, Livigni stopped by the Cahokia National store. As manager, he was authorized to check and supervise the operation of the store even during off-duty hours. He was intoxicated at the time of this visit, which was a violation of National rules. Livigni observed a young man urinating on the store wall outside the east exit doors. He hollered at the young man and followed the fleeing youth to the parked vehicle of Diana Bryant.

Livigni pulled four-year-old Farris Bryant from the automobile. Farris, however, was not the boy who had been urinating. The boy Livigni had followed was Donya Jackson, age 10, who was also in the back seat of the Bryant automobile. Livigni shouted racial epithets and profanity at Diana Bryant and the children, and he attacked four-year-old Farris, throwing the child through the air.

Farris was taken to Centreville Township Hospital's emergency room for medical treatment. Farris was admitted to the hospital and released after four days. He was released from all medical treatment approximately one month after the battery.

In a multicount complaint, Farris Bryant and Diana Bryant sued both Livigni and his employer, National. Mark Livigni was accused of committing battery and invasion of privacy. National was alleged to be liable based upon theories of *respondeat superior* (for Livigni's actions), negligent retention of Livigni as an employee, and willful and wanton retention of Livigni as an employee.

At trial, Livigni's supervisor testified that during Livigni's 17-year tenure with National, Livigni had been a good employee. This supervisor never received any reports from customers or employees that Livigni had "violent-related" problems, although he was aware of a report that Livigni threw an empty milk crate which struck a co-worker.

Evidence was offered of two batteries committed by Livigni prior to his attack of Farris. In 1980, Livigni had a disagreement with a subordinate employee resulting in Livigni throwing an empty milk crate at the employee striking him on the arm and necessitating medical treatment. At the time of this battery, Livigni was an assistant store manager. A workers' compensation claim was filed against National by the injured employee. A short time after the workers' compensation claim was resolved, Livigni was promoted to store manager by National in spite of this incident.

The second battery occurred in 1985 when Livigni, while disciplining his 13-year-old son, threw the boy into a bed causing the boy to sustain a broken collar bone. In June 1986 Livigni pleaded guilty to aggravated battery to a child and was sentenced to two years' probation. He was still on probation at the time he attacked Farris.

Livigni testified at trial that he had not told any of his supervisors at National about the battery of his son. He admitted to telling employees of equal or lesser positions than himself about the battery. He considered these people to be his friends.

Plaintiffs sought to prove that Livigni's district manager and supervisor, Ben Rodell, was aware of Livigni's battery of his son by presenting the testimony of Gary Kapecchi, a former National employee. Kapecchi testified, over a hearsay objection, that he had discussed Livigni with Carl Oller, the man who replaced Livigni as the Cahokia store manager after Livigni was fired. Kapecchi testified that after the Bryant battery, Oller related to Kapecchi a prior conversation Oller had with Livigni and Rodell. This conversation concerned Livigni's battery of his son and how the incident had been blown out of proportion. Kapecchi testified that the conversation took place prior to March 18, 1987. Thus, it would have

			Verdicts*		
	Plaintiff	Defendant	Theory of recovery	Type of damages sought	Prevailing party and verdict
Count I	Farris	Livigni	Battery	Compensatory	Plaintiff-$20,000
Count II	Farris	Livigni	Battery	Punitive	Plaintiff—$0
Count III	Farris	National	Battery (*respondeat superior*)	Compensatory	Plaintiff—$20,000
Count IV	Farris	National	Battery (*respondeat superior*)	Punitive	Plaintiff—$0
Count V	Farris	National	Negligent Retention	Compensatory	Plaintiff—$15,000
Count VI	Farris	National	Willful & Wanton Retention	Punitive	Plaintiff—$115,00
Count VII	Diana	Livigni	Invasion of Privacy	Compensatory	Plaintiff—$2,000
Count VIII	Diana	National	Invasion of Privacy (*respondeat superior*)	Compensatory	Plaintiff—$2,000
Count IX	Diana	National	Invasion of Privacy/ Negligent Retention	Compensatory	Plaintiff—$8,000

*The numbering of the counts on the verdict forms differed from the numbering in the complaint. The numbers indicated in the chart are those used on the verdict forms

predated the Farris Bryant attack and shown that Livigni's supervisor had notice of the incident involving Livigni's son. Oller, Rodell, and Livigni all denied that any such conversation took place.

At the close of the evidence, the following counts were submitted to the jury for consideration and the following verdicts were returned.

Judgment was entered for Farris Bryant in the sum of $20,000 for compensatory damages and $115,000 in punitive damages. Judgment was entered for Diana Bryant in the sum of $8,000 for compensatory damages.

National's post-trial motion was denied, and this appeal ensued. Mark Livigni is not a party to this appeal.

I.

National first contends that the trial court erred in denying its motion for a directed verdict and further erred in refusing to grant its motion for a judgment notwithstanding verdict on the claims of both Farris and Diana Bryant (counts V and IX) which alleged National was negligent in retaining Mark Livigni as an employee. This allegation of error is premised on the proposition that insufficient evidence was presented to show that National knew or should have known that Livigni was not fit to serve as a store manager. National also argues that there was no showing that its retention of Livigni was a proximate cause of the plaintiff's injuries.

In considering this issue, we must remain cognizant of the relief sought by National. National is not seeking a new trial. Rather the only relief requested is a judgment notwithstanding the verdict. The standard of review in such a case is that set forth in *Pedrick v. Peoria & Eastern R.R. Co.* (1967), 37 Ill.2d 494, 510, 229 N.E.2d 504, 513-14:

> "verdicts ought to be directed and judg-
> ments *n.o.v.* entered only in those cases
> in which all of the evidence, when viewed
> in its aspect most favorable to the
> opponent, so overwhelmingly favors movant
> that no contrary verdict based on that
> evidence could ever stand."

According to National, there was no evidence that it knew or had reason to know that Livigni was anything other than "an excellent store manager, fit for his position." To support this argument, National claims that there was conflicting evidence regarding the 1980 incident where Livigni threw a milk crate at a co-worker causing injury. It argues that the 1980 incident was of uncertain origin since differing versions of the incident and its cause were presented in the evidence. It asserts that due to this conflicting evidence the incident could not form the basis for a negligent retention claim.

[1] We find this argument to be unpersuasive. Under the *Pedrick* standard, we must refrain from resolving conflicts in the testimony. That is the function of the jury. Our charge is to view the evidence in the light most favorable to the party which prevailed below. In doing so, we believe that the evidence regarding this incident was sufficient to present an issue for the jury to resolve on the question of whether National was negligent in retaining Livigni in its employ.

Rather than disciplining Livigni after he injured a subordinate employee in an unprovoked attack, National promoted him following the resolution of the injured employee's workers' compensation claim.

National further argues that it had no knowledge of the incident involving Livigni's son that resulted in Livigni's felony conviction for aggravated battery of a child. Relying upon *Campen v. Executive House Hotel, Inc.* (1982), 105 Ill.App.3d 576, 61 Ill.Dec. 358, 434 N.E.2d 511, National points to the general rule which states that to impute knowledge of this occurrence to National a showing was required that an agent or employee of National had notice or knowledge of the incident and that the knowledge concerned a matter within the scope of the agent's authority. According to National, evidence of such knowledge was lacking.

This claimed ignorance of the facts is premised upon several different bases. Each will be considered in order.

[2] National first admits that Livigni told employees of equal or lesser rank within the corporation about the battery involving his son. However, it claims that this is insufficient notice to the corporation. It argues that the people Livigni told were his "friends" and that as mere "co-workers" of equal or subordinate position no notice could legally be imputed to National. We disagree.

The issue is not whether these persons were Livigni's friends. The law does not recognize a "friendship privilege." Being his friends did not diminish the duty of loyalty owed to the corporation (Restatement (Second) of Agency sec. 39, at 130 (1958)), nor did this friendship relieve them of the duty to act in the best interest of the corporation. The duty is clearly set forth in comment *a* to Section 39 of the Restatement as follows:

> "Authority is conferred to carry out the purposes of the principal and not those of someone else. These purposes, as manifested to the agent, constitute the benefit for which, as the agent should realize, the agency is created. In business enterprises, an agent normally has no authority… to conduct his principal's

business with a mind to the benefit of
others." Restatement (Second) of Agency,
sec. 39, comment *a*, at 130 (1958).

The friendship enjoyed by Livigni and his co-workers
does not constitute a reason to ignore the information
admittedly learned by these co-workers, nor does it
constitute a reason to forego imputing this information
to National.

Assuming for the sake of discussion that no National
employees in a position to supervise Livigni knew he
had battered his son, we believe that the knowledge
admittedly held by the co-workers must still be imputed
to National. *St. Paul Mercury Insurance Co. v. Statistical
Tabulating Corp.* (1987), 155 Ill.App.3d 545, 550, 108
Ill.Dec. 272, 275, 508 N.E.2d 433, 436.

[3]   There is no rule which requires this knowledge
held by co-workers to be disregarded unless the co-worker
possessing the knowledge "outranks" the employee in
question. To the contrary, whether the agent obtaining
the knowledge is in a subordinate or a superior position
in the corporation, that knowledge is still chargeable
to the corporation if the information concerns a matter
within the scope of the agent's authority. *St. Paul
Mercury Insurance Co.*, 155 Ill.App.3d at 550, 108 Ill.Dec.
at 275, 508 N.E.2d at 436.

[4]   In this case, the defendant tendered, as its
instruction No. 2, IPI No. 50.06 (Illinois Pattern Jury
Instruction, Civil, No. 50.06 (2d ed. 1971)), and it
was given by the court. The instruction explained to
the jury the concept of agency and what was required
for the jury to conclude that a matter was within the
scope of an agent's authority. Viewing the evidence in
the light most favorable to the plaintiff (*Pedrick*), we
believe that a reasonable jury could have concluded that
the information concerning the battery of Livigni's son,
learned by these co-workers, was within the scope of
their authority to act upon. Whether reported to higher
authorities or not, the information still constitutes
"corporate knowledge." (*Campen*, 105 Ill.App.3d at 586,
61 Ill.Dec. at 364, 434 N.E.2d at 517.) In such a case,
their knowledge is chargeable to National.

National claims that evidence that Livigni's supervisor was aware of this incident involving his son was improperly admitted. It claims this evidence was hearsay, should not have been admitted over its objection, and as such could not constitute notice of the incident.

Specifically, National complains that a former National employee, Gary Kapecchi, was allowed to testify over its hearsay objection about a conversation he had with Carl Oller, a National store manager. According to Kapecchi, Oller told him that before the incident involving Farris Bryant, National management knew of the incident involving Livigni's son. Kapecchi stated that Oller told him of a meeting attended by Oller, Livigni, and Ben Rodell, Livigni's supervisor. At this meeting, there was allegedly a discussion about the incident involving Livigni's son.

[5, 6] This objection has been waived. National has not requested a new trial. It has requested a judgment *n.o.v.*, an outright reversal. In such a case, error in the admission or rejection of evidence is not preserved for review. (*Chiribes v. Bjorvik* (1968), 100 Ill.App.2d 150, 241 N.E.2d 626: *Gundich v. Emerson-Comstock Co.* (1960), 21 Ill.2d 117, 171 N.E.2d 60.) Despite this waiver, we believe that any error in the admission of this evidence was at most cumulative and harmless. Our previous discussion demonstrates that knowledge of this incident is imputable to National through the knowledge of other employees.

Finally, we have considered the claim that National's retention of Livigni in its employ was not a proximate cause of the plaintiff's injuries. In its brief, National devotes only two sentences to this allegation and never states why it believes proof of proximate cause is lacking. The issue is waived. Supreme Court Rule 341(3)(7)(134 Ill.2d R. 341(e)(7)).

[7] We conclude that the circuit court did not err in denying National's motion for a directed verdict, nor did it err in refusing to grant a judgment *n.o.v.* on plaintiff's claim of negligent retention. Viewing the evidence in the light most favorable to the plaintiff, we cannot state that the evidence so overwhelmingly favored National that this verdict cannot stand.

II.

National next claims that the circuit court should have directed a verdict in its favor or granted a judgment *n.o.v.* on the plaintiff's punitive damages claim. This count alleged that National's retention of Mark Livigni as a management employee constituted willful and wanton misconduct (count VI).

[8] National reiterates its claim that there was insufficient evidence to prove negligence. It, therefore, states that if the evidence was insufficient to prove simple negligence, then as a matter of law the evidence must be insufficient to prove willful and wanton misconduct. It relies upon *In re Application of Busse* (1984), 124 Ill.App.3d 433, 79 Ill.Dec. 747, 464 N.E.2d 651, as authority for this argument. As an abstract principle of law, we agree that if the evidence was insufficient to prove negligence it must necessarily also be insufficient to prove willful and wanton misconduct. However, we have already found that sufficient evidence was presented to create a jury issue on the negligent-retention claim. The jury resolved that issue against National. Now we must decide whether sufficient evidence was presented to support the punitive damages award.

The Restatement (Second) of Torts, Section 909, at 467 (1977) provides:

> "Punitive damages can properly be awarded against a master or other principal because of an act by an agent if, but only if,
>
> * * * * * * * * * * * * * *
>
> (b) the agent was unfit and the principal or a managerial agent was reckless in * * * retaining him."

This count did not seek to impose liability upon the defendant vicariously. Rather, the plaintiff's cause of action alleged wrongful conduct on the part of National itself. Section 909(b) of the Restatement (Second) of Torts speaks directly to the issue under discussion. So too does the case of *Easley v. Apollo Detective Agency,*

Inc. (1979), 69 Ill.App.3d 920, 26 Ill.Dec. 313, 387 N.E.2d 1241.

[9, 10] *Easley* recognized that it is settled law that a cause of action exists in Illinois for negligent hiring of an employee, and that if the defendant's conduct could properly be characterized as willful and wanton then punitive damages are recoverable. (*Easley*, 69 Ill.App.3d at 931, 26 Ill.Dec. at 320, 387 N.E.2d at 1248.) We see little difference between a punitive damages claim for willfully and wantonly hiring an employee in the first instance and a claim for willfully and wantonly retaining an unfit employee after hiring. In both instances, the interest to be protected is the same. Employers that wrongfully (whether negligently or willfully and wantonly) hire or retain unfit employees expose the public to the acts of these employees. In such cases it is not unreasonable to hold the employer accountable when the employee causes injury or damage to another. The principle at issue is not *respondeat superior*, although that may also be implicated. Rather, the cause of action is premised upon the wrongful conduct of the employer itself. (*Easley*, 69 Ill.App.3d at 931, 26 Ill.Dec. at 320, 387 N.E.2d at 1248.) For this reason, the cause of action is distinguishable from the situation in *Mattyasovszky v. West Towns Bus Co.* (1975), 61 Ill.2d 31, 330 N.E.2d 509, where the plaintiff sought to hold the employer responsible for an employee's acts based upon principles of vicarious liability.

[11] Punitive damages are not awarded as compensation. They serve the purpose of punishing the offender and deterring others from committing like conduct. (*Loitz v. Remington Arms Co.* (1990), 138 Ill.2d 404, 150 Ill.Dec. 510, 563 N.E.2d 397.)

> "It has long been established in this State that punitive or exemplary damages may be awarded when torts are committed with fraud, actual malice, deliberate violence or oppression, or when the defendant acts willfully, or with such gross negligence as to indicate a wanton disregard of the rights of others." *Kelsay v. Motorola, Inc.* (1978), 74 Ill.2d 172, 186, 23 Ill.Dec. 559, 565, 384 N.E.2d 353, 359.

[12, 13] Citing *Kritzen v. Flender Corp.* (1992), 226 Ill.App.3d 541, 168 Ill.Dec. 509, 589 N.E.2d 909, National argues that punitive damages require proof beyond that necessary for the basic action. We agree that proof of misconduct more egregious than simple negligence is required. Nevertheless, a party may plead in alternative counts that certain conduct constitutes either negligence or willful and wanton misconduct. (*Alley v. Champion* (1979), 75 Ill.App.3d 878, 31 Ill.Dec. 533, 394 N.E.2d 735.) The question then becomes one for the jury to determine, whether the conduct amounted to simple negligence or rose to the level of willful and wanton misconduct. *Giers v. Anten* (1978), 68 Ill.App.3d 535, 24 Ill.Dec. 878, 386 N.E.2d 82.

[14, 15] Because the cause of action pleaded in count VI was not based upon *respondeat superior*, there was no need to demonstrate that the wrongfully retained employee was in a managerial position. (*Cf. Kemner v. Monsanto Co.* (1991), 217 Ill.App.3d 188, 160 Ill.Dec. 192, 576 N.E.2d 1146 (dealing with punitive-damage claims based upon *respondeat superior*).) Rather, focus of the inquiry was on whether the employer committed willful and wanton misconduct by retaining the unfit employee in the service of the business. In other words, the position of those persons with the power to make decisions regarding the employment status of the unfit employee must be examined, not the position occupied by the person alleged to be unfit.

This does not mean that the position in the corporate chain of command of the unfit employee is irrelevant. An unfit supervisor is frequently in a position to do more harm than a lower-level employee. Also, as the authority of the unfit employee increases, the likelihood that his acts will be deemed corporate policy increases. There is a distinct qualitative difference between acts committed by a corporate official with authority to set policy and acts committed by an employee with no management authority. (*Kemner*). But, given proper circumstances, an employee with little or no authority, who was wrongfully retained, could cause an employer to be subjected to a punitive award if the employer was willful and wanton in its retention of the employee.

[16] The jury heard evidence that Livigni attacked a fellow employee in 1980 and was then promoted. He injured his own son, he was convicted of aggravated battery in a criminal proceeding, and members of National's management admittedly knew of that incident. National took no action. Then while a store manager, in an intoxicated state, he attacked a four-year-old child and threw him through the air, resulting in his hospitalization. National itself characterizes this attack on young Farris as outrageous. We cannot say the jury was unjustified in concluding the same and also concluding that retaining this man as a managerial employee constituted willful and wanton misconduct.

[17] Since a motion for judgment notwithstanding the verdict presents only a question of law, it will be granted only if there is a total failure to prove any essential element of the plaintiff's case. (*Baier v. Bostitch* (1993), 243 Ill.App.3d 195, 202, 183 Ill.Dec. 455, 460, 611 N.E.2d 1103, 1108.) Under *Pedrick*, we must view the evidence of this case in the light most favorable to the plaintiff. There is no claim that the jury did not receive proper instruction, and given this standard of review, we cannot say that there was a total failure of proof on any element, nor can we say that the evidence so overwhelming favored the defendant that the verdict cannot stand. There was sufficient credible evidence presented to lead the jury to render the verdict that it did. We will not disturb that determination. The circuit court did not err in refusing to grant a directed verdict, nor did it err in refusing to grant a judgment *n.o.v.* on plaintiff's punitive damages claim (count VI).

III.

[18] Finally, National asks that a judgment *n.o.v.* be entered in its favor on the plaintiff's *respondeat superior* claims counts III and VIII). National argues in its brief that:

> "Mark Livigni's conduct on March 18, 1987, in arriving at the store in an intoxicated state, contrary to known store policy, shouting racial epithets

> at a customer and her young children,
> and throwing and attacking an innocent
> four-year-old child, was unexpectable and
> outrageous and thus did not come within
> the scope of Livigni's employment."

We agree with National that the conduct was outrageous. We disagree that this precludes a judgment against National based upon principles of *respondeat superior*.

National appears to argue that because the conduct was reprehensible that it should be insulated from liability. Relying upon Section 245 of the Restatement (Second) of Agency (Restatement (Second) of Agency sec. 245, at 537 (1958)), and *Harrington v. Chicago Sun-Times* (1986), 150 Ill.App.3d 797, 104 Ill.Dec. 69, 502 N.E.2d 332, which cites Section 245 with approval, National argues that we should find that Livigni's conduct was "unexpectable" in view of his duties.

Plaintiff cites the case of *Metzler v. Layton* (1939), 373 Ill. 88, 25 N.E.2d 60, in opposition. In *Metzler*, an employer was found to be responsible based upon *respondeat superior* principles for the act of an office manager who mistakenly shot an innocent bystander after a robbery. The court quoted with approval the following language from a Texas decision:

> "The master who puts a servant in a place
> of trust or responsibility, or commits to
> him the management of his business or the
> care of his property, is justly held
> responsible when the servant, through lack
> of judgment or discretion, or from infir-
> mity of temper, or under the influence
> of passion aroused by the circumstances
> in the occasion, goes beyond the strict
> line of his duty or authority and inflicts
> an unjustifiable injury on a third per-
> son." *Metzler*, 373 Ill. at 91-92, 25N.E.2d
> at 61, quoting *Central Motor Co. v. Gallo*
> (Tex.Civ.App.1936), 94 S.W.2d 821, 822.

We believe that the foregoing language is fully applicable to this case.

We also believe that the *Harrington* case relied on by National is distinguishable. National contends in its brief that *Harrington* involved a shooting by a newspaper delivery man of a person suspected of stealing newspapers. However, contrary to the claims of National, in *Harrington* the delivery man specifically *denied* suspecting the victim of stealing newspapers. (*Harrington*, 150 Ill.App.3d at 800-02, 104 Ill.Dec. at 71-72, 502 N.E.2d at 334-35.) Under these circumstances, it would be difficult to claim that the delivery man was trying to further the employer's interest or protect the employer's property.

[19] Moreover, we believe that the criteria for imposition of liability suggested by Section 245 of the Restatement (Second) of Agency have been satisfied. Given the evidence of Livigni's prior acts that preceded the Farris Bryant attack, it was not "unexpectable" that an incident of violence might take place.

In order to impose liability upon National, it was not necessary that Livigni be motivated *solely* by a desire to further National's interest. It is sufficient if his actions were prompted only *in part* by a purpose to protect store property or further the employers business. (*Wilson v. Clark Oil & Refining Corp.* (1985), 134 Ill.App.3d 1084, 1089, 90 Ill.Dec. 40, 43, 481 N.E.2d 840, 843.) The evidence was sufficient to justify such a conclusion by the jury.

Finally, the actions of Livigni in attacking Farris were committed within the constraints of the authorized time and location of his employment, thus bolstering a finding that the battery occurred within the course and scope of his employment. *Sun-seri v. Puccia* (1981), 97 Ill.App.3d 488, 493, 52 Ill.Dec. 716, 721, 422 N.E.2d 925, 930.

Accordingly, we conclude that while the actions of Livigni were obviously not the normal actions of a store manager, we cannot say the jury was unjustified in concluding that Livigni was acting within the course and scope of his employment at the time of the Farris Bryant incident.

CONCLUSION

For the foregoing reasons, the judgment of the circuit court of St. Clair County is affirmed.

Affirmed.

CHAPMAN, P.J., concurs.

Justice WELCH, concurring in part and dissenting in part:

I concur with the majority's opinion with respect to, and would affirm the judgment of the circuit court on the jury verdict against National Food Stores on, the *respondeat superior* counts of plaintiffs' complaint. With respect to the majority's opinion concerning the judgment against National on the negligent and wilful and wanton retention counts of plaintiffs' complaint, however, I must respectfully dissent.

A cause of action for negligent hiring or retention must establish that a "particular unfitness" of an applicant or employee creates a danger of harm to third parties of which the employer knew, or should have known, when he hired, placed, or continued the person in employment where he could injure others. (*Huber v. Seaton* (1989), 186 Ill.App.3d 503, 508, 134 Ill.Dec. 285, 288, 542 N.E.2d 464, 467.) Under plaintiffs' complaint alleging wilful and wanton retention, plaintiffs were further required to show a deliberate intention to harm or an utter indifference to or conscious disregard for the safety of others in the retention of defendant Livigni as an employee. (*Gregor v. Kleiser* (1982), 111 Ill.App.3d 333, 336, 67 Ill.Dec. 38, 41, 443 N.E.2d 1162, 1165.) In the instant case plaintiffs failed to establish by the two incidents occurring, respectively, two and seven years earlier that defendant Livigni had a violent propensity that could manifest itself in Livigni's dealings with those persons in the zone of foreseeable risk created by Livigni's employment.

As the majority notes, Livigni had been employed at National for 17 years at the time of the incident and was a good employee; National had never received any reports from customers or employees that Livigni had any violence-related problems. The 1980 incident in which a subordinate employee was struck by an empty milk crate thrown by Livigni and the 1985 incident in which he injured his 13-year-old son while disciplining him involved persons over whom Livigni was trying to assert authority, not the general public. As such, National was not put on notice by these two incidents that Livigni's continued employment created a danger of harm to the store customers or others with whom Livigni, as store manager, would have come in contact. In order to prove the negligent and the wilful and wanton retention counts of the complaint to the jury, plaintiffs were required to establish a causal relationship between the particular unfitness and the negligent act of the agent. (*Huber*, 186 Ill.App.3d at 508, 134 Ill.Dec. at 288, 542 N.E.2d at 467.) In my opinion, plaintiffs failed to show a connection between the nature of Livigni's prior acts of violence and the conduct of Livigni towards the plaintiffs in the instant case.

From a practical standpoint, the majority's opinion sends a message to all employers that in order to insulate themselves from liability for negligent or wilful and wanton retention any employee who has ever had an altercation on or off the workplace premises must be fired. Moreover, the majority opinion places an unreasonable investigative burden upon the employer by forcing the employer to discover, retain, and analyze the criminal records of its employees. Is not the majority's opinion then at cross-purposes with the established public policy and laws of Illinois protecting the privacy of citizens and promoting the education and rehabilitation of criminal offenders? See Ill.Rev.Stat.1991. ch. 68, par. 2-103 (making it a civil rights violation to ask a job applicant about an arrest record); see also Ill.Rev.Stat.1991. ch. 38, par. 1003-12-1 *et seq.* (concerning correctional employment programs whose function is to teach marketable skills and work habits and responsibility to Illinois prisoners).

I would have granted defendant National Food Stores' motion for judgment *non obstante veredicto* on the negligent retention count of plaintiffs' complaint, because "no contrary verdict based on that evidence could ever stand" (*Pedrick*, 37 Ill.2d at 510, 229 N.E.2d at 514), in that plaintiffs failed to sustain their burden of proof. Similarly, since the theory of negligence was not proved, plaintiffs also failed to prove that the alleged conduct or omission showed a deliberate intention to harm or an utter indifference or conscious disregard for the safety of others. Accordingly, I would have also granted defendant National's motion for judgment *non obstante veredicto* on the wilful and wanton retention count of plaintiff's complaint.

(Modified for the purposes of this text)

Melody PERKINS, Appellant,
v.
Thomas S. SPIVEY, Appellee.
Melody PERKINS, Appellant,
v.
GENERAL MOTORS CORPORATION
OF AMERICA, Appellee.
Melody PERKINS, Appellant,
v.
GENERAL MOTORS CORPORATION
OF AMERICA, Appellee.
Melody PERKINS, Appellee,
v.
Thomas S. SPIVEY, Appellant.
Nos. 89-1833WM, 89-2136WM,
89-2137WM.
United States Court of Appeals,
Eighth Circuit.
Submitted March 12, 1990.
Decided July 24, 1990.
Rehearing and Rehearing En Banc Denied
Aug. 29 and Aug. 31, 1990.

Female employee brought action against her employer, alleging that employer breached its duty under Kansas common law to maintain a workplace free from sexual harassment, and that employer had violated Title VII. The United States District Court for the Western District of Missouri, 709 F.Supp. 1487, D. Brook Bartlett, J., entered summary judgment in favor of employer on claim that employer breached its duty under Kansas common law to maintain a workplace free from sexual harassment and, following a bench trial, rejected as not credible allegations that employer had violated Title VII. Employee brought action against employer's superintendent, alleging battery, assault and intentional infliction of emotional distress. The District Court, Scott O. Wright, Chief Judge, granted summary judgment in favor of superintendent on collateral estoppel grounds, and denied superintendent's motion to impose sanctions against employee and her attorney. Appeals were taken. The Court of Appeals, Magill, Circuit Judge, held that: (1) a cause of action exists under Kansas common law when employers negligently hire or retain employees they

know or should know are incompetent or dangerous when
another employee is physically injured by the dangerous
employee or is emotionally harmed such that immediate
physical injury is the result; (2) employee could not
assume the risk of rape; (3) judge did not abuse his
discretion in refusing to recuse himself; (4) judge's
failure to join legal claim against superintendent to
mixed legal and equitable claims against employer did
not violate employee's right to a jury trial; and (5)
trial court findings supporting decision not to impose
sanctions on employee and her attorney were supported
by the evidence.

Affirmed in part, reversed and remanded in part.

Gwen G. Caranchini, Kansas City, Mo., for appellant.

Michael J. Gallagher, Kansas City, for appellee
Spivey.

Paul Scott Kelly Jr., Kansas City, for appellee
General Motors Corp. of America.

Before JOHN R. GIBSON and MAGILL, Circuit Judges,
and DUMBAULD,[1] Senior District Judge.

MAGILL, Circuit Judge.

In these consolidated appeals, Melody Perkins challenges two orders of Judge Bartlett[2] and one order of
Chief Judge Wright.[3] First, Perkins appeals Judge Bartlett's entry of summary judgment in favor of her
employer, General Motors (GM), on her claim that GM
breached its duty under Kansas common law to maintain
a workplace free from sexual harassment. She alleges
that GM negligently hired and retained employees, including a superintendent, who physically and/or psychologically assaulted her over a period of several years.[4]
On appeal, she argues that Judge Bartlett erred (1) in
finding that GM owed no common law duty to prevent sexual

---

[1] THE HONORABLE EDWARD DUMBAULD, Senior United States District Judge
  for the Western District of Pennsylvania, sitting by designation.
[2] The Honorable D. Brook Bartlett, United States District Judge for
  the Western District of Missouri.
[3] The Honorable Scott O. Wright, Chief Judge, United States District
  Court for the Western District of Missouri.

harassment in the workplace, and (2) in alternatively finding that she assumed the risk of harm. We affirm Judge Bartlett's entry of summary judgment to the extent that he found Kansas common law does not impose a duty on GM to provide a workplace free from the purely psychological harms of sexual harassment. However, we reverse the entry of summary judgment in favor of GM insofar as Perkins alleges that her emotional harm was accompanied by or resulted in physical injuries, and remand for further proceedings consistent with this opinion.

Second, Perkins appeals Judge Bartlett's entry of judgment following a thirty-day bench trial on her claim that GM violated Title VII. 709 F.Supp. 1487. At the conclusion of the trial, Judge Bartlett rejected as not credible her allegations of sexual assault and harassment. Three months later he issued his formal findings of fact and conclusions of law. On appeal, Perkins argues that Judge Bartlett erred when he denied her motion for recusal, which she filed nearly three months after the conclusion of the bench trial. Because we find that Judge Bartlett did not abuse his discretion when he denied her motion to recuse, we affirm his entry of judgment in favor of GM on Perkins' Title VII claim.

Perkins also appeals Chief Judge Wright's entry of summary judgment in favor of GM superintendent, Thomas Spivey, on her claim that Spivey was liable for battery, assault and intentional infliction of emotional distress. In granting Spivey's motion for summary judgment, Chief Judge Wright found that because of Judge Bartlett's factual findings, Perkins was collaterally estopped from relitigating whether Spivey had psychologically and physically attacked her. On appeal, Perkins argues that Chief Judge Wright erred in granting summary judgment in favor of Spivey on collateral estoppel grounds, and, in so doing, denied her right to a jury trial on her claim against Spivey. We affirm.

Finally, in his cross-appeal Spivey argues that Chief Judge Wright erred when he refused to impose sanctions against Perkins and her attorney. We affirm.

---

[4] *Perkins v. General Motors*, Nos. 86-0665-CV-W-9, 87-0048-CV-W-9, slip op. at 1-2, 1988 WL 125706 (W.D.Mo. Oct. 11, 1988) (order dismissing case No. 86-0665).

I.

On July 24, 1978, GM hired Perkins as a sixth-level production supervisor in the body shop of its plant in Kansas City, Kansas. With a few exceptions,[5] her employment at this plant continued uninterrupted until January 1986 when she filed a charge with the Equal Employment Opportunity Commission (EEOC) and a claim for workers' compensation. Perkins alleged that while employed at the plant, she was subjected to sexual harassment. She testified that several GM employees told sexual jokes, and directed leering gestures and catcalls toward her and other women. She testified that despite her complaints, lodged early in her employment at the plant, the harassment grew worse and she gradually became immune to the conduct.

Perkins further alleged that Spivey, superintendent of the body shop and/or acting general superintendent at the plant, raped her at least twice a month during much of the time she worked at the plant. The alleged rapes occurred at places other than the plant. She maintains that Spivey accomplished these rapes by threatening her with death or termination of employment and, on several occasions, by brandishing a knife or gun at her.

Perkins filed three separate suits. In the first suit, filed in federal court on May 27, 1986, she alleged that GM breached its Kansas common law duty to maintain a safe workplace free from sexual harassment by negligently hiring and retaining various employees whom GM knew or should have known would physically and/or psychologically injure her. After receiving her notice of right to sue from the EEOC, Perkins filed her second suit in federal court on May 29, 1986, alleging that GM had violated Title VII. Shortly thereafter, however, the negligence and Title VII claims were consolidated for

---

[5] Perkins' employment was interrupted for periods of training, layoff, transfer, special assignment and disability. For example, on May 1, 1980, GM laid off Perkins. Several months later, she was hired at GM's plant in Bowling Green, Kentucky. She worked at this plant until laid off on February 28, 1982. Shortly before her layoff, Perkins and Spivey discussed the possibility of her returning to Kansas City plant. In fact, Perkins initiated some of these discussions. With Perkins' encouragement, Spivey helped her return to the plant, effective May 10, 1982.

discovery and trial. On October 28, 1988, Judge Bartlett granted GM's motion for summary judgment on Perkins' negligence claim, finding that no duty existed under Kansas common law for employers to provide a workplace free from sexual harassment. In the alternative, Judge Bartlett found that even if such a duty did exist, Perkins had assumed the risk of harm through her conduct.

The bench trial on Perkins' Title VII claim commenced on November 1, 1988. Three weeks later, Judge Bartlett informed the parties that his law clerk had just accepted employment to commence at the conclusion of her clerkship with a law firm occasionally retained by GM, but not involved in the case. Although Judge Bartlett offered to remove his law clerk from the case, counsel for both parties stated they had no objection to her continued participation.

On January 11, 1989, after a thirty-day bench trial, Judge Bartlett ruled in favor of GM on Perkins' Title VII claim. Formal entry of judgment was deferred pending consideration of proposed findings of fact and conclusions of law, which were ultimately issued on April 10, 1989. Judge Bartlett rejected as not credible Perkins' claims of sexual harassment and assault by Spivey and other employees. He also found that Perkins participated in a consensual relationship with Spivey in the hope of advancing her career and that her allegations of threats and violence were the result of a fertile and twisted imagination. He also found Perkins' allegation that the plant environment was sexually hostile to be more creative than factual. He found that she was not only generally encouraged and welcomed such conduct but was an active and encouraging participant in the sexually explicit conversations. On the few occasions when Perkins did lodge a complaint, either the conduct ceased immediately or GM dealt with it appropriately. *Perkins v. General Motors*, 709 F.Supp. 1487, 1488-98 (W.D.Mo.1989) (findings of fact, conclusions of law and order granting judgment for GM). Perkins does not challenge these factual findings on appeal as clearly erroneous.

On February 9, 1989, nearly one month after the bench ruling, partners in the firm of Gage & Tucker, which represented GM in the Title VII action, and partners in the firm of Lathrop, Koontz & Norquist, were informed for the first time that their management committees were engaged in preliminary discussions concerning the possibility of a merger between the firms.[6] Judge Bartlett's wife, a labor attorney, was a partner in the firm of Lathrop, Koontz & Norquist. On April 3, 1989, Perkins filed a motion requesting Judge Bartlett recuse himself on three grounds: (1) the preliminary merger discussions between his wife's firm and Gage & Tucker, (2) the future employment of his law clerk with a firm which occasionally represents GM; and (3) his wife's position as a labor attorney. In rejecting Perkins' motion for recusal, Judge Bartlett found in part that her motion "was a last ditch, desperate attempt to delay the issuance of findings and conclusions in [the Title VII] case until after [Perkins'] case against Thomas Spivey [could] be tried in another division of this court." *Perkins v. General Motors*, 129 F.R.D. 655, 672 (W.D.Mo.1989) (order denying Perkins' motion to recuse).

In that case, filed originally in Missouri state court on May 29, 1986, Perkins alleged that Spivey assaulted, battered, and engaged in outrageous conduct against her. GM and Perkins conducted an extensive amount of discovery on her two claims against GM throughout most of 1987. With only a few exceptions, the depositions were not cross-noticed in the state and federal actions, and Spivey's counsel did not attend depositions in the federal case against GM. Furthermore, Perkins sought and obtained a protective order from the Missouri state court restricting Spivey from disclosing her medical and psychiatric records to GM, in order to prevent GM from using them against her in either the Title VII or negligence actions. Perkins' counsel subsequently wrote a letter to counsel for Spivey and GM warning them that violation of the state court order restricting disclosure of Perkins' medical records would force her to take strong steps. Finally, on numerous occasions, Perkins stated that no new parties would be added in the federal case against GM.

On November 27, 1987, Perkins moved to join her claim against Spivey with the negligence and Title VII claims asserted against GM which were pending in federal court. Judge Bartlett denied her motion citing undue delay and

---

[6] No formal or informal agreement to merge was ever concluded and after the filing of this appeal, the law firms terminated negotiations.

:ejudice to GM and Spivey. He then ordered that no
irther discovery be allowed without leave of court on
≥r claims against GM, finding that the amount of
iscovery was out of proportion to the claims asserted
ı the case.

On March 10, 1988, after successfully moving to
ismiss her state court action against Spivey without
:ejudice, Perkins refiled her claims against Spivey in
≥deral court. On April 10, 1989, Judge Bartlett issued
is formal findings of fact in the Title VII action. On
ɔril 19, 1989, Chief Judge Wright granted Spivey's
ɔtion for summary judgment on the basis of the collateral
ȝtoppel effect of Judge Bartlett's findings. *Perkins*
. *Spivey*, No. 88-0213-CV-W-5 (W.D.Mo. Apr. 19, 1989).

II.

In order to best resolve the numerous issues raised
ɣ Perkins, we will first examine whether Judge Bartlett
:operly granted GM's motion for summary judgment on
≥rkins' common law claim of negligence. Second, we will
≥termine whether Judge Bartlett committed reversible
:ror during the course of the Title VII suit against
₄ when he refused to recuse himself.[7] Third, we will
ːamine whether Judge Bartlett abused his discretion in
≥fusing to grant Perkins' motion for leave to join her
ːtion against Spivey to her GM claims.[8] Finally, we
ːll examine Spivey's claim that Chief Judge Wright erred
ıen he refused to impose sanctions on Perkins and her
ːtorney.

Perkins argues not only that we should reverse the judgment entered
because Judge Bartlett refused to recuse himself, but that his
refusal to do so infringed her seventh amendment right to a jury
trial on her claims against Spivey. She argues that if Judge Bartlett
had recused himself, his formal findings of fact and conclusions of
law would never have been issued. Therefore, Chief Judge Wright
would not have entered summary judgment on the basis of collateral
estoppel and she would have received a jury trial on her claims
against Spivey. Because we hold that Judge Bartlett did not abuse
his discretion when he denied Perkins' motion to recuse, we need
not reach her seventh amendment argument. It is well settled that
an otherwise proper ruling is not erroneous merely because it has
the incidental effect of precluding a jury trial. *See, e.g., Fidelity
& Deposit Co. of Md. v. United States*, 187 U.S. 315, 319-21, 23
S.Ct. 120, 121-22, 47 L.Ed. 194 (1902) (summary judgment does not
violate seventh amendment).

A.

Perkins first argues that Judge Bartlett erred when he granted GM's motion for summary judgment on her negligence claim. In granting summary judgment, Judge Bartlett found that no duty exists under Kansas common law which requires GM to provide a workplace free of sexually offensive conduct. In the alternative, he held that even if such a duty exists, Perkins assumed the risk as a matter of law because she voluntarily returned to the Kansas plant in 1982 with the full knowledge of the alleged sexual harassment.

Our review of a grant of summary judgment is governed by the same standard used by the court below. The entry of summary judgment is appropriate only when there is no genuine issue of material fact and the moving party is entitled to judgment as a matter of law. Fed.R.Civ.P. 56; *Kegel v. Runnels*, 793 F.2d 924, 926 (8th Cir.1986). In reviewing the validity of Judge Bartlett's decision to grant GM's motion for summary judgment, we review the facts and the inferences reasonably drawn therefrom in the light most favorable to Perkins. *Id.*

GM first argues that an employer has no duty under Kansas common law to provide a workplace free from sexual harassment. We agree, but only in part. In *EEOC v. General Motors Corp.*, 713 F.Supp. 1394 (D.Kan.1989), the court confronted one of the issues raised by Perkins. The plaintiff-employee brought suit against GM alleging

---

[8] Perkins argues that the effect of Judge Bartlett's refusal to grant her motion for leave to join her action against Spivey to her claims against GM was to violate her seventh amendment right to a jury trial. She argues that if Spivey had been joined, her claims would have presented mixed legal and equitable issues. As a result, Perkins would have received a jury trial on her legal claim against Spivey before a bench trial on her Title VII claims against GM. Her seventh amendment argument therefore hinges on whether Judge Bartlett abused his discretion when he refused to grant Perkins' motion for leave to join her action against Spivey to her claims against GM. Because we hold that Judge Bartlett did not abuse his discretion, Perkins' claims against Spivey and GM properly remained independent causes of action. Therefore, under *Parklane Hosiery Co., Inc. v. Shore*, 439 U.S. 322, 99 S.Ct. 645, 58 L.Ed.2d 552 (1979), discussed *infra* at 35-36, Chief Judge Wright's entry of summary judgment on the basis of collateral estoppel was proper and did not violate Perkins' seventh amendment right to a jury trial on her legal claim against Spivey.

negligence and intentional infliction of emotional distress stemming from a coemployee's alleged sexual harassment. The employee argued that GM breached its common
law duty to provide a workplace "free of sexual assault,
harassment, intimidations and advances of a verbal and
physical nature." *Id.* at 1396. The district court, in
a well-reasoned opinion, held that Kansas common law
does not impose upon an employer a duty to provide a
workplace free from sexual harassment. *Id.*

[1,2] We defer to the state law rulings of federal
district courts which sit in the state whose law is
controlling. We may refuse to follow the federal district
court ruling only if we find that it is "'fundamentally
deficient in analysis, without a reasonable basis, or
contrary to reported state-court opinion.'" *Pony Exp.
Cab & Bus., Inc. v. Ward*, 841 F.2d 207, 209 (8th Cir.1988)
(per curiam) (quoting *Economy Fire & Cas. Co. v. Tri-
State Ins. Co.*, 827 F.2d 373, 375 (8th Circ.1987)).
Because we do not find any of these conditions to exist,
we adopt the Kansas district court's holding that Kansas
common law does not impose a duty on its employers to
provide a workplace free from sexual harassment.[9] In so
doing, we note that we are obliged to apply, not amend,
existing state law. *Tidler v. Eli Lilly & Co.*, 851 F.2d
418, 426 (D.C.Cir.1988).

[3] However, as counsel for GM admitted in oral
argument and as the court in *EEOC v. General Motors*
explicitly noted, the lack of a Kansas common law duty
to provide a workplace free from sexual harassment was
based upon the district court's conclusion that the only
injuries alleged were psychological, and not physical,
in nature. *EEOC v. General Motors*, 713 F.Supp. at 1396
(citing *Hoard v. Shawnee Mission Medical Ctr.*, 233 Kan.
267, 662 P.2d 1214, 1219-20 (1983)). The Kansas federal
district court explained:

> It appears from the relevant case law
> that the traditional application of the
> employer's common law duty to provide a
> safe work place was intended to protect
> employees from unsafe work places where
> they could suffer physical injury. At

---

[9] We note that Perkins' counsel does not attack, nor even discuss,
the district court's holding in *EEOC v. General Motors*.

> common law, the employer had a limited tort liability to its employees. *Kansas case law exhibits a general concern for the physical safety of employees. [Plaintiff's] alleged injuries here were purely psychological. Generally, Kansas law does not allow recovery for emotional distress caused by negligence absent some bodily injury. Hoard v. Shawnee Mission Medical Center, 233 Kan. 267, 274, 662 P.2d 1214, 1219-20 (1983) (Kansas does not recognize the tort of negligent infliction of emotional distress absent bodily injury). This court is not at liberty to arbitrarily expand the common law of this state beyond those boundaries set by the Kansas Supreme Court. Since no authority exists to establish a cause of action for failure to provide a safe work place arising out of incidents of sexual harassment on the job, the court has no choice but to grant summary judgment in favor of GM on count III of [plaintiff's] complaint.*

Id. (emphasis added) (citations omitted). Therefore, .ile GM does not have a duty under Kansas common law maintain a workplace free from the psychological harm sexual harassment, it does have a duty not to retain superintendent it knows or should have known emotionally .rming his coemployees in such a manner that they suffer .ysical injuries. *See id.; Hoard,* 662 P.2d at 1219-20, 23; *Plains Resources Inc. v. Gable,* 235 Kan. 580, 682 2d 653, 663, 665 (1984).

[4] Perkins' claims against GM based upon the sexual .rassment within the plant do not state a cause of tion under Kansas common law. As Perkins concedes, the .rassment was "offensive, bothersome, embarrassing, sgusting, shocking and infuriating…[but she eventually came] immune" to the conduct. Such harms are purely ychological in nature and, therefore, Judge Bartlett operly granted GM's motion for summary judgment based . these claims.

[5,6] However, Perkins also argues that she suffered both purely physical harm and emotional harm accompanied by physical injuries, when Spivey, taking advantage of his position as superintendent, raped her repeatedly over a period of several years. Perkins alleges that Spivey threatened her with termination, physical harm, and death if she did not submit. She claims that GM, through its managers, knew or should have known that Spivey was subjecting her to repeated rapes throughout the course of her employment with GM. She maintains GM was negligent because the company, despite its actual or constructive knowledge, continued to employ Spivey and failed to take any action against him, thereby subjecting her to his continued attacks. To this extent, Perkins' negligence claim is based upon both emotional harms, which are accompanied by or result in immediate physical injuries, and physical injury alone. Therefore, insofar as her claim was so based, the entry of summary judgment in favor of GM was improper. A cause of action does exist under Kansas common law when employers negligently hire or retain employees they know or should know are incompetent or dangerous when another employee is physically injured by the dangerous employee or is emotionally harmed such that immediate physical injury is the result. *See Plains Resources, Inc.*, 682 P.2d at 665 (citing Restatement (Second) of Torts § 909, comment b (1977)).

[7] Because a duty does exist under Kansas common law, we must examine GM's argument that Perkins assumed the risk of physical harm by voluntarily returning to the Kansas City plant to continue her work. GM relies on the Kansas Supreme Court decision in *Gabbard v. Sharp*, 167 Kan. 354, 205 P.2d 960 (1949), to support its position that Perkins assumed this risk. In *Gabbard*, the plaintiff-employee, a waitress, was assaulted by a fellow employee who was alleged to be drunk and upset, about his thwarted advances toward her. In her suit, the waitress also alleged that the employer knew of her assailant's disposition toward her and other women and his habits in and about the premises during the hours that she worked. She claimed that the employer owed her a duty to provide a safe workplace, which it neglected, thereby causing and contributing to her injuries. The Kansas Supreme Court held in part that because the waitress

knew of her coemployee's propensity for violence, she assumed the risk of harm. *Id.* 205 P.2d at 964.

After a thorough review of Kansas law, we have found that this decision, published forty-one years ago, no longer represents, in significant part, the common law of Kansas. First, we note that *Gabbard* has not been cited favorably in thirty-three years. *See, e.g., Murray v. Modoc State Bank*, 181 Kan. 642, 313 P.2d 304, 311 (1957). Given the tremendous social changes since 1957 in the evolution of society's attitudes towards women and abuse, it is difficult to believe Kansas courts would continue to hold to such archaic views. *See, e.g.,* Kan.Stat.Ann. § 44-1001, *et seq.* (Kansas Acts Against Discrimination). More importantly, however, two recent Kansas Supreme Court cases directly conflict with the underlying holdings of *Gabbard.*

First, the *Gabbard* court stated that the waitress did not state a cause of action. The waitress claimed that because her employer negligently retained a coemployee known to have a vicious disposition, her employer was liable for injuries she suffered when her coemployee assaulted her. The reasoning behind the court's conclusion is that an assault is an intentional act which is not committed within the scope of one's employment. *Gabbard,* 205 P.2d at 963-64. The idea that an employer is not directly liable for negligently retaining an employee who assaults another person because an assault is not within the scope of employment has since been emphatically rejected in Kansas. In *Plains Resources, Inc.,* the Kansas Supreme Court stated:

> The negligent hiring and/or retention doctrine recognizes that an employer has a duty to use reasonable care in the selection and retention of employees. This duty requires that an employer hire and retain only safe and competent employees. An employer breaches this duty when it hires or retains employees that it knows or should know are incompetent. Liability exists under either of these doctrines relating to negligent hiring or retention despite the fact that the direct cause of injury to the injured

> person is the negligent or intentional
> acts of an employee *acting outside the
> scope of his employment.*

Plains Resources, Inc., 682 P.2d at 662 (emphasis added)
(citation omitted) (distinguishing cause of action stemming from indirect liability, *respondeat superior*, and
direct liability, negligent hiring and retention of
incompetent employee; "'[i]n a respondeat superior
action, it must be proved that the employee was acting
within the course and scope of his employment and that
he was negligent, which negligence was the proximate
cause of the plaintiff's injuries. In a negligent hiring
action, there is no need to prove that the employee's
act was committed within the course and scope of
employment.'") (quoting 2 Am.Jur. Proof of Facts 2d,
Negligent Hiring § 3, p.616); *compare id.* (employer
directly liable for employees' assaults under doctrine
of negligent hiring and retention) *with Williams v.
Community Drive-in Theater, Inc.*, 214 Kan. 359, 520 P.2d
1296, 1301-02 (1974) (employer not liable under doctrine
of respondeat superior for employee's assault when not
committed in furtherance of employer's interest). Therefore, one of the primary rationales underlying *Gabbard*
has since been rejected by the Kansas Supreme Court.[10]

Second, the *Gabbard* court stated that if "an employee
knows that another employee is incompetent, or habitually
negligent, and continues his work without objection, and
without being induced by his employer to believe that
a change will be made or that his employment will be made
less dangerous he assumes the risk and hazards, of which
he has full and complete knowledge, incident to such
employment." *Gabbard*, 205 P.2d at 964. Although we have
no quarrel generally with the continued validity of this
common law rule, we note that its application to the facts
of *Gabbard*, which involved an intentional criminal act
against a coemployee, does not survive in Kansas.

---

[10] The fact that the injured party in *Plains Resources, Inc.* was not
another employee is irrelevant. The common law doctrine as developed
in Kansas focuses on the direct liability of the employer for its
negligence in hiring and retaining an unfit employee who harms
another person. We find no case in which the identity of the person
harmed is important. *See generally, Plains Resources, Inc.*, 682 P.2d
at 661-67.

[8]   Under Kansas common law, the assumption of the
risk doctrine is based upon a theory implied contract
whereby an employee impliedly contracts to assume both
the ordinary and at times the extraordinary risks of
employment. In *Jackson v. City of Kansas City*, 235 Kan.
278, 680 P.2d 877 (1984), the Kansas Supreme Court stated
that:

> "The doctrine of assumption of risk is
> still viable in Kansas though its appli-
> cation is limited to cases...where a mas-
> ter-servant relationship is involved....
>
> Assumption of risk, in the law of
> master and servant, is a phrase commonly
> used to describe a term or condition in
> the contract of employment, either ex-
> press or implied from the circumstances
> of the employment, by which the employee
> agrees that certain dangers of injury,
> while he is engaged in the service for
> which he is hired, shall be at the risk
> of the employee. Assumption of risk
> generally bars recovery by an employee
> who knows of the danger in a situation
> but nevertheless voluntarily exposes him-
> self to that danger.
>
> [A]ssumption of risk arises through
> implied contract of assuming the risk of
> a known danger; the essence of it is
> venturousness; it implies intentional
> exposure to a known danger, it embraces
> a mental state of willingness; ... it
> defeats recovery because it is a previous
> abandonment of the right to complain if
> an accident occurs.'

Id. 680 P.2d at 897 (citations omitted) (quoting *Borth
v. Borth*, 221 Kan. 494, 561 P.2d 408, 412-13 (1977));
see also id.680 P.2d at 891 ("'reduced to its last
analysis, the doctrine of assumed risk must rest for
its support upon the express or implied agreement of
the employe [sic] that, knowing the danger to which he
is exposed, he agrees to assume all responsibility for
resulting injury'") (quoting *Blackmore v. Auer*, 187 Kan.
434, 357 P.2d 765, 773 (1960)).

Despite the fact that the doctrine is based upon contract principles, GM asks us to find that under Kansas common law, Perkins assumed the risk of harm by entering into an implied contract under which she assumed the risk of an illegal act, namely rape. Given the state's strong policy against such violent acts, *see* Kan.Stat.Ann. §§ 21-3301, 21-3502, we do not believe Kansas courts would sanction a contract, whether implied or express, with such an illegal subject matter. *See e.g., Southern American Ins. v. Gabbert-Jones, Inc.*, 13 Kan.App.2d 324, 769 P.2d 1194, 1198 (1989); *cf. Wagstaff v. Peters*, 203 Kan. 108, 453 P.2d 120, 126 (1969) (because contract never challenged as illegal, agreement valid). Therefore, we hold that Perkins did not, and could not, assume the risk of rape whatever her knowledge of the risk.

[9,10] We conclude that while summary judgment was properly granted as to Perkins' claim that GM breached its common law duty to provide workplace free from sexual harassment, it was not properly granted as to Perkins' claim that GM breached its common law duty by retaining Spivey. In *Lytle v. Household Mfg., Inc.*, — U.S. —, 110 S.Ct. 1331, 108 L.Ed.2d 504 (1990), the Supreme Court refused "to accord collateral estoppel effect to a district court's determinations of issues common to equitable and legal claims first solely because it erroneously dismissed the legal claims." *Id.* 110 S.Ct. at 1338. Any other result, the Court explained, would infringe the plaintiff's seventh amendment right to a jury trial. *Id.* Perkins' claims against GM under Kansas common law and Title VII were consolidated for discovery and trial, and as such were pursued in the same action. Therefore, because the entry of summary judgment on behalf of GM was partially in error, Perkins is entitled to a jury trial on her erroneously dismissed claim notwithstanding the intervening bench trial on her equitable claim under Title VII.[11] As a result, we must remand to the district court for proceedings consistent with this opinion. If the district court finds that a genuine issue as to material fact exists as to Perkins' allegations against GM concerning its negligent retention of Spivey, then Perkins must be given a jury trial.

B.

Perkins next argues that Judge Bartlett erred when he refused to recuse himself because (1) his law clerk who assisted in the case had accepted employment to commence at the conclusion of her clerkship with a law firm occasionally retained by GM but not involved in the Title VII case; (2) his wife specializes in labor law practice and attended portions of the trial; and (3) rumors surfaced nearly one month after the bench ruling that indicated the firm at which Judge Bartlett's wife was a partner was engaged in preliminary merger discussions with the firm representing GM in this case.

[11, 12] We review Judge Bartlett's decision declining to recuse himself only for abuse of discretion. *Gray v. University of Arkansas*, 883 F.2d 1394, 1398 (8th Cir. 1989). Disqualification is appropriate only if the facts would provide an objective, knowledgeable member of the public with a reasonable basis for doubting a judge's impartiality. *United States v. DeLuna*, 763 F.2d 897, 907 (8th Cir.), *cert. denied sub nom. Thomas v. United States*, 474 U.S. 980, 106 S.Ct. 382, 88 L.Ed.2d 336 (1985). However, any grounds for recusal must be asserted promptly. *Oglala Sioux Tribe v. Homestake Min. Co.*, 722 F.2d 1407, 1414 (8th Cir.1983). Perkins does not once argue that Judge Bartlett abused his discretion. She appears to urge upon us a *de novo* standard of review, forgetting that our inquiry is not merely whether we would have made the same decision as Judge Bartlett, but whether we believe that he abused his discretion.

[13] We cannot say that Judge Bartlett abused his discretion based upon the record in this case. First, on November 21, 1988, Judge Bartlett disclosed to counsel

---

[11] GM argues that we cannot allow Perkins to raise a violation of her seventh amendment jury right on appeal because she pursued a trial strategy whereby she attempted, although unsuccessfully, to waive her right to a jury trial. We reject GM's argument. First, it was GM who prevented Perkins' from waiving her right to a jury trial by objecting to Perkins' attempted waiver. To that extent, GM pursued a trial strategy of seeking a jury trial. Under GM's own argument, it should not be allowed to complain because it will now get exactly what it had originally sought, a jury trial by objecting to Perkins' attempted waiver. Second, Perkins correctly claims and GM admits that the waiver of her jury trial right was never accepted. Therefore, it was not effectively waived.

for Perkins and GM that his law clerk had accepted an offer of employment with a law firm which occasionally represented GM. The judge asked both parties if they had any concerns about her continued participation in the case. Perkins responded, "No sir, I don't," and her counsel stated, "None whatsoever, Your Honor." Grounds for disqualification under 28 U.S.C. § 455(a) may be waived after "full disclosure on the record." By electing to proceed and failing to seek Judge Bartlett's recusal, Perkins waived any such claim she might have had regarding the law clerk's future employment.

[14] Second, Perkins' claim that Judge Bartlett should have recused himself because his wife was an attorney specializing in labor law and management is not only legally insufficient, but it was not asserted in a timely fashion. Merely because his wife is a labor attorney does not mean that Judge Bartlett must recuse himself from all labor cases. Otherwise, every judge married to an attorney would be forced to recuse himself or herself from every case involving matters in which the spouse specializes. In fact, a judge whose spouse is a general practitioner would have to recuse himself or herself in almost every case.

[15] Furthermore, Judge Bartlett had been assigned to the Title VII case for over two years prior to Perkins' motion for recusal. During the interim, discovery had taken place, the pretrial orders had been issued, the negligence claim dismissed and a thirty-day bench trial held. If Perkins had wanted to seek Judge Bartlett's recusal on the basis of his wife's position as an attorney specializing in labor law, she should have asserted it before such a great expenditure of resources. Therefore, her motion was not asserted promptly. *See Oglala Sioux Tribe*, 722 F.2d at 1414.

[16] Third, Judge Bartlett did not abuse his discretion when he refused to recuse himself on the basis of the preliminary merger discussions underway between the firm representing GM and the firm in which his wife was a partner. The discussions were merely preliminary. Therefore, his wife did not have a reasonable future interest in the firm representing GM. Second, the merger discussions did not begin and the judge and his wife did not learn about them until approximately one month after the judge

had already ruled against Perkins. To that extent, there could be no conflict. Although the precise wording of the factual findings and conclusions of law had not yet been determined, Perkins had already lost and GM had won. The parties with the most significant interest in the precise wording of the findings were Perkins and Spivey because of the possible collateral estoppel effect on Perkins' subsequent claim against Spivey. Spivey was not represented by GM's law firm, Gage & Tucker.

[17] Perkins also argues that Judge Bartlett should have recused himself because his wife was present in the courtroom on several occasions. The mere presence of his wife in the courtroom has no independent significance. Perkins' argument regarding the wife's presence is only relevant in the context of the merger and the specialization arguments. Because, as we have found, Judge Bartlett did not abuse his discretion in refusing to recuse himself because of his wife's labor and management specialization, he could not have abused his discretion merely because his wife observed some of the testimony in open court.[12] Furthermore, at the time his wife attended small portions of the trial, her firm had not even begun preliminary negotiations with Gage & Tucker. The announcement of the talks were made over one month later. Therefore, there could not have been a connection between her attendance and the discussions.

C.

Perkins further argues that Chief Judge Wright's failure to join her legal claim against Spivey to her mixed legal and equitable claims against GM violated her right to a jury trial, because if they had been joined, the seventh amendment would have required that she receive a jury trial on her legal claims. The cases Perkins cites in support of her seventh amendment argument hold that where a single action involves both legal and

---

[12] Perkins' claim that Judge Bartlett schemed to subvert her claim against Spivey by refusing to recuse himself is frivolous and without any basis whatsoever. In fact, if anyone schemed, it was Perkins' counsel whose apparent motivation behind filing the motion to recuse was to delay Judge Bartlett's completion of the findings of fact until after Perkins' jury trial against Spivey. *Perkins v. General Motors*, 709 F.Supp. 1487, 1489-90 (W.D.Mo.1989) (order denying Perkins' motion for recusal).

equitable claims, the court must conduct a jury trial of the legal claims first. *See Beacon Theatres, Inc. v. Westover*, 359 U.S. 500, 79 S.Ct. 948, 3L.Ed.2d 988 (1959); *Dairy Queen, Inc. v. Wood*, 369 U.S. 469, 82 S.Ct. 894, 8 L.Ed.2d 44 (1962); *Meeker v. Am. Ambassador Oil Corp.*, 375 U.S. 160, 84 S.Ct. 273, 11 L.Ed.2d 261 (1963) (per curiam). These cases are inapposite because Perkins' legal claim against Spivey was not joined in the same action with her equitable claim against GM. Because Perkins' seventh amendment claim hinges on whether Judge Bartlett abused his discretion when he refused to grant her motion for joinder, the claim necessarily fails unless we find Judge Bartlett in fact abused his discretion. *See infra* text at 34-36.

[18, 19] We review the district court's decision to grant or deny leave to amend for abuse of discretion. *See Thompson-El v. Jones*, 876 F.2d 66, 67 (8th Cir.1989). Leave to amend can be properly denied for several reasons, such as undue delay, bad faith or dilatory motive by the movant, undue prejudice to the opposing party, and futility of amendment. *Id.* Despite her allegations of error, Perkins does not even allege that Judge Bartlett abused his discretion. After a thorough review of the record, we hold that Judge Bartlett did not abuse his discretion when he refused to join Perkins' claim against Spivey to the same action with her claims against GM.

Perkins waited eighteen months after filing the federal and state cases before seeking leave to amend her complaint and join her claim against Spivey to her claims against GM. She finally admitted that her effort to join Spivey was motivated solely by a last minute change in strategy. However, she did not originally inform Judge Bartlett of her true motivations because, as she subsequently informed him, she did not believe he would appreciate her candor.

In denying her motion to join Spivey, Judge Bartlett found: (1) in light of the extensive discovery already conducted, most of it without Spivey's participation, it would be "contrary to the overall purposes of the federal rules to promote efficiency and cost saving in litigation to permit amendments that appear to run a very real risk of extensive additional discovery;" (2) joinder and amendment would unduly prejudice both GM and

Spivey at trial, where evidence admissible against only
one defendant would be heard by a single jury deciding
both cases (i.e., Spivey would be prejudiced by alleged
misconduct of other GM employees and GM by alleged
threats and assaults not made on GM premises); and (3)
because Perkins asserted no single claim against both
GM and Spivey, there was a good chance that even if
joinder were granted, separate trials would be ordered
to avoid prejudice and confusion. *Perkins v. Spivey*, No.
88-0213-CV-W-5, slip op. at 5-6 (W.D.MO. June 9, 1988).
We cannot find based upon this record that Judge Bartlett
abused his discretion when he refused to permit joinder.
Therefore, Judge Bartlett's valid ruling does not become
erroneous merely because it has the incidental effect
of precluding a jury trial on Perkins' claim against
Spivey. *See Fidelity & Deposit Co.*, 187 U.S. at 319-21,
23 S.Ct. at 121-22.

[20] Because Judge Bartlett did not abuse his discretion, only two avenues are left to Perkins to challenge
Chief Judge Wright's entry of summary judgment on her
claim against Spivey. In examining these avenues, the
first issue we must resolved is whether, apart from
Perkins' right to a jury trial under the seventh
amendment, Spivey, who was not a party to the prior
judgment, may nevertheless use that judgment defensively
to prevent Perkins from relitigating issues in the
subsequent suit which were resolved against her in the
earlier proceeding. The Supreme Court in *Blonder-Tongue
Lab, Inc. v. University of Illinois Foundation*, 402 U.S.
313, 91 S.Ct. 1434, 28 L.Ed.2d 788 (1971), conclusively
resolved this issue. The Court held that a party who
had fully litigated a claim and lost in a prior action
could be collaterally estopped from relitigating the
same claim in a subsequent action against a different
defendant unless the party did not have a full and fair
opportunity to litigate her claim in the prior suit.
Id. at 328-29, 91 S.Ct. at 1442-43. Not only did Perkins
fully litigate her Title VII claim in front of Judge
Bartlett, but Perkins conceded during oral argument that
the ensuing findings of fact resolved all relevant factual
disputes in her claim against Spivey. Because Perkins
had a full and fair opportunity to litigate her claims
of harassment and assault in her Title VII suit, she is
collaterally estopped from relitigating these fact issues
in her subsequent action against Spivey. *Id.; see also*

Parklane Hosiery Co., Inc. v. Shore, 439 U.S. 322.
329-33, 99 S.Ct. 645, 650-52, 58 L.Ed.2d 552 (1979).

The issue which remains is whether the use of
defensive collateral estoppel in this case violated
Perkins' seventh amendment right to a jury trial. In
Parklane, the Supreme Court stated:

> The question that remains is whether,
> notwithstanding the law of collateral
> estoppel, the use of offensive collateral
> estoppel in this case would violate the
> petitioners' Seventh Amendment right to
> a jury trial.
>   '[T]he thrust of the [Seventh] Amend-
> ment was to preserve the right to jury
> trial as it existed in 1791.' At common
> law, a litigant was not entitled to have
> a jury determine issues that had been
> previously adjudicated by a chancellor
> in equity.
>
>   • • • •
>
>   Thus if, as we have held, the law of
> collateral estoppel forecloses the pe-
> titioners from relitigating the factual
> issues determined against them in the
> [equitable] action, nothing in the Sev-
> enth Amendment dictates a different re-
> sult [if the subsequent action is at
> law], even though because of lack of
> mutuality there would have been no col-
> lateral estoppel in 1791.

Id. at 333, 337, 99 S.Ct. at 652, 655 (citations omitted).
Parklane is fatal to Perkins' argument. The seventh
amendment is not violated although Chief Judge Wright's
entry of summary judgment on collateral estoppel grounds
foreclosed Perkins from relitigating factual issues
determined against her in the prior equitable action.

Perkins attempts to distinguish *Parklane* in several
ways. First, Perkins argues she had a procedural avenue
(jury trial) in the potential Spivey trial which was
not available to her in the Title VII trial. This argument
does no more than restate the issue and as such does
not distinguish *Parklane*.

Second, Perkins argues that *Parklane* is inapposite because it involved the offensive use of collateral estoppel, while Spivey's case involved its defensive use. Her argument is without merit because the distinction is not relevant for purposes of her claim under the seventh amendment. The Court's holding in *Parklane* that the seventh amendment is not violated when a party is barred from relitigating the factual issues determined against it in a prior equitable action does not turn on whether collateral estoppel was used offensively or defensively. Because Judge Bartlett's ruling in the Title VII action was without error, his findings of fact preclude trial on Perkins' separate intentional tort claim.[13]

D.

[21] In his cross-appeal, Spivey argues that Chief Judge Wright erred when he refused to impose sanctions on Perkins and her attorney. In support of this argument, Spivey points to the following: (1) Perkins' continued prosecution of the Spivey claim after Judge Bartlett ruled against her on January 11, 1989; (2) Perkins' continued prosecution of the case against Spivey after formal findings were issued; and (3) misleading statements made to Chief Judge Wright. We review the factual findings under the clearly erroneous standard and the

---

[13] Perkins argues that *Lytle v. Household Mfg., Inc.,*— U.S. —, 110 S.Ct. 1331, 108 L.Ed.2d 504 (1990), requires a different result. We disagree. First, *Lytle* does not overrule *Parklane* but cites it favorably. *See id.* 110 S.Ct. at 1334, 1336, 1337. Second, the holding of *Lytle* is limited to a case where the party brought both equitable and legal claims *in the same action,* but the district court erroneously dismissed the legal claim. *Id.* at 1334. In the case at bar, the legal and equitable claims were not brought in the same action and the district court properly refused to grant Perkins leave to join them. Therefore, *Parklane* controls, not *Lytle*

Our affirmance of Chief Judge Wright's grant of Spivey's motion for summary judgment is based upon our finding that Chief Judge Wright properly relied upon Judge Bartlett's factual findings in the Title VII proceeding. Our affirmance should not be read as stating that we believe Judge Bartlett's findings were necessarily correct. We affirm on the basis of the procedural tool of collateral estoppel. Our partial reversal of Judge Bartlett's grant of GM's motion for summary judgment on Perkins' common law claims is based upon the seventh amendment. Our partial reversal should not be read as stating that we believe that Perkins' version of the facts was more credible.

court's legal conclusion that the rule was not violated
de novo. EEOC v. Milavetz & Assocs., 863 F.2d 613, 614
(8th Cir. 1988).

28 U.S.C. § 1927 provides for the imposition of
costs and attorneys fees against an attorney "who so
multiplies the proceedings in any case unreasonably and
vexatiously." The standard under § 1927 and Rule 11 is
whether the attorney's conduct "viewed objectively,
manifests either intentional or reckless disregard of
the attorney's duties to the court." *Braley v. Campbell*,
832 F.2d 1504, 1512 (10th Cir. 1987) (en banc).

In a well-reasoned opinion, Chief Judge Wright refused
to impose sanctions, holding that: (1) he could not say
that a reasonable attorney would conclude that Judge
Bartlett's indication through his bench ruling of the
probable final order to be issued, no matter the degree
of certainty indicated, constituted final judgment for
the purposes of collateral estoppel; (2) Perkins could
not reasonably be expected at the January 11, 1989
hearing to hypothesize with absolute certainty the
specific findings of April 10, 1989; and (3) he could
not find that Perkins' actions between April 10, 1989
and April 19, 1989 were so unreasonable as to merit
sanctions. After a thorough review of the record and
the parties' arguments, we find no error.

III.

In conclusion, we hold that Judge Bartlett properly
granted GM's motion for summary judgment on Perkins'
negligence claim based upon his finding that no duty
exists under Kansas common law for an employer to provide
a workplace free from sexual harassment. However, GM's
motion for summary judgment was erroneously granted to
the extent Perkins' negligent retention claim was based
upon physical injury to Perkins and/or emotional harm
accompanied by immediate physical injury. Therefore, we
reverse on this one point alone, and remand for further
proceedings consistent with this opinion. We further
hold that Judge Bartlett did not abuse his discretion
when he failed to recuse himself and when he denied
Perkins' motion for leave to join her claim against
Spivey to her claims against GM. Furthermore, we hold
that Chief Judge Wright properly granted Spivey's motion

for summary judgment based on the collateral estoppel effect of the findings made in the prior Title VII bench trial. Perkins' seventh amendment right to a jury trial on her legal claim against Spivey was not infringed. Finally, we hold that Chief Judge Wright properly denied Spivey's motion for sanctions.

chapter eight

Negligent supervision

"So much of what we call management consists in making it difficult for people to work."

Peter Drucker

"The minute you read something you can't understand, you can almost be sure it was drawn up by a lawyer."

Will Rogers

The theory of negligent supervision has been gaining momentum in various courts in the United States. Under this theory, the company may assume liability for their management team members or agents where a management person fails to properly supervise an employee who ultimately inflicts harm on fellow employees or co-workers. In the vast majority of negligent supervision cases, the proximate cause issues are the primary focus of the courts in finding or failing to find negligent supervision. In general, the courts appear to have limited the scope of this type of extended liability to employers and companies to identifiable causal links between the failure to supervise or manage and the actual incident (i.e., "Would the workplace violence incident have occurred in any event?"). However, there are numerous variations in the approach courts are taking with regard to this issue.

In the case of *St. Paul Fire and Marine Insurance v. Knight,*[1] the issue of negligent supervision, as well as negligent hiring and negligent retention, was brought before the court. In this public sector case, a claim was made that an adolescent stress center had improperly hired, supervised, and retained an employee who sexually assaulted a young patient. The particular incident occurred off premises and the individual was aware that the meeting with the ex-supervisor was not part of the center's "after care" program. The court reversed an earlier decision finding no evidence that the employer should have known of its employee's sexual activities and that the incident

---

[1] 764 S.W. 2d 601 (Ark. 1989).

did not arise out of the employment since the employer had specific policies prohibiting contact with former patients.

This theory of negligent supervision is gaining ground in the courts. The general rule is that the employer has an affirmative duty to appropriately supervise its employees in the workplace. Where an employer fails to properly supervise or take appropriate actions which could ultimately prevent workplace violence, the potential of liability under this theory exists. This theory, in most circumstances, would be applied in combination with the negligent hiring and negligent retention theories.

The theory of negligent supervision has been extended to new and unique areas. In *Geise v. Phoenix Company of Chicago*,[2] the Illinois Supreme Court extended this theory to circumstances involving sexual harassment. In this case, the court found that sexual harassment by a manager was a foreseeable hazard in a manager's job function and the company or employer could violate its duty to exercise ordinary care in the hiring of employees by failing to make a "reasonable examination" of the manager's history of sexual harassment.

It should be noted that negligent supervision, as well as the other theories of negligence discussed in this text, is often added to wrongful discharge or statutory discrimination actions by legal counsel to expand the cause of action. In *Geise v. Phoenix Company of Chicago*,[3] the court found no prohibition to the use by the plaintiff of parallel causes of action for statutory sexual harassment and negligent supervision within the same action. The benefit to the plaintiff for this type of parallel action is an extended statute of limitations, absence of punitive damage caps, and direct liability implications. For example, under Title VII of the Civil Rights Act, the plaintiff would be required to file a claim with the Equal Employment Opportunity Commission within 180 days. This period can often be extended to 300 days through a filing with the individual human rights commission. A claim filed after this statutory period of time would be barred from being filed. Thus with the parallel action being permitted, the statute of limitations for a negligence-type action may be 2 to 5 years, thus permitting a longer period of time to initiate the action.

Conversely, Title VII of the Civil Rights Act of 1991, the Americans with Disabilities Act, and other statutory laws limit punitive damages up to $300,000, depending on the number of employees. Negligence-type claims do not place limits on the amount a judge or jury can award against the employer or company in the area of punitive (or punishment) damages.

The defenses which may be available to a claim of negligent supervision (also called "negligent entrustment") are virtually the same as with other negligence-based claims. Prudent professionals should establish policies and

---

[2] 246 Ill. App. 3d 441, 615 N.E. 2d 1179 (2d Dist. 1993), *reversed on other grounds*, 159 Ill. 2d at 507, 639 N.E. 2d at 1273 (1994). *Also see, Bryant v. Livigni*, 250 Ill. App. 3d 303, 619 N.E. 2d 550 (5th Dist. 1993).

[3] Id.

procedures (see Chapter 3) to identify and address circumstances with members of management where liability can be vicariously attached. Companies and employers are responsible for the actions or inactions of their supervisors, managers, and agents. To avoid the potential liability within this growing area, prudent professionals must be diligent in the assessment of their supervisors, managers, and agents to identify and address actions which may be negligent in nature, however keeping a keen eye so as not to trod upon the individual privacy rights of the supervisor, manager, or agent.

Companies and employers often state that this situation is a "damned if you do and damned if you don't" situation. Such situations are often difficult; however, statutory law and case law have provided guidance in this emerging area. Prudent professionals must be knowledgeable in these important areas in order to make an informed and appropriate judgment in these difficult situations. Careful evaluation, avoiding snap judgments, and appropriate research or guidance often removes the murk of uncertainty and provides a clear path of decision.

(Modified for the purposes of this text)

The STATE of Nevada, DEPARTMENT
OF HUMAN RESOURCES, DIVISION
OF MENTAL HYGIENE AND MENTAL
RETARDATION, Appellant,
v.
Julie JIMENEZ, as Guardian Ad Litem
for John Doe, a Minor,
Respondent.
No 26021.
Supreme Court of Nevada.
March 27, 1997.

Juvenile brought action against state to recover damages for sexual assaults committed against him while he was in group home for adolescent sex offenders. The Second Judicial District Court, Washoe County, Peter I. Breen, J., awarded damages to juvenile, and state appealed. The Supreme Court held that: (1) state waived its sovereign immunity for intentional torts committed by employees in scope of their employment; (2) group home leader's sexual assaults of juvenile were committed within scope of group home leader's employment; (3) evidence supported damage awards on nine counts of sexual assault; and (4) allowing juvenile damages for negligent supervision amounted to impermissible double recovery.

Ordered accordingly.

Frankie Sue Del Papa, Attorney General, and Cynthia A. Pyzel, Senior Deputy Attorney General, Carson City, for Appellant.

Durney and Brennan, Reno; Calvin R.X. Dunlap, Reno, for Respondent.

OPINION

PER CURIAM:

The instant appeal arises from allegations of sexual assaults of a minor involuntarily placed in a State agency for adolescent sex offenders. After the program's

supervisor, Mike Peters, had been relieved of his duties
for reasons unrelated to the allegations in this lawsuit,
the minor revealed that he had been repeatedly sexually
assaulted by Peters. Julie Jimenez, mother and guardian
ad litem of the minor (hereinafter referred to as John
Doe), sued the State for negligent supervision of Peters
and for the resulting false imprisonment, battery, and
sexual assault of her son.

The district court found the State liable for nine
counts of sexual assault and one count of negligent
supervision and awarded Jimenez the statutory maximum
allowed for each count, $50,000. We conclude that the
district court properly found liability and assessed
damages on the counts for sexual assault. We also conclude
that the district court properly found liability on the
count for negligent supervision but that the award of
damages for that count amounted to a double recovery
for a single injury and must be reverted.

FACTS

In 1990, the State of Nevada Department of Human
Resources, Division of Mental Hygiene and Mental Retardation (the State), operated the Northern Nevada Child
and Adolescent Services (NNCAS). NNCAS ran a variety of
programs ranging from residential to outpatient programs
and programs for very young children to adolescents. Two
of the programs, including the one in which John Doe
was enrolled, were based on a nationally recognized
behavioral program, the Family Learning Home Model. The
State had been running family learning home programs
since 1975.

As an agency of the State of Nevada, NNCAS's hiring
and employment practices are required to meet the State's
laws and regulations. These include reference and fin-
gerprint checks and a probationary evaluation period for
new employees. If an employee successfully completes
probation, he or she is considered a permanent employee.
As employees of a state agency, individuals are also
subject to an annual review.

In 1987, Michael Peters was hired as a Mental Health
Technician by the Nevada Mental Health Institute (NMHI),
an agency also under the control of the State's Division

of Mental Hygiene and Mental Retardation. After three months of reviews indicating quality work performance, Peters laterally transferred to a position in the NNCAS with the Adolescent Treatment Center, a twenty-four hour awake supervision facility for the treatment of adolescents with mental disorders. For this position, Peters filled out another employment application, provided references from his supervisor at NMHI, and was fingerprinted so that the FBI could do a background check for any criminal history.

In late 1989, NNCAS was reviving its sexual offender treatment home at Desert Hills and was looking for a professional teaching partner. Until 1989, the Desert Hills program had operated with a teaching couple living at the home. After the last couple left the program, the State could not find replacements and was forced to terminate the program. The State then chose to use "teaching partners" instead of the teaching couple in order to facilitate the hiring of better qualified individuals and to enable continuation of the program if one of the teaching partners were to leave.

In December of 1989, Peters commenced work at the Desert Hills program as the administrator. He received training from nationally recognized experts in the treatment of adolescent sexual offenders and spent time with the former supervisor of the program. At that time, no problems or concerns were identified regarding Peters' ability to perform the job or about his ability to work with adolescents. Peters was initially brought over to the Children Behavioral Services program to learn the Family Learning Home Model upon which Desert Hills was based.

Around April of 1990, two other professional teaching partners, Noel Cullen and Deborah Henson, joined Peters at Desert Hills. Desert Hills reopened in June of 1990 with Peters designated the home supervisor of the program. Peters' co-workers began to have problems with the manner in which Peters was supervising them. The State claims that the complaints did not address how Peters dealt with the children, but only how he treated his co-workers. Henson and Cullen felt Peters was belittling them and was not letting them share equally in the duties and responsibilities.

The complaints culminated in two meetings between Peters and his supervisor, Les Gruner. At the second meeting on July 24, 1990, it became clear to Gruner that Peters had not listened to him and had not implemented any changes in his supervisory methods. That afternoon, Peters was relieved of his duties at Desert Hills and placed on administrative leave. The State contends that up to this point there had been no indication that Peters might be sexually assaulting any of the children. However, evidence was presented at trial that, prior to his removal, Peters had demonstrated inappropriate behavior at Desert Hills, specifically that he had sexually suggestive magazines at the group house and that complaints had been lodged with Gruner regarding this behavior.

John Doe was the third child brought into the program. He was brought in on June 25, 1990, when he was fourteen. He had been diagnosed since birth as learning disabled and mildly retarded. For the preceding four years, John Doe had been subjected to sexual assaults by a maternal uncle and his cousins. Around 1988 or 1989, John Doe began to sexually assault his younger brother. In February of 1990, his mother caught him having anal intercourse with his younger brother. He was put on juvenile probation; however, he continued to sexually assault his brother, and the court ultimately put him in the Desert Hills program for treatment.

On July 28, 1990, approximately one month after John Doe's admission to the program and four days after Peters was relieved of his duties, allegations of Peters' sexual assaults of John Doe were made by John Doe. The staff decided to question John Doe in response to his behavior after Peters had left the program. John Doe stated that Peters would come into his room, talk to him, and then subject him to anal intercourse. John Doe said that to the best of his recollection, the number of times Peters sexually assaulted him was in the "double figures." John Doe testified at trial that he did not report the assaults because Peters threatened him and because Peters' assaults on him made him feel like "trash." He also testified he had looked up to Peters like a father.

Following these allegations, Cullen and Henson notified law enforcement, child protective services, and their supervisors. The State claims this was the first time any allegation of impropriety toward clients was made about Peters. Law enforcement officials interviewed the other children in the program who had been around Peters and uncovered no other allegations of impropriety. However, at trial, a minor client of Children's Behavior Services, the agency Peters had worked for prior to going to Desert Hills, testified that Peters had also sexually assaulted him and subjected him to anal intercourse "over one-hundred times" during a six-month period.

A medical examination of John Doe revealed that he had sustained sexual trauma, but the State claimed that due to John Doe's extensive history of prior abuse, no conclusions could be reached regarding the allegations against Peters. Peters was placed on administrative leave for the duration of the investigation and then laterally transferred back to NMHI.

Julie Jimenez, John Doe's mother, was appointed as his guardian ad litem. She sued the State for negligent supervision of Peters and for the resulting false imprisonment, battery, and sexual assault of her son. The district court found that Peters scheduled the employee rotations so that he would be the sole staff member on the premises during the nighttime shifts. His duties on the night shift including checking the boys' rooms when they were sleeping. The district court also found that "on nine separate and distinct occasions, when Mike Peters entered [John Doe's] room while performing his assigned employment duties, he sodomized [John Doe] without the consent and against the will of [John Doe]."

The district court concluded by finding the State liable for nine counts of sexual assault and one count of negligent supervision. The court found that John Doe suffered physical and emotional injuries and would continue to suffer such injuries. Accordingly, the district court awarded $50,000 for each of the nine separate acts of sexual assault, along with the additional sum of $50,000 for the State's failure to reasonably supervise the program and its staff, for a total of $500,000 in damages. We conclude that the district court properly awarded damages for the nine counts of sexual assault but that the award of damages for the negligent supervision is tantamount to a double recovery for the same injury and must be reversed.

DISCUSSION

The State waived immunity from liability for Peters' sexual assaults

The State claims that sovereign immunity is not waived for intentional torts and therefore the district court improperly found the State liable for Peters' actions of sexual assault.[1] The State has not challenged the district court's finding of liability on the claim of negligent supervision.

> NRS 41.031(1) states:
> The State of Nevada hereby waives its immunity from liability and action and hereby consents to have its liability determined in accordance with the same rules of law as are applied to civil actions against natural persons and corporations, except as otherwise provided in NRS 41.032 to 41.038, inclusive, 485.318, subsection 4 and any statute which expressly provides for governmental immunity....

In *State v. Silva*, 86 Nev. 911, 914, 478 P.2d 591, 593 (1970), this court discussed the statutory codification and the interpretation of sovereign immunity in this state. We stated:

> Before the enactment of the statutory waiver of immunity, Nevada case law on the viability of the doctrine of sovereign immunity was uncertain and in flux. The trend was toward the judicial abolition of that doctrine. It is only fair to assume that the 1965 Legislature

---

[1] We find this argument puzzling because in its opening brief to this court, the State conceded that it had waived its immunity for the torturous acts of its employees who acted within the course and scope of employment. Despite this concession, we will still reach this issue.

> reacted to that trend, and elected to
> waive immunity within limits and impose
> a ceiling upon the recovery allowable to
> a claimant, rather than await further
> judicial action upon the subject. *The*
> *apparent legislative thrust was to waive*
> *immunity and, correlatively, to strictly*
> *construe limitations upon that waiver.*

Id. (citations omitted; emphasis added). Strict construction of NRS Chapter 41 indicates that the Legislature intended to waive the State's sovereign immunity
generally and then reinstate that immunity in certain
limited circumstances.

[1] NRS 41.0334(2)(a) states that the immunity
granted to the State in NRS 41.0334(1)[2] will not apply
to *any* action for injury "intentionally caused" by a
State employee. We construe this provision to mean that
the State can be held liable for the intentional torts
of its employees. However, this liability is subject to
the restrictions of NRS 41.03475, which states:

> No judgment may be entered against the
> State of Nevada or any agency of the
> state or against any political subdivi-
> sion of the state for any act or omission
> of any present or former officer, em-
> ployee, immune contractor, member of a
> board or commission, or legislator which
> was outside the course and scope of his
> public duties or employment.

Based on this language, and employing the *Silva*
standard of strictly construing any limitation on the
waiver of immunity, we conclude that the State can be
held liable for the intentional torts of its employees

---

[2] NRS 41.0334(1) states: Except as otherwise provided in subsection 2,
no action may be brought under NRS 41.031 or against an office or
employee of the state or any of its agencies or political subdivisions
for injury, wrongful death or other damage sustained in or on a
public building or public vehicle by a person who was engaged in
any criminal act proscribed in NRS 202.810, 205.005 to 205.080,
inclusive, 205.220, 205.225, 205.235, 205.240, 205.245, 205.271 to
205.2741, inclusive, 206.310, 206.330, 207.210, 331.200 or 393.410,
at the time the injury, wrongful death or damage was caused.

provided that those torts are committed within the course and scope of employment.

[2]   The rule of law in Nevada permits civil actions to be filed to collect damages for injuries resulting from sexual assaults. Therefore, based on NRS 41.031(1), 41.0334(1), and 41.03475, the State can be held liable for injuries resulting from sexual assaults which occurred during the course and scope of employment. We conclude that the district court did not err in finding that the State waived its immunity for Peters' intentional actions provided that those actions were within the course and scope of Peters' employment.

The State was properly found liable for both the sexual assaults and the negligent supervision

1.   Peters acted within the scope of his employment

[3]   The State argues that Peters' actions were not within the course and scope of his employment and therefore it is not liable for Peters' actions. The district court, however, found that Peters' duties and the unconditional discretion vested in him by the State gave him unfettered control of and access to John Doe, making the State liable under the doctrine of *respondeat superior*.

NRS 41.031(1) provides that when the State has waived immunity, the State will have its "liability determined in accordance with the same rules of law as are applied to civil actions against natural persons and corporations." This court has held that in order for an employer to be liable for the intentional tort of an employee, that tort must occur within the scope of the task assigned to that employee. *Prell Hotel Corp. v. Antonacci*, 86 Nev. 390, 391, 469 P.2d 399, 400 (1970). "[I]f the employee's tort is truly an independent venture of his own and not committed in the course of the very task assigned to him, the employer is not liable." *Id.* at 391, 469 P.2d at 400.

[4]   We take this opportunity to further refine our explanation of "course and scope of employment" from *Prell* because that explanation gives no substantial guidance as to what is included within the scope of employment. In *Prell*, we declined to follow the "moti-

vation" test which states generally that there will be no cause of action for respondeat superior if the employee's actions were not primarily motivated by the desire to serve his other employer. We are mindful of the fact that many jurisdictions followed the "motivation" test. *See Andrews v. United States*, 732 F.2d 366, 370 (4th Cir.1984) (applying South Carolina law); *Hoover v. University of Chicago Hospitals*, 51 Ill.App.3d 263, 9 Ill.Dec. 414, 418, 366 N.E.2d 925, 929 (1977); *Cosgrove v. Lawrence*, 214 N.J.Super. 670, 520 A.2d 844, 846-848 (Law Div.1986). However, we will continue not to require that the employee's actions be motivated by a desire to serve the employer, and in doing so, we join several other jurisdictions. *See Doe v. Samaritan Counseling Center*, 791 P.2d 344, 346-47 (Alaska 1990) (reversing and remanding a grant of summary judgment in favor of the employer on the issue of vicarious liability for a sexual assault committed on a patient by an employee counselor); *Mary M. v. City of Los Angeles*, 54 Cal.3d 202, 285 Cal.Rptr. 99, 102, 814 P.2d 1341, 1344 (1991) (reversing and remanding a court of appeals decision reversing a jury verdict imposing liability on the city of Los Angeles for a police officer's rape of a woman in custody); *Samuels v. Southern Baptist Hospitals*, 594 So.2d 571, 573 (La.Ct.App.1992) (affirming a district court decision finding a hospital vicariously liable for an employee nurse's sexual assault of a patient); *Marston v. Minneapolis Clinic of Psychiatry*, 329 N.W.2d 306, 309-10 (Minn.1982) (reversing and remanding a lower court decision that the employer was not vicariously liable for a sexual assault committed on a patient by an employee doctor).

We now adopt a new test which maintains *Prell's* basic tenet of not following the "motivation" test and which employs a "rationale that the employer's liability should extend beyond his actual or possible control over the employees to include risks inherent in or created by the enterprise because [the employer], rather than the innocent injured party, is best able to spread the risk through prices, rates or liability insurance." *Rodgers v. Kemper Construction Co.*, 50 Cal.App.3d 608, 124 Cal.Rptr. 143, 148 (1975). The test we not adopt was first employed by the California Court of Appeals, and states:

> One way to determine whether a risk is inherent in, or created by, an enterprise is to ask whether the actual occurrence was a generally foreseeable consequence of the activity. However, "foreseeability" in this context must be distinguished from "foreseeability" as a test for negligence. In the latter sense "foreseeable" means a level of probability which would lead a prudent person to take effective precautions whereas "foreseeability" as a test for *respondeat superior* merely means that *in the context of the particular enterprise an employee's conduct is not so unusual or startling that it would seem unfair to include the loss resulting from it among other costs of the employer's business.*

Id., 124 Cal.Rptr. at 148-49 (emphasis added). This test was later used by the California Supreme Court in *Mary M. v. City of Los Angeles*, 54 Cal.3d 202, 285 Cal.Rptr. 99, 102, 814 P.2d 1341, 1344 (1991), and *Perez v. Van Groningen & Sons, Inc.*, 41 Cal.3d 962, 227 Cal.Rptr. 106, 108, 719 P.2d 676, 678 (1986).

[5] The policy objectives underlying respondeat superior are "(1) to prevent recurrence of the torturous conduct; (2) to give greater assurance of compensation for the victim; and (3) to ensure that the victim's losses will be equitably borne by those who benefit from the enterprise that gave rise to the injury." *Mary M.*, 814 P.2d at 1343. We believe that these three policy objectives underlying respondeat superior would be achieved by applying the doctrine in this case.

First, requiring the State to pay for Peters' sexual assaults would encourage the State to take preventative measures to ensure that such egregious behavior does not occur again. Second, requiring the State to pay would give greater assurance of compensation to the Jimenez family. As was stated earlier, the Legislature has recognized that the imposition of vicarious liability on the State is an appropriate method to ensure that victims are compensated for State actions committed within the course and scope of employment. *See* NRS

41.03475; *Mary M.*, 285 Cal.Rptr. at 106, 814 P.2d at
1348. Finally, the appropriateness of spreading the risk
of loss among the beneficiaries of the enterprise favors
imposition of vicarious liability against the State.
Because the community derived a substantial benefit from
the lawful exercise of Peter's authority, i.e., the sex
offenders were separated from society while they were
being rehabilitated and the rehabilitated sex offenders
would no longer pose a threat to society, the community
should bear the cost of Peters' misuse of power. See
Mary M., 285 Cal.Rptr. at 107, 814 P.2d at 1349 (stating
that the cost resulting from misuse of police power
should be borne by the community because of the substantial benefit that the community derived from the
lawful exercise of police power.)

[6] In cases concerning course and scope of employment, no single factor is necessarily controlling, and
each case must be determined on its own particular facts
and circumstances. *Chapin v. United States*, 258 P.2d
465, 467 (9th Cir.1958). "In each case involving scope
of employment all of the relevant circumstances must be
considered and weighed in relation to one another." *Loper
v. Morrison*, 23 Cal.2d 600, 145 P.2d 1, 3 (1944).

"Ordinarily, the determination whether an employee
has acted within the scope of employment presents a
question of fact; it becomes a question of law, however,
when the facts are undisputed and no conflicting inferences are possible." *Mary M.*, 285 Cal.Rptr. at 105, 814
P.2d at 1347 (quoting *Perez* 227 Cal.Rptr. at 109, 719
P.2d at 679). We conclude that in light of the new test
adopted today, the factual circumstances of this case
did not present a question of law. Therefore, because
the judge's conclusion that Peters acted within the scope
of his employment was not clearly erroneous, we will
not set it aside. *Hermann Trust v. Varco-Pruden Buildings*,
106 Nev. 564, 566, 796 P.2d 590, 592 (1990).

In employing this new test, this court must first
determine whether, given the context of Peters' employment, his conduct was so "unusual or startling" that it
would seem unfair to include the loss resulting from it
in the costs of the employer's business. On this issue,
we find the analysis from *Mary M.* persuasive. *Mary M.*

involved a police officer who raped a woman in custody, and the court in *Mary M.* stated that because of the great control that police officers have over criminal suspects, "it is neither startling nor unexpected that on occasion an officer will misuse that authority by engaging in assaultive conduct."[3] *Mary M.*, 285 Cal.Rptr. at 108, 814 P.2d at 1350.

In the case at bar, Peters had extensive control over the children he supervised, although it was a different type of control than that which a police officer has over a detainee. As the group home supervisor, Peters had control over almost every aspect of the children's lives. As their counselor, the children looked to him for guidance; John Doe even stated that he looked up to Peters as a father. Furthermore, the children Peters supervised were often the products of troubled backgrounds, vulnerable, and in the case of John Doe, mentally retarded. We therefore conclude that given this type of control, it is extremely unfortunate, but not "startling or unexpected," that Peters, or someone in his position, would misuse his authority by engaging in such behavior.

---

[3] The *Mary M.* court was careful to note that its decision flowed from the unique authority vested in police officers and that employees who do not have such authority and who commit sexual assaults may be acting outside of the scope of employment as a matter of law. *Mary M. v. City of Los Angeles*, 54 Cal.3d 202, 295 Cal.Rptr. 99, 108 n. 11, 814 P.2d 1341, 1350 n. 11 (Cal.1991).

An example of conduct that a court found "startling and unusual" was presented in *Alma W. v. Oakland Unified School District*, 123 Cal. App.3d 133, 176 Cal.Rptr. 287 (1981). In *Alma W.* a janitor at an elementary school sexually assaulted a young female student. The student's mother filed a lawsuit against the school district on a theory of respondeat superior, and the complaint was dismissed by the trial court. In affirming the trial court's action, the Court of Appeal stated that the plaintiff's case stretched the *Rodgers* "foreseeability" standard far beyond its logical limits. *Id.*, 176 Cal.Rptr. at 291. The court stated:

Thus, while it might be foreseeable for a school custodian to become involved in a dispute over the manner in which he swept the floors or cleaned a classroom and for the dispute to end in someone being hit with a mop, the same statement cannot be made with reference to rape. There is no aspect of a janitor's duties that would make sexual assault anything other than highly unusual and very startling.

Id. at 291-92.

However, there are exceptions to this new test, and in evaluating such exceptions, this court must consider whether this was a situation where an employee "substantially depart[ed] from his duties for purely personal reasons" such that liability will not attach. *John R. v. Oakland Unified School Dist.*, 48 Cal.3d 438, 256 Cal.Rptr. 766, 771, 769 P.2d 948, 953 (1989). When considering this issue, the proper inquiry is not "'whether the wrongful act itself was authorized but whether it was committed in the course of a series of acts of the agent which were authorized by the principal.'" *Perez*, 227 Cal.Rptr. at 110, 719 P.2d at 680 (quoting *Fields v. Sanders*, 29 Cal.2d 834, 180 P.2d 684, 688 (1947)). In *Mary M.*, the court concluded that the police officer was acting within the scope of his employment when he detained the victim, performed a field sobriety test on the victim, and ordered the victim into his car. The court then stated that the officer misused his authority when he sexually assaulted the victim, but viewed as a whole, it could not be said as a matter of law that the officer had substantially departed from his duties for purely personal reasons when he sexually assaulted the victim. *Mary M.*, 285 Cal.Rptr. at 109, 814 P.2d at 1351.

In the case at bar, we conclude that when viewing the incidents as a whole, it cannot be said as a matter of law that Peters had substantially departed from his duties for purely personal reasons when he sexually assaulted John Doe because the sexual assaults were committed during a series of acts authorized by the State. Peters was acting in the scope of his employment when he counseled the children and conducted bedchecks. During the bedchecks and the discussions that sometimes occurred during the bedchecks, Peters misused his authority as a counselor and sexually assaulted John Doe. As such, a conclusion that Peters did not substantially depart from his duties for purely personal reasons is supported by substantial evidence.

We therefore, conclude that the judge's determination that Peters' conduct was within the course and scope of Peters' duties was not clearly erroneous and, therefore, that determination will not be overturned.

2. *The supervising of the employee was an operational* act

[7] Under NRS 41.032, the State can only be held liable for operational and not discretionary acts. NRS 41.032(2) states:

> Except as provided in NRS 278.0233 no action may be brought under NRS 41.031 or against an ... employee of the state ... which is:
>
> ...
>
> 2. Based upon the exercise or performance or the failure to exercise or perform a discretionary function or duty on the part of the state or any of its agencies or political subdivisions or of any officer, employee or immune contractor of any of these, whether or not the discretion involved is abused.

In *Hagblom v. State Dir. of Motor Vehicles*, 93 Nev. 599, 604, 571 P.2d 1172, 1176 (1977), this court stated:

> When the State qualifiedly waived its immunity from liability and consented to civil actions, it did so to provide relief for persons injured through negligence in performing or failing to perform non-discretionary or operational actions. It *did not* intend to give rise to a cause of action sounding in tort whenever a state official or employee made a discretionary decision injurious to some persons.

[8, 9] A discretionary act is an act which requires "the exercise of personal deliberation, decision and judgment." *Travelers Hotel v. City of Reno*, 103 Nev. 343, 345-46, 741 P.2d 1353, 1354 (1987). In a close case between whether an act is discretionary or operational, this court will favor "a waiver of immunity and accommodate the legislative scheme," and only when the court concludes "that discretion alone is involved may [it] find immunity from suit." *State v. Silva*, 86 Nev. 911, 914, 478 P.2d 591, 593 (1970).

In *Silva*, this court subjected the State to liability for negligent supervision after a prisoner at an honor camp escaped and sexually assaulted a woman. In subjecting the State to liability, the court stated:

> Although the selection of inmates for honor camp may primarily be a discretionary act, the manner in which the camp is supervised and controlled is mainly operational in nature. Indeed, the very fact that such inmates are not released from prison to roam at will, but remain under state control for work assignment and honor camp living, establishes state recognition that control and supervision is essential.

Id.

As in *Silva*, we conclude that while the creation of Desert Hills may have been a discretionary function, the State's suspension of Peters was an operational function. Therefore, even if the daily operation of the program and supervision of Peters are not clearly operational functions, it is a very close case between discretionary and operational and warrants our favoring a waiver of immunity. Accordingly, the district court properly found that the supervision of Peters was an operational act and that the State was liable for negligent supervision.

3. *The district court did not err in finding the State liable on nine separate counts of sexual assault in addition to one count of negligent supervision.*

[10] In its findings of fact, the district court stated:

> As the entity operating the Desert Hills program, [the State] had the obligation and the opportunity to structure the program and supervise the staff to protect the vulnerable clients from physical, mental, and emotional harm while they were enrolled in the program.

The district court concluded that "[a]s a direct and proximate result of the negligence of the [State] and sexual molestation and false imprisonment by [Peters], [John Doe] suffered severe physical and emotional injuries, and will continue to suffer consequent injuries in the future." Although the district court acknowledged John Doe could not recover damages exceeding $50,000 for a single action sounding in tort, he could recover for each separate and distinct tort. The district court stated that each act of sodomy, in addition to the one instance of ongoing negligent supervision, constituted a separate and distinct tort, thus allowing ten recoveries, the maximum amount allowed under the statute.

[11] This court has stated that when determining whether a particular situation constitutes a single occurrence or multiple occurrences for the purpose of an insurance contract, it will use the "causal approach" employed by the majority of the jurisdictions. *Bish v. Guaranty Nat'l Ins.*, 109 Nev. 133, 135, 848 P.2d 1057, 1058 (1993). While this case does not involve an insurance contract, the discussion from *Bish* is instructive to determine whether Peters' actions resulted in one or nine "occurrences" of negligent supervision and sexual assault for which the State is liable. Under the "causal approach," "the inquiry is focused on whether there was one or more than one cause which resulted in all of the injuries or damages." *Id.; see also Welter v. Singer,* 126 Wis.2d 242, 376 N.W.2d 84, 87 (1985) (stating that the proper inquiry when employing the "causal approach" analysis is whether "a single, uninterrupted cause results in all of the injuries and damage").

This case involves two separate and distinct theories of liability—negligent supervision and respondeat superior. The analysis of each theory of liability under the "causal approach" is very different, and we take special care not to combine that analysis. We will apply the "causal approach" analysis to the negligent supervision claims first.

[12] The district court determined that as a direct and proximate result of the State's negligent supervision, John Doe suffered damages. This court recently addressed the issue of whether negligent supervision

which results in multiple child sexual assaults constitutes one occurrence or multiple occurrences of negligent
supervision. *Washoe County v. Transcontinental Ins.*, 110
Nev. 798, 878 P.2d 306 (1994). In that case, parents of
children who were sexually assaulted at a county licensed
day care center sued Washoe County for negligently
investigating and monitoring the day care center in
conjunction with its licensing process. Employing the
"causal approach," this court held Washoe County liable
for only one count of negligence, stating that Washoe
County's liability stemmed from its failure to adequately
perform an ongoing duty to investigate and monitor the
day care center.

Similarly, in the instant case, the State breached
its duty to properly supervise Peters; the only issue
is whether the State was liable for nine counts of
negligent supervision (one count for each sexual assault)
or only one count. We conclude that based on *Washoe
County*, the district judge properly found that the State
was liable for only one count of negligent supervision
because its liability on that theory stemmed from a
single, ongoing cause—the continuous failure to supervise
Peters. *See State Farm Fire & Cas. v. Elizabeth N.*, 12
Cal.Rptr.2d 327 (Ct.App.1992).

[13] However, just because the State was liable for
only one count of negligent supervision does not mean
that it was also liable for only one count of sexual
assault on the respondeat superior theory. Under a theory
of respondeat superior, "the act of the servant within
the scope of his employment must be considered the act
of the master, for which he is liable to the same extent
as though he had performed the act in person." 30 C.J.S.
Employer-Employee § 216(1992); *see also Chaney Bldg. Co.
v. City of Tucson*, 148 Ariz. 571, 716 P.2d 28, 30-31
(1986). Therefore, it is proper to determine which torts
Peters would have been liable for had he been sued
directly and then impute any of Peters' liability to
the State.

Initially, we note that the issue of negligent
supervision is irrelevant to the discussion of the State's
liability for Peters' intentional torts. *See Samuels v.
Southern Baptist Hosp.*, 594 S.2d 571, 574
(La.Ct.App.1992) (stating that "vicarious liability is

imposed upon the employer without regard to his own negligence or fault"). Had Peters been sued directly, he would have been liable for nine separate counts of sexual assault regardless of whether he had been negligently supervised or not. Peters could not have successfully employed the "causal approach" to prove that the nine instances of sexual assault stemmed from one proximate cause of negligent supervision, and therefore, that he was liable for only one count of sexual assault.[4]

Furthermore, the situation presented in this case is not similar to that presented in *Bish v. Guaranty National Insurance Company*, 109 Nev. 133, 848 P.2d 1057 (1993), a negligence case involving a woman who accidently twice ran over a child, first by backing over the child, then after realizing what she had done and panicking, by putting the car in forward gear and driving over the child again. In *Bish*, this court concluded that there was only one "occurrence" for purposes of liability because the injuries were the result of one proximate, uninterrupted and continuing cause: the driver's negligence. *Id.* at 136-37, 848 P.2d at 1059. In the case at bar, John Doe was subject to nine separate and distinct sexual assaults over a period of approximately one month, and these assaults cannot be considered to be "'so closely linked in time and space as to be considered by the average person as one event.'" *Id.* at 139, 948 P.2d at 1058 (quoting *Welter v. Singer*, 126 Wis.2d 242, 376 N.W.2d 84, 87 (1985)). Therefore, Peters was liable for nine counts of sexual assault, and pursuant to the rules of respondeat superior, the State was also liable for nine counts of sexual assault.

The respondent presented sufficient evidence of damages, but the award of damages for negligent supervision must be reversed because it is tantamount to an improper double recovery

[14] The State argues that Jimenez failed to prove the fact of damages and the amount thereof, and as a result, the damages granted to Jimenez were necessarily excessive. In assessing alleged excessive damage awards, this court will not substitute its judgment on the issue

---

[4] In any criminal action on these same charges, Peters could have been charged with nine separate counts of sexual assault.

of damages for the decision of the fact finder. *Automatic Merchandisers, Inc. v. Ward*, 98 Nev. 282, 284-85, 646 P.2d 553, 555 (1982). This court will only reverse a verdict when the award is flagrantly improper, indicating passion, prejudice, or corruption. *Id.* at 285, 646 P.2d at 555.

[15] In cases where the damages are for extreme emotional distress, such distress must be manifested by physical injury or illness, something more than just embarrassment. In *Branda v. Sanford*, 97 Nev. 643, 648, 637 P.2d 1223, 1227 (1981), this court intimated that severe emotional distress could be manifested through such symptoms as hysteria and nervousness, nightmares, great nervousness, and bodily illness and injury.

[16] Initially, evidence of physical injury was presented at trial in the form of medical testimony which proved that John Doe had suffered sexual trauma as a result of being subjected to anal intercourse. Furthermore, John Doe testified that Peters' assaults made him feel "like trash." John Doe also stated: "Sometimes I couldn't even go to sleep at night because I was afraid he would come to my room or do something." Wilford W. Beck, the former agency director for NNCAS, testified that with each rape, John Doe could become more and more ashamed of himself, and that it was his impression that when John Doe testified, he seemed ashamed. Beck also stated: "[I] will say that [sexual assault is] a terrible thing that happens. Whether it... makes [improvement] less possible, I don't know."

This medical evidence and the testimony of emotional distress, along with the multiple criminal acts committed upon John Doe, support a $50,000 award to John Doe for each of the nine acts of sexual assault. However, despite the fact that the district court properly concluded that the State was liable for one count of negligent super-vision, we believe that it was improper to award damages to John Doe on that count. John Doe was properly compensated for the nine acts of sexual assault, and the damages proved by John Doe were all caused by these sexual assaults. To permit further recovery on the basis of negligent supervision is tantamount to awarding the victim an improper double recovery for a singly injury. *See General Electric Co. v. Bush*, 88 Nev. 360, 367-68,

498 P.2d 366, 371 (1972) (explaining that an award of damages for a wife's claim for loss of consortium after her husband had been injured and compensated was not an improper double recovery for the same injury to the husband), *abrogated on other grounds by Motenko v. MGM Dist., Inc.,* 112 Nev. 1038, 921 P.2d 933 (1996). Therefore, while the State was liable on the theory of negligent supervision, we conclude that the district court erroneously awarded damages on that claim when John Doe was fully compensated on the theory of respondeat superior.

The State is not entitled to a setoff for the cost of treatment respondent's son received

[17] The State claims it is entitled to a $130,000 setoff because John Doe was a ward of the court at Colorado Boy's Ranch from September 10, 1992, through March 30, 1994. The State claims that it is entitled to this setoff under NRS 42.020,[5] which states:

1. In any action for damages for personal injury against any provider of health care, the amount of damages, if any, awarded in the action must be reduced by the amount of any prior payment made by or on behalf of the provider of health care to the injured person or to the claimant to meet reasonable expenses of medical care, other essential goods or services or reasonable living expenses.

2. As used in this section, "provider of health care" means a physician ... [or] licensed psychologist....

This statute has not been interpreted by this court. However, we need not reach the issue of its construction because the State failed to pursue a ruling on the issue of the setoff in the district court, thus waiving the issue on appeal. In *Cerminara v. District Court,* 104 Nev. 663, 765 P.2d 182 (1988), the district court failed to rule on a motion for a new trial after granting a judgment notwithstanding the jury's verdict. The defendant failed to object to this and failed to take any action to compel a ruling on that motion until after this court had rendered a decision on appeal. *Id.* at

---

[5] This statute was amended in 1995 and now contains different provisions.

665, 765 P.2d at 184. This court concluded the defendant had abandoned its motion for a new trial. *Id.; see also Old Aztec Mine, Inc. v. Brown*, 97 Nev. 49, 52-53, 623 P.2d 981, 983-84 (1981) (refusing to consider issues not precisely raised below).

We conclude that the State is not entitled to a setoff in this case because it failed to pursue a ruling on this issue from the district court.

CONCLUSION

The district court properly found that the State had waived its immunity for Peters' actions and concluded that the State was liable for Peters' sexual assaults because Peters was acting within the course and scope of his employment when he sexually assaulted John Doe. Furthermore, we conclude that the district court properly found the State liable on all nine counts of sexual assault and awarded the maximum statutory amount of damages allowed on each count for a total of $450,000. Additionally, we conclude that while the district court also properly found that the State was liable for the one count of negligent supervision of Peters because such supervision was an operational rather than a discretionary act, the award of damages to John Doe on that count amounted to a windfall or double recovery and must be reversed. Finally, the State is not entitled to a setoff for the amount spent on John Doe while at the Colorado Boy's Camp. Accordingly, we reverse the district court's award of damages on the count of negligent supervision and affirm the remainder of the district court's decisions.[6]

---

[6] The Honorable A. William Maupin, Justice, did not participate in the decision of this matter.

(Modified for the purposes of this text)

Kimberly M. WALSH, Individually and as
Parent and Guardian of Johnathon E.
Walsh, an Infant, Respondent,
v.
CITY SCHOOL DISTRICT OF
ALBANY, Appellant.
Supreme Court, Appellate Division,
Third Department.
March 13, 1997.

Action was brought against school district under theories of premises liability and negligent supervision in connection with accident in which first-grader's finger was caught in bathroom door. The Supreme Court, Albany County, Keegan, J., denied district's motion for summary judgment, and it appealed. The Supreme Court, Appellate Division, Casey, J., held that: (1) district was not liable under negligent supervision theory, and (2) no evidence established that bathroom door constituted dangerous condition, or that district had actual or constructive notice that any defective or dangerous condition existed.

Reversed; motion granted.

Carter, Conboy, Case, Blackmore, Napierski & Maloney (Joseph T. Johnson, of counsel), Albany, for appellant.

McClung, Peters & Simon (Christine M. Legorius, of counsel), Albany, for respondent.

Before CARDONA, P.J., and MERCURE, CASEY, SPAIN and CARPINELLO, JJ.

CASEY, Justice.

Appeal from an order of the Supreme Court (Keegan, J.), entered April 22, 1996 in Albany County, which denied defendant's motion for summary judgment dismissing the complaint.

Johnathon E. Walsh, a first-grade student at Public School No. 27 in the City of Albany, and a classmate were permitted to go to the bathroom together. While there, Johnathon's finger became caught in the bathroom door, resulting in the amputation of his fingertip. Subsequently, his fingertip was surgically reattached. Plaintiff commenced this action in negligence, individually and on behalf of Johnathon, alleging that defendant was negligent in failing to adequately supervise Johnathon and that the hydraulic device installed on the bathroom door created a dangerous condition, which defendant failed to adequately warn against. Following joinder of issue and some discovery, defendant moved for summary judgment. Supreme Court, not reaching the issues of dangerous condition or failure to warn, denied defendant's motion finding that a question of fact existed regarding the adequacy of defendant's supervision. This appeal ensued.

[1, 2] There must be a reversal. Schools are under a duty to adequately supervise students and can be liable for foreseeable injuries proximately related to the lack of adequate supervision (*see, Mirand v. City of New York*, 84 N.Y.2d 44, 49, 614 N.Y.S.2d 372, 637 N.E.2d 263; *Matter of Kimberly S.M. v. Bradford Cent. School*, — A.D.2d —, —, 649 N.Y.S.2d 588, 589-590, *lv. denied* — N.Y.2d — [Feb. 7, 1997]). Schools are not, however, insurers of the safety of their students (*see, Mirand v. City of New York, supra*, at 49, 614 N.Y.S.2d 372, 637 N.E.2d 263; *Ceglia v. Portledge School*, 187 A.D.2d 550, 590 N.Y.S.2d 228; *James v. Gloversville Enlarged School Dist.*, 155 A.D.2d 811, 548 N.Y.S.2d 87). Although Johnathon and his classmate engaged in horseplay while in the bathroom and defendant was aware that such inappropriate activity occurred, there is no indication that Johnathon's injury was caused during any horseplay.

[3, 4] There is testimony indicating that Johnathon's classmate may have shut the door on his finger because the classmate did not see Johnathon there. Accepting this testimony as true, this would constitute an intervening act of a third party relieving defendant of liability (*see, Ceglia v. Portledge School, supra*). Plaintiff further asserts that regardless of how the injury occurred, the presence of a supervisor could have kept Johnathon and/or his classmate attentive and the injury would have been prevented. Schools are not under a duty to guarantee safety and, therefore, we find this assertion unpersuasive. Moreover, even conceding the

fact that first-grade children are energetic, physical and easily distracted, the manner in which Johnathon's injury occurred could have happened even if he had been supervised. As such, no liability can be imposed (*see, Tomlinson v. Board of Educ. of City of Elmira*, 183 A.D.2d 1023, 1024, 583 N.Y.S.2d 664; *Swaitkowski v. Board of Educ. of City of Buffalo*, 36 A.D.2d 685, 686, 319 N.Y.S.2d 783).

[5] Furthermore, we find no evidence that the bathroom door constituted a dangerous condition. In any event, there is no evidence that defendant had actual or constructive notice that any defective or dangerous condition existed. As such, no liability can be imposed (*see, Warren v. Wilmorite Inc.*, 211 A.D.2d 904, 905, 621 N.Y.S.2d 184). We conclude that in these circumstances defendant's motion for summary judgment should be granted.

ORDERED that the order is reversed, on the law, with costs, motion granted, summary judgment awarded to defendant and complaint dismissed.

CARDONA, P.J., and MERCURE, SPAIN and CARPINELLO, JJ., concur.

chapter nine

Negligent training

"The man who graduates today and stops learning tomorrow is uneducated the day after."

Newton T. Baker

"The teaching profession is the only profession that has no definition for malpractice."

Merimon Cuninggim

The theory of negligent training is exactly what it is called — namely failing to train and educate or improperly or negligently training or educating employees or management team members. This theory can involve situations where the employers failure to provide necessary or required education or training to employees or members of management, or conversely where the employer provided training but the training was improper or wrong information was provided to the employees or management team members. This theory has limited application and is primarily based or focused on the specific facts of the situation. For example, an employer has an affirmative duty to train employees under the Occupational Safety and Health Act (OSHA) before entering a confined space area. If the employer fails to provide this mandatory training in violation of the OSHA standard and the employee enters the confined space area and becomes injured, the employer may be liable not only under the Occupational Safety and Health Act, but also when the individual state's workers' compensation act and under this theory of negligence. Although this type of action may be barred under state workers' compensation laws or willful negligence required, the basic elements are in place to proceed with this type of action:

Duty
- The employer has an affirmative duty under the Occupational Safety and Health standards.

Breach
- The employer has a duty to train and failed to provide this training.

Causation
- The failure to train the employees was the direct and proximate reason for the employee's injuries.

Damages
- What were the compensatory (i.e., medical, lost wages, etc.) damages, and how much does the jury want to punish the employer (i.e., punitive damages)?

Although this particular scenario may be covered under workers' compensation, several states have permitted tort recovery outside of workers' compensation in areas where the accident was caused by the willful negligence of the employer.[1]

The theory of negligent training has also surfaced in the public sector in dealing with firearm safety for police officers, public safety, and related areas. The principal elements in a negligent training action usually involve the employer's duty to provide appropriate training to employees. For example, in *Meistinsky v. City of New York*,[2] a police officer shot a hostage in an attempt to stop a robbery. The police officer was not trained in "double action" firing and his lack of training resulted in his fire tending to go to the right, resulting in the plaintiff being shot. In this case the court found the failure to train the police officer was negligent and assessed liability to the city. In the similar case of *Strachan v. Kitsap County*,[3] the plaintiff, a deputy sheriff, was shot by a city police officer when the officers were engaging in "horseplay." The court found that the county could be liable for failure to provide proper weapons training to the frequently deputized officer.

Conversely, several courts have denied liability for inadequate training. For example, in *Martin v. Garlotte*,[4] the police chief was found not to be liable for inadequate training when he had lectured officers on firearms safety and the department had work rules forbidding horseplay. In this case, the plaintiff was accidentally shot by another officer during an incident of horseplay.

As with virtually all of the theories of negligence which could be brought against an employer, the plaintiff has the burden of proving each of the elements of negligence (i.e., duty, breach, causation, and damages). The standard of care under the duty and breach elements is often a major area of contention, and expert witnesses are often utilized to provide testimony as to the standard of care to be utilized by the jury. This is an especially important element in a negligent training case. For example, in *District of Columbia v. Davis*,[5] the plaintiff brought a negligent training action against the district where an off-duty officer shot himself in a negligent manner. The

---

[1] For example, some states such as West Virginia permit a secondary recovery beyond workers' compensation in cases of willful negligence.
[2] 285 A.D. 1153, 140 N.Y.S.2d 212 (1955).
[3] 27 Wash. App. 2d 217, 616 P.2d 1251 (1980).
[4] 270 So. 2d 252 (La. App. 1972).
[5] 442 A.2d 159 (D.C. App. 1982).

plaintiff failed to call an expert to testify as to the standard of care, which resulted in the dismissal of the action. The judge in this case stated the standard of care issue is "so distinctly related to some science, profession, business, or occupation...beyond the ken of the average laymen" that the jury was not able to make a determination in this case.

The cause of action for negligent training is often included within other causes of action such as Section 1983 actions in the public sector and wrongful death actions in the private sector. Prudent employers should pay special attention where an affirmative duty is created, such as with the OSH Act, EPA or state laws, where mandatory training is required by law, and ensure that all training is properly performed and thoroughly documented. Employers should also evaluate any special relationships which have been created, such as with contractors, where failure to train may affect third parties or the general public.

(Modified for the purposes of this text)

Gail MANDY, Plaintiff,

v.

MINNESOTA MINING AND
MANUFACTURING, a/k/a
3M, Defendant.
Civil No. 4-95-774.
United States District Court,
D. Minnesota,
Fourth Division.
Sept. 26, 1996.

Female former employee brought action against her former employer alleging sex discrimination and retaliation in violation of Title VII and Minnesota Human Rights Act, and for negligent training, retention, and supervision. On employer's motion to dismiss, or for summary judgment, adopting in part and rejecting in part magistrate judge's report and recommendation, the District Court, Tunheim, J., held that: (1) incidents of sexual harassment, which occurred within statute of limitations period, could be part of series of related acts of discrimination for employee's Minnesota Human Rights Act claim to be timely under continuing violation doctrine, precluding summary judgment for employer on statute of limitations grounds; (2) employee failed to demonstrate any alleged acts of sexual harassment which occurred during 300-day limitations period before she filed charge of discrimination under Title VII; (3) employee stated claims of negligent retention and supervision under Minnesota law; and (4) preemption of those claims under Minnesota's Workers' Compensation Act could not be decided without further development of factual record and additional briefing.

Motion granted in part and denied in part.

Jeffrey Robert Anderson, Teresa Kathleen Patton, Reinhardt & Anderson, St. Paul, MN, for plaintiff.

Thomas Patrick Kane, Kathleen Mary Mahoney, David M. Wilk, Oppenheimer, Wolff & Donnelly, St. Paul, MN, for defendant.

MEMORANDUM OPINION AND ORDER ADOPTING IN PART AND REJECTING IN PART REPORT AND RECOMMENDATION

TUNHEIM, District Judge.

Plaintiff Gail Mandy was employed by defendant Minnesota Mining and Manufacturing ("3M") as a laborer from approximately September 12, 1991 to November 1993. Mandy filed a complaint against 3M on August 3, 1995, alleging sex discrimination and retaliation in violation of Title VII of the Civil Rights Act of 1964, 42 U.S.C. §§ 2000e, *et seq.* ("Title VII") and the Minnesota Human Rights Act, Minn.Stat. §§ 363.03, *et seq.* ("MHRA") and claims of negligent training, retention, and supervision. Defendant moved to dismiss or, in the alternative, for summary judgment as to plaintiff's Title VII and MHRA sex discrimination claims and her negligence claims. Defendant argues that plaintiff's discrimination claims are barred by the statue of limitations, that her negligent supervision and retention claims are preempted by state statutes, and that Minnesota law does not recognize a cause of action for negligent training. Defendant also moved to strike paragraph 15 of plaintiff's complaint. Defendant's motion was referred to Magistrate Judge John M. Mason for a Report and Recommendation pursuant to 28 U.S.C. § 636(b)(1)(B).

Magistrate John Mason filed his Report and Recommendation on April 23, 1996, and recommended that the Court: (1) grant defendant's motion for summary judgment and dismiss plaintiff's sex discrimination and negligence claims; and (2) strike paragraph 15 from plaintiff's complaint pursuant to Rules 8 and 12(f) of the Federal Rules of Civil Procedure. The matter is before the Court on plaintiff's objections to the Magistrate Judge's recommendation that the Court dismiss her sex discrimination and negligence claims.[1] The Court has reviewed *de novo* plaintiff's objections to the Report and Recommendation on this dispositive pretrial matter, pursuant to 28 U.S.C. § 636(b)(1)(C) and D.Minn. LR 72.1(c)(1). The Court agrees with the Magistrate Judge's conclusion that plaintiff's Title VII sex discrimination claim and her negligent training claim should be dismissed, but rejects the Magistrate Judge's recommendation as to plaintiff's sex discrimination claim under the MHRA and her negligent retention and supervision claims.

I. SEX DISCRIMINATION CLAIMS

[1] Defendant argues that plaintiff's claims are barred because she did not file a charge of discrimination within 300 days of the alleged discrimination as Title VII requires, or within 365 days as the MHRA requires. *See* 42 U.S.C. § 2000e-5(e)(1); Minn.Stat. § 363.06, subd. 3. Given that plaintiff filed a charge of discrimination on March 21, 1994, plaintiff's discrimination claims are timely only if the alleged sexual harassment occurred on or after March 21, 1993 for purposes of her MHRA claim, and on or after May 25, 1993 for her Title VII claim.

In support of her argument that her claims are timely, plaintiff relies on her complaint, a signed statement she gave 3M investigators on April 8, 1993, and an affidavit filed with her opposition to defendant's motion.[2] Plaintiff's complaint states that she was sexually harassed by her supervisor, William Palmer, beginning on April 22, 1992, and continuing through March 22, 1993. Plaintiff's statement to 3M's investigators

---

[1] Plaintiff did not object to the Magistrate Judge's recommendation that paragraph 15 of her complaint be stricken pursuant to Federal Rules of Civil Procedure 8 and 12(f), and there is little consequence to the Magistrate Judge's recommended ruling absent a Rule 12(e) motion for a more definite statement from defendant. Although defendant's motion to strike paragraph 15 of plaintiff's complaint was premised solely on its argument that the claim was time-barred, the Magistrate Judge recommended that even if the claim was timely filed, the Court should strike the paragraph as "impertinent and scandalous" material unnecessary to the complaint, apparently because the factual allegations describe the sexual comments and advances plaintiff allegedly experienced.

The Court will adopt the Magistrate Judge's recommendations on this issue because it was not objected to, but notes that a plaintiff in a sexual harassment case has the burden of stating a claim showing entitlement to relief with sufficient specificity to survive a Rule 12(e) motion. A prima facie case of sexual harassment may include allegations that the plaintiff was subjected to unwelcome sexual conduct which interfered with a term or condition of her employment. The Court hesitates to restrict the manner in which a plaintiff may plead such a case in the absence of evidence of an improper purpose or lack of evidentiary support, which are adequately addressed by the obligations imposed by Fed.R.Civ.P. 11.

and her affidavit contain many specific allegations supporting her claim that Palmer subjected her to repeated and ongoing unwelcome conduct of a sexual nature throughout the time period alleged in her complaint. The allegations included repeated sexual comments about plaintiff's body, sexual advances toward plaintiff, and other unwelcome attention such as telephone calls, cards and gifts. Plaintiff also alleges that Palmer repeatedly referred to his control over her employment status.

Plaintiff's statement to 3M describing specific incidents of unwelcome conduct includes an incident which occurred on Friday, March 19, 1993 at 10:00 a.m. in which Palmer grabbed her sweatshirt, looked down her shirt, commented on her breasts and put his hand on her neck. Plaintiff states that on the next work day, Monday, March 22, 1993, she reported Palmer's behavior. Palmer then came to her area, grabbed her arm, and, as plaintiff described the incident:

> [Palmer] said, "What are you trying to do to me, get me fired? I didn't say anything. He said, "I want you to go tell them guys that what I did was nothing" again. I said nothing. We went to the office, and the 5 of us were talking about what had happen[ed] and Bill [Palmer] said it was nothing, grabbing at my shirt trying to show what he did. I said, "that's not the way it was." He said that he didn't want anything that was said to leave the office. He really didn't want Butch to know. Bill said he was really scared. (punctuation added).

Plaintiff argues that this March 22 incident, which occurred within the statute of limitations period, was part of a series of related acts of discrimination based on plaintiff's sex and that her claims therefore are timely pursuant to the continuing violation doctrine.

---

[2] Defendant's motion is to dismiss or, in the alternative, for summary judgment. Given the Magistrate Judge's recommendation that the Court grant summary judgment on plaintiff's sex discrimination claims, the Court must take all evidence produced by the parties into consideration in its review.

The Magistrate Judge concluded that the March 22 incident was not an incident of discrimination based on sex and therefore did not consider whether it was part of a continuing violation. The Magistrate Judge noted only the part of the incident in which Palmer grabbed plaintiff's arm; he apparently did not consider Palmer's later action of grabbing at plaintiffs shirt to demonstrate what he claimed he had done the previous Friday. The Magistrate Judge concluded that the arm-grabbing incident happened to plaintiff, not because she was female, but because she complained about sexual harassment. He therefore characterized the incident as one of retaliation, and not harassment, as a matter of law. Because the incident was not harassment, the Magistrate Judge specifically declined to consider whether it was part of a continuing violation which included the prior conduct of incidents of discrimination based on sex.

[2, 3] The Court disagrees with the Magistrate Judge's analysis for the following reasons. First, in determining whether the incident within the statute of limitations period constituted sexual harassment, the Court must consider "'the record as a whole' and 'the totality of the circumstances, such as the nature of the sexual advances and the context in which the alleged incidents occurred.'" *Meritor Sav. Bank, FSB v. Vinson*, 477 U.S. 57, 69, 106 S.Ct. 2399, 2406, 91 L.Ed.2d 49 (1986) (citing EEOC Guidelines). The Court must look at "the nature, frequency, intensity, location, context, duration, and object or target" of the language and conduct. *Klink v. Ramsey County*, 397 N.W.2d 894, 901 (Minn.App.1986), *rev. denied* (Minn. Feb. 13, 1987) (citing *Meritor*). As the Magistrate Judge noted, the Eighth Circuit Court of Appeals has made it clear that incidents which are not of an explicitly sexual nature may still constitute sexual harassment:

> We have never held that sexual harassment
> or other unequal treatment of an employee
> or group of employees that occurs because
> of the sex of an employee must, to be
> illegal under Title VII, take the form
> of sexual advances or of other incidents
> with clearly sexual overtones. And we
> decline to do so now. Rather, we hold
> that any harassment or unequal treatment

of an employee or group of employees
that would not occur but for the sex of
the employee or employees may, if sufficiently patterned or pervasive, comprise an illegal condition of employment
under Title VII.

Hall v. Gus Constr. Co., Inc., 842 F.2d 1010, 1014 (8th
Cir.1988); *see also Kopp v. Samaritan Health System,
Inc.*, 13 F.3d 264, 269 (8th Cir.1993) (reaffirming *Hall*).[3]
The Court stated that "[i]ntimidation and hostility
towards women because they are women can obviously result
from conduct other than explicit sexual advances." *Hall*,
842 F.2d at 1014. The Court also noted with approval
other Circuit Court opinions holding that threats of
physical violence and acts of physical aggression were
properly considered sexual harassment. *Id.*

[4, 5] Against the backdrop of the definition of
sexual harassment, the Court must also, apply the
continuing violation doctrine. Under that doctrine, there
may be redress for unlawful discriminatory acts which
occurred prior to the statute of limitations period if
they are related to violative acts which occurred within
the statutory period. A plaintiff may challenge incidents
which occurred outside the statute of limitations period
if the various acts of discrimination constitute a
continuing pattern of discrimination. *Hukkanen v. Int'l
Union of Operating Eng'rs, Hoisting & Portable*, 3 F.3d
281, 285 (8th Cir.1993) (when Title VII violations are
continuing in nature, the limitations period does not
begin to run until the last occurrence of discrimination);
see also Delaware State College v. Ricks, 449 U.S. 250,
257-58, 101 S.Ct. 498, 503-04, 66 L.Ed.2d 431 (1980);
United Air Lines, Inc. v. Evans, 431 U.S. 553, 558-60,
97 S.Ct. 1885, 1888-90, 52 L.Ed.2d 571 (1977). Minnesota
courts have adopted the continuing violation theory in

---

[3] Defendant argues that *Hall*, a case applying Title VII to a sexual
harassment claim, does not apply to a sexual harassment claim under
the MHRA due to differences in the language of the two statutes.
The Minnesota Supreme Court, however, has repeatedly held that
principles developed under Title VII may be applied to cases brought
under the MHRA because of the similarities between the two statutes.
See Sigurdson v. Isanti County, 448 N.W.2d 62, 67 (Minn. 1989);
Sigurdson v. Isanti County, 386 N.W.2d 715, 719 (Minn. 1986); *Hubbard
v. United Press Int'l, Inc.*, 330 N.W.2d 428, 441 (Minn. 1983).
Defendant has not cited any contrary authority.

discrimination cases. *See Hubbard v. United Press Int'l, Inc.*, 330 N.W.2d 428, 441 n. 11 (Minn.1983); *Sigurdson v. Isanti County*, 448 N.W.2d 62, 68 (Minn.1989).

Federal courts have recognized two types of continuing violations, a series of related acts, one or more of which falls within the limitations period, or the maintenance of a discriminatory system both before and during the limitations period. *Jenson v. Eveleth Taconite Co.*, 824 F.Supp. 847, 877 (D.Minn.1993); *Laffey v. Independent School Dist. No. 625*, 806 F.Supp. 1390, 1400 (D.Minn.1992), *aff'd sub nom.*, 994 F.2d 843 (8th Cir.1993), *cert. denied*, 510 U.S. 1054, 114 S.Ct. 715, 126 L.Ed.2d 680 (1994). The issue in the instant motion is whether defendant has shown that the acts plaintiff alleges occurred within the limitations period are, as a matter of law, not part of a series of related acts. In the incident which occurred within the statute of limitations period, the alleged harasser physically grabbed the plaintiff in a threatening manner because he learned she had complained about his sexual advances toward her on the previous work day, which was outside the limitations period. He also touched her shirt in the presence of other employees to demonstrate what he claimed he had done to her on the previous work day. This touching occurred at his initiative, not as part of any investigation of plaintiff's complaint. The issue of the Court is whether Palmer's actions on March 22 could be construed as part of a series of related acts.

Defendant argues that a bright-line distinction can be drawn in this case between sexually harassing behavior and retaliatory behavior, that Palmer's actions on March 22 were motivated only by retaliation, and that these actions therefore are unrelated to Palmer's previous actions. Defendant's interpretation of the March 22 incident must be rejected because it would require drawing an inference adverse to plaintiff when all inferences must be drawn in plaintiff's favor for purposes of this motion.

[6] Moreover, even if the Court accepted defendant's characterization of the events on March 22 as motivated solely by reprisal that would not bar the Court from also considering them to be acts of sexual harassment under Minnesota law. The Minnesota Court of Appeals

specifically rejected the argument that acts of reprisal cannot constitute sexual harassment in *Giuliani v. Stuart Corp.*, 512 N.W.2d 589, 595-96 (Minn.Ct.App.1994). The defendant in *Giuliani* argues that acts of reprisal cannot constitute continuing sexual harassment violation because the reprisals were merely "consequences" of discriminatory acts. The court found that the retaliatory conduct flowed directly from the sexual harassment and stated:

> A "consequence," according to the [Minnesota Supreme Court], is merely a "continuing effect." [The] reprisals were not only continuing effects but were discriminatory acts in and of themselves. [Plaintiff's] claim of sexual harassment is thus not barred by the statute of limitations.

Id. at 596. This analysis seems particularly appropriate where the act which defendant urges the court to consider only as an act of reprisal involves physical touching of the plaintiff, and where it "flows directly" from an incident of alleged unwelcome touching which occurred on the previous workday and which could not have been motivated by retaliation.

[7] The Court views the acts which occurred within the statute of limitations period on March 22, 1993, as closely related to those which occurred on March 19, 1993 and as part of a series of related acts of alleged harassment which would not have occurred but for plaintiff's sex. The Court's view is guided by reference to federal cases applying the continuing violation doctrine to hostile environment sexual harassment claims. Other federal courts have shared this Court's view that to properly evaluate whether the acts within the limitations period are part of a series of related acts, it must consider all of the alleged related acts. As the court stated in *Jenson*:

> In the arena of sexual harassment, particularly that which is based on the existence of a hostile environment, it is reasonable to expect that violations are continuing in nature: a hostile environment results from acts of sexual

harassment which are pervasive and continue over time, whereas isolated or
single incidents of harassment are insufficient to constitute a hostile environment. Accordingly, claims based on
hostile environment sexual harassment
often straddle both sides of an artifi-
cial statutory cut-off date.

Moreover, evidence of what occurred
prior to the beginning of the statutory
period is relevant evidence which may be
considered in determining whether a hos-
tile environment existed during the rel-
evant period. This view is implied in
Meritor: The Supreme Court considered
acts of harassment that had occurred over
a four year period. At no time did the
Court mention that some of these events
occurred outside the relevant time period.

Jenson, 824 F.Supp. at 877-78 (emphasis added); *see also*
Robinson v. Jacksonville Shipyards, Inc., 760 F.Supp.
1486, 1494-95 (M.D.Fla.1991) (considering incidents outside limitations period to provide context for actionable
incidents in determining whether more recent conduct was
a part of the work environment or an aberration). Such
an approach is consistent with the Supreme Court's
directive that courts consider the nature, context, and
frequency of incidents of sexual harassment in considering hostile environment claims. Although the Court is
mindful that plaintiff's claim is timely only if a
violation exists during the limitations period, a hostile
environment claim by its nature is comprised of a series
of incidents. Any one incident occurring within the
limitations period, if temporally and substantively
related to those preceding the limitations period,
supports a claim that the environment continued to be
hostile into the limitations period. In this case, the
incidents within the limitations period occurred one
work day after those outside the period, were initiated
by the same alleged harasser, and were similar in nature
to previous acts of threats and intimidation alleged by
plaintiff to have occurred on a regular basis throughout
the immediately preceding year.

[8] Were the Court to employ the more mechanical analytical framework for applying the continuing violation theory enunciated in *Berry v. Bd. of Supervisors of L.S.U.*, 715 F.2d 971, 981 (5th Cir.1983), and adopted by several other circuit courts, the Court would reach the same result. Three factors were set out in *Berry* to determine whether a continuing violation exists. *Id.* The first was similarity of subject matter. This factor is easily satisfied in this case where all alleged acts were committed by the same person and all involved sexual advances, unwelcome touching or implied threats.

[9] The second factor, the frequency of the acts, also points toward a continuing violation in this case. Plaintiff alleged an unbroken string of incidents of sexually motivated behavior by Palmer during an approximately one-year period. The last two incidents, which straddle the limitations period cut-off, occurred on subsequent work days. Noting the implications of the Supreme Court's recognition of hostile environment sexual harassment claims in *Meritor* for the continuing violation analysis it set out in *Berry*, the Fifth Circuit Court of Appeals has held that in a hostile environment, an individual feels constantly threatened even in the absence of constant harassment, and that courts should therefore avoid mechanical calculations of the frequency of harassment in such cases. *Waltman v. Int'l Paper Co.*, 875 F.2d 468, 476 (5th Cir.1989). Courts should instead review the pattern and frequency of harassment and determine whether a reasonable person would feel that the environment was hostile throughout the period alleged. *Id.* Based on the nature and frequency of the acts alleged in the instant case, the Court cannot conclude as a matter of law that a reasonable woman would not find plaintiff's work environment to have been hostile on March 22, 1993.

The third *Berry* factor, permanence, involves inquiry into whether an act outside the limitations period should have triggered the plaintiff's awareness and duty to assert her rights. This factor is difficult to apply in the context of allegations of recurrent incidents of harassment. *See* Ramona L. Paetzold & Anne M. O'Leary-Kelly, *Continuing Violations and Hostile Environment Sexual Harassment: When is Enough, Enough?*, 31 Am. Bus. L.J. 365, 390-92 (November 1993) (noting difficulty of applying third factor of *Berry* test to hostile environment claims). The Court is unable to distinguish any particular alleged incident prior to the limitations period which, as a matter of law, should have spurred plaintiff to assert her rights.

[10] For the reasons stated above, defendant has failed to demonstrate that it is entitled to judgment as a matter of law on the statute of limitations with respect to plaintiff's sex discrimination claims under the MHRA. The Court rejects the Magistrate Judge's recommendation and denies defendant's motion to dismiss or for summary judgment as to plaintiff's MHRA sex discrimination claim. The Court agrees with the Magistrate Judge that plaintiff failed to demonstrate any alleged acts of sexual harassment which occurred on or after May 25, 1993; therefore, the Court accepts the Magistrate Judge's recommendations that defendant's motion for summary judgment on plaintiff's Title VII sex discrimination claim be granted.

II.   NEGLIGENCE CLAIMS

[11, 12] The Magistrate Judge recommended dismissing plaintiff's state law claims of negligent supervision and negligent retention because he concluded they were preempted by the MHRA and Minnesota's Workers Compensation Act (WCA). The tort of negligent retention has been recognized in Minnesota law for over 100 years. *Dean v. St. Paul Union Depot Co.*, 41 Minn. 360, 43 N.W. 54, 55(1889)(employer "had no more right ... to knowingly and advisedly employ, or allow to be employed, ... a dangerous and vicious man, than it would have to keep and harbor a dangerous and savage dog....."). Liability for negligent retention is:

> predicated on the negligence of an employer in placing a person with known propensities, or propensities which should have been discovered by reasonable investigation, in an employment position in which because of the circumstances of the employment, it should have been foreseeable that the hired individual posed a threat of injury to others.

Ponticas v. K.M.S. Investments, 331 N.W.2d 907, 911 (Minn.1983). *See also Restatement (Second) Agency* § 213. The Minnesota Court of Appeals has adopted the following definition:

> Negligent retention ... occurs when, during the course of employment, the employer becomes aware or should have been aware of problems with an employee that indicated his unfitness, and the employer's failure to take further action such as investigating, discharge, or reassignment.

Yunker v. Honeywell, Inc., 496 N.W.2d 419, 423 (Minn.Ct.App.1993), *rev. denied* (Minn. April 20, 1993) (quoting *Garcia v. Duffy*, 492 So.2d 435, 438-39 (Fla.Dist.Ct.App.1986 citations omitted). Negligent retention imposes liability on the employer for the employee's intentional torts; thus, the employee's acts are almost always outside the scope of employment. *Id.* at 422.

[13, 14] The Minnesota Supreme Court has also recognized the tort of negligent supervision. Minnesota courts look to *Restatement (Second) of Torts* § 317 and *Restatement (Second) Agency* § 213 in defining the parameters of the tort. In *Semrad v. Edina Realty, Inc.*, 493 N.W.2d 528, 534 (Minn.1992), the court described the tort of negligent supervision as follows:

> [T]he entire thrust of § 317 is directed at an employer's duty to control his or her employee's physical conduct while on the employer's premises or while using the employer's chattels, even when the employee is acting outside the scope of the employment, in order to prevent intentional or negligent infliction of personal injury.

Id. The duty imposed is unambiguously limited to preventing an employee from inflicting personal injury upon a third person on the master's premises or inflicting bodily harm by use of the employer's chattels. *Id.* at 534. The tort has also been described as "the failure of the employer to exercise ordinary care in supervising the employment relationship so as to prevent the fore-seeable misconduct of an employee causing harm to others." *Fletcher v. St. Paul Pioneer Press*, 1995 WL 379140 at *4 (Minn.Ct.App. June 27, 1995), *rev. denied* (Minn. Aug. 30, 1995), *rev. denied* (Minn. Oct. 10, 1995) (unpublished opinion citing *Cook v. Greyhound Lines, Inc.*, 847 F.Supp. 725, 732 (D.Minn. 1994).

[15, 16] The Court first considers whether plaintiff has stated a claim of negligent retention or supervision under Minnesota law. Plaintiff has alleged that she complained to Palmer's supervisor in August 1992 about Palmer's harassment, that nothing was done in response to her complaints, and that the harassment became worse after her complaints and continued through March 1993. Plaintiff also alleges that Palmer touched her in a sexual manner on more than one occasion after she complained about his behavior. Plaintiff has therefore alleged that in the course of her employment, her employer became aware of Palmer's alleged unfitness and failed to take further action. Plaintiff has adequately pleaded a claim of negligent retention. *See Yunker*, 496 N.W.2d at 423. She has also stated a claim for negligent supervision by alleging that after she reported Palmer's sexual harassment, his continued harassment of her was foreseeable to defendant and that defendant had a duty to protect her from further sexual harassment.

Defendant argues that the torts of negligent retention and supervision apply only to violent acts causing physical injury. Some courts have suggested that redress for the tort of negligent retention under Minnesota law is confined to cases involving physical injury. *Leidig v. Honeywell, Inc.*, 850 F.Supp. 796, 807 (D.Minn.1994) ("Although Minnesota courts have never directly addressed this issue, the case law strongly suggests that this duty not to retain employees with 'known dangerous proclivities' is limited to circumstances involving a threat of physical injury."). Minnesota courts have, however, recognized claims of negligent retention involving allegations of sexual harassment. In *Kresko v. Rulli*, 432 N.W.2d 764 (Minn.Ct.App.1988), *rev. denied* (Minn. Jan. 31, 1989), the court considered a negligent retention claim in a case where a sexual harassment claim was also brought and affirmed the trial court's dismissal of the negligent retention claim only because the plaintiff provided no evidence that the employer knew or should have known of the harassment. *Id.* at 769-70. In an unpublished case, the Minnesota Court of Appeals has explicitly held that "the tort of negligent supervision or retention is not limited to negligence that results in physical injuries; it also applies to negligent retention or supervision of a sexual harasser." *DeRochemont v. D & M Printing of Minneapolis*, 1994 WL 510153 at *1 (Minn.Ct.App. Sept. 20, 1994). Minnesota law is thus unclear as to whether the tort of negligent retention is limited to claims involving physical injury or the threat of physical injury.

Even if the tort of negligent retention is limited to cases involving physical injury, however, plaintiff has alleged physical injury by alleging that she was subjected to physical acts of sexual misconduct after she complained of sexual harassment. It is this Court's view that allegations of such physical conduct are sufficient to satisfy any requirement of a threat of physical injury. While Minnesota law on the parameters of the tort of negligent retention are not well defined, *see Thompson v. Campbell*, 845 F.Supp. 665, 676 (D.Minn.1994), they are sufficiently clear to enable the Court to determine that defendant has not demonstrated it is entitled to judgment as a matter of law on plaintiff's negligent retention claim on the grounds that she failed to allege an act of violent aggression.

[17] With respect to negligent supervision, it is even less clear that Minnesota law requires an allegation of violent aggression. *See Semrad*, 493 N.W.2d at 534 (duty is to control employee's *physical conduct* to prevent intentional or negligent infliction of *personal injury*). Plaintiff has alleged that Palmer's physical conduct subjected her to personal injury and that defendant's failure to control Palmer's behavior was the proximate cause of her injury. Defendant has also failed to demonstrate that it is entitled to judgment as a matter of law on plaintiff's negligent supervision claim for failure to allege an act of violent aggression.

[18] Having determined that plaintiff has stated a claim of negligent retention or supervision, the Court addresses defendant's argument that plaintiff's negligence claims are preempted by the MHRA and the WCA. The

Magistrate Judge concluded that the claims were preempted based solely on the holdings in an unpublished Minnesota Court of Appeals case, *Wise v. Digital Equipment Corp.*, 1994 WL 664973 (Minn.Ct.App. Nov 29, 1994), *rev. denied* (Minn. Jan. 25, 1995). The *Wise* court held that prior to the MHRA there was no duty imposed on an employer to take prompt remedial action to respond to sexual harassment. *Wise,* 1994 WL 664973 at *2. The court held that because the duty alleged in *Wise* was created by the MHRA, the plaintiff's negligence claim was preempted by the exclusivity provision of the MHRA found in Minn.Stat.§356.11.[4] *Id.* The Court also held that Wise's negligence claim was preempted by the exclusivity provision of the WCA because the injury claimed arose out of and in the course of employment. *Id; see* Minn.Stat.§ 176.031.

The Minnesota Supreme Court has never addressed the issue of preemption of negligent retention or supervision claims by the MHRA or the WCA. The *Wise* case is unpublished and contains little analysis of these important state law issues. Although *Wise* appears to hold that negligent supervision or retention claims for failure to respond to sexual harassment are preempted by the MHRA, another unpublished Minnesota Court of Appeals case has held that claims may be brought for negligent supervision or retention of a sexual harasser. *See DeRochemont v. D & M Printing of Minneapolis*, 1994 WI 510153 (Minn.Ct.App. Sept. 20, 1994). In yet another unpublished Court of Appeals case, *Huffman v. Pepsi-Cola Bottling Co. of Minneapolis and St. Paul*, 1995 WL 434467 at *5 (Minn.Ct.App. July 25, 1995), the court held that threats of physical violence by a co-worker which occurred during a period of alleged sexual harassment fall outside the MHRA and could support a separate negligent retention or supervision claim. Other courts applying Minnesota law have also allowed plaintiffs to proceed with both harassment and negligent supervision or retention claims, although there is no discussion of preemption in these cases. *See Fletcher v. St. Paul Pioneer Press*, 1995 WL 379140 (Minn.Ct.App. June 27, 1995, *rev. denied* (Minn. Aug. 30, 1995), *rev. denied* (Minn. Oct. 10 1995); *Thompson*

---

[4] The Minnesota Court of Appeals case of *Vaughn v. Northwest Airlines, Inc.*, 546 N.W.2d 43 (Minn.Ct.App.1996) is consistent with this holding.

v. Campbell 845 F.Supp. 665 (D. Minn.1994). Under these circumstances, it is unclear how the Minnesota Supreme Court would resolve the MHRA preemption issue in this case.

There is a similar lack of clarity in Minnesota law with respect to preemption by the Workers Compensation Act. Although *Wise* appears to hold that any negligent retention or supervision claim involving allegations of sexual harassment is preempted by the WCA, the Minnesota Court of Appeals has held in a published case that a battery claim based on an unwelcome kiss which occurred in the workplace was not barred by the WCA because the kiss had no association with the job itself and therefore did not arise out of the employment, Johnson v. Ramsey County, 424 N.W.2d 800, 804 (Minn.Ct.App.1988). In Huffman, another panel of the Minnesota Court of Appeals concluded in an unpublished case that sexual comments and physical contact which occurred at plaintiff's workplace were not associated with her job and that her negligent retention or supervision claims therefore were not barred by the WCA. *Huffman*, 1995 WL 434467 at *5-6 (citing *Johnson*).

[19] These cases as well as a review of cases from other jurisdictions,[5] compel the conclusion that the Court cannot decide the WCA preemption issue presented in this case without further development of the factual record and additional briefing. Under Minnesota law, to

---

[5] Although some courts have held that their state worker's compensation statutes preempt negligent retention or supervision claims regarding alleged sexual harassment, courts in other jurisdictions have reflected the same preemption arguments. *Compare Fields v. Cummins Employees Fed. Credit Union*, 540 N.E.2d 631 (Ind. Ct. App. 1989). (preemption where plaintiff conceded the harassment arose in the course of employment); *Brooms v. Regal Tube Co.*, 881 F.2d 412, 421 (7th Cir.1989) (preemption under Illinois law), with *Byrd v. Richardson-Greenshields Securities, Inc.*, 552 So.2d 1099 (Fla.1989) (Florida worker's compensation statute did not preempt negligent hiring and retention claims in sexual harassment case); *Hogan v. Forsyth Country Club Co.*, 79 N.C.App. 483, 340 S.E.2d 116 (1986), rev.denied (N.C. July 2, 1986) (negligent retention claim based on sexual harassment not barred by North Carolina worker's compensation statute as emotional injury suffered not natural and probable consequence of employment); *Cox v. Brazo*, 165 Ga.App. 888, 303 S.E.2d 71 (1983), *aff'd sub nom*, 251 Ga. 491, 307 S.E.2d 474 (1983) (negligent retention claim based on knowledge of employee's sexually offensive conduct toward female employees not barred by Georgia worker's compensation statute).

be compensable under the WCA, the plaintiff's injury must arise out of the employment and must not come within the "assault exception" to the definition of personal injury in the WCA. *Foley v. Honeywell, Inc.*, 488 N.W.2d 268, 271 (Minn.1992). Neither party has briefed the issues of the nature of the injuries in this case, whether they occurred "in the course of" the employment, or whether the assault exception applies. Defendant has not demonstrate that, accepting plaintiff's allegations as true and drawing all favorable inferences therefrom, plaintiff's alleged injuries are exclusively compensable under the WCA.

In conclusion, the Court finds defendant has failed to demonstrate that it is entitled to judgment as a matter of law at this stage of these proceedings. The Court rejects the Magistrate Judge's recommendation that it grant summary judgment for defendant on plaintiff's negligent retention or supervision claims at this time, but the Court may entertain a motion for partial summary judgment on the preemption issues when the record is more fully developed.

[20] With respect to plaintiff's negligent training claim, the Court agrees with the Magistrate Judge's conclusion that Minnesota does not recognize such a cause of action. See *M. L. v. Magnuson*, 531 N.W.2d 849, 856 (Minn.Ct.App.1995), *rev. denied* (Minn. July 20, 1995) ("Minnesota recognizes three causes of action where a claimant sues an employer in negligence for injuries caused by one of its employees; negligent hiring, negligent retention, and negligent supervision."); Fletcher, 1995 WL 379140 at *4 (no case law recognizing an independent cause of action for negligent training). Plaintiff did not object to the Magistrate Judge's recommendation that her negligent training claim be dismissed. The Court therefore adopts the Magistrate Judge's Report and Recommendation with respect to the negligent training claim.

ORDER

Based on the submissions of the parties, the arguments of counsel and the entire file and proceedings herein, IT IS HEREBY ORDERED that the Magistrate Judge's Report and Recommendation [Docket No. 14] is ADOPTED as to plaintiff's Title VII sex discrimination claim and plaintiff's negligent training claim and REJECTED as to plaintiff's sex discrimination claim under the Minnesota Human Rights Act and plaintiff's negligent retention and supervision claims. Count One (Title VII Sex Discrimination) and the Negligent Training claim in Count Five of plaintiff's Complaint are hereby DISMISSED.

chapter ten

Negligent security

"All business proceeds on beliefs, on judgments of probabilities, and not on certainties."

Charles Williams Eliot

"Take calculated risks. That is quite different from being rash."

General George S. Patton

The theory of negligent security is often invoked in areas where the employer had an affirmative duty to safeguard employees or the general public. This legal theory is most applicable in situations where the landowner, company, or employer has an affirmative duty set forth in the law or has actual or constructive knowledge of past incidents and fails to safeguard the public, employees, or other individuals.

A duty to safeguard employees or the public is often created for the employer by federal laws, such as the Occupational Safety and Health Act, through state laws or statutes, or through knowledge because of past incidents. The pivotal issues in a negligent security case normally involve the issue as to whether a duty was created and whether that duty was breached, rather than on issues of causation and damages.

For example, an employer has a large employee parking lot where employees are required to park their vehicles. The employee parking area has substandard lighting, and the employer knew that there had been several attempted assaults in the parking area. A female employee, working the late shift, leaves the facility and walks to her vehicle where she is sexually assaulted. Utilizing this example, the employer knew of past assaults, and the issue is whether this knowledge created a duty to safeguard this employee in the parking area. Was this duty breached when the employer failed to provide adequate lighting or security for the female employee leaving the plant? Was the failure to provide adequate security or lighting the cause of the sexual assault? The only remaining issue would be the extent of the damages and whether the particular workers' compensation statute applied to the parking lot areas.

The issue of whether a duty exists or is created for the employer is often the pivotal question in cases of negligent security. In assessing the basic elements of a potential negligent security issue, prudent professionals may want to consider the following:

Duty
1. Has a duty to safeguard employees been created by federal law? State law? Local ordinance?
2. Has the workplace experienced any attempts, threats, or actual incidents in the recent past?
3. Does the employer know of any attempts, threats, or actual incidents?
4. Does the employer have second-hand knowledge of an incident?
5. Should the employer have known of attempts, threats, or actual incidents?

Breach
1. Did the employer fail to act on the knowledge of the risk?
2. Did the employer breach its duty to create and maintain a safe workplace?
3. If the employer reacted, were the safeguards initiated adequate under the circumstances?

Causation
1. Was the injury incurred the result of the employer's actions?
2. Was the injury incurred the result of the employer's inactions?

Damages
1. How much were the individual's medical costs?
2. How severe were the individual's physical injuries? Psychological injuries?
3. How much time did the individual lose from work?
4. How much were the other compensatory damages?
5. How much does the judge or jury want to punish the employer for not safeguarding the employee (i.e., punitive damages)?

With negligent security cases, the elements of whether a duty was present for the employer and whether this duty was breached are normally the heart of the case. For employers in the public sector, this duty is usually provided by statute and often extends to the general public. For private sector employers, a careful assessment of the federal and state laws can identify any duty which may be applicable. More often in negligent security cases in the private sector, the employer's duty is created when knowledge is acquired or should have been acquired of a risk or incident to an individual or group of individuals. Once knowledge is acquired or should have been acquired and the situation or risk of harm is permitted to exist by the employer without response or with inadequate response, the second or subsequent incident could result in a duty being created and subsequently breached.

Another area where the theory of negligent security may be applicable is the issue of domestic violence filtering into the workplace. Does the employer have an affirmative duty to safeguard employees from the potential of outside violence from family members or significant others? In many circumstances, the employer does have a duty to safeguard the individual involved as well as other employees when knowledge of the threat becomes known to the employer. The potential of domestic violence transferring to the workplace is present; however, employers must also be cognizant of individual privacy rights and involvement in activities outside of the workplace. This can often create a dilemma for employers, especially where all individuals involved work for the employer. Prudent employers may want to address these potential issues in their personnel procedures and policies beforehand and provide strict parameters to be followed by management team members. Additionally, employers may want to reduce the potential risk to employees by nonemployees or outside individuals by providing additional physical security and limiting access to the workplace.

A substantial number of workplace violence incidents in recent years involved ex-employees who had been terminated from their employment who returned to the workplace seeking revenge or for other motives. Does the employer have an affirmative duty to provide security against ex-employees returning to the workplace? As a general rule, most employers limit access to the workplace by ex-employees except under specific circumstances, such as grievance hearings, insurance issues, and related issues. With a substantial number of workplace violence incidents, the focus of the violence is the ex-employee's supervisor, personnel manager, or related member of management. Given this general knowledge, as well as specific incident knowledge, it can be assumed that employers in the United States have been placed on notice of the potential risks in this area. Employers should be cognizant of the potential risk of ex-employees returning to the workplace and address this potential through appropriate policies, procedures, training, and physical security measures.

In order to prevent or minimize the potential of liability in the area of negligent security, employers may wish to consider the following:

- Assess the potential risks and issues in the particular workplace.
- Develop company policies and procedures addressing the potential risks.
- Develop training programs addressing proper methods of terminating employees, nonviolent response, and conflict resolution.

- Assess physical security measures, including but not limited to, use of surveillance cameras in entrance areas, lighting in parking areas, drop boxes in money handling areas, and other measures.
- Establish policies and procedures to address reports of incidents or threats and an assessment/response procedure for management.
- If an incident should occur, take proactive measures to prevent recurrence.
- Include workplace violence risks, within the required emergency and disaster preparedness plans.

The level of security for the individual workplace will vary depending upon size, location, and various other factors; however, management should be cognizant of the potential risks of workplace violence and take necessary and appropriate security actions in order to safeguard the employees in the workplace. The level and degree of security appear to be dependent upon the past history and type of workplace as well as the employer's actual or constructive knowledge that a risk exists.

Prudent employers should not place their "heads in the sand" with regard to the potential of workplace violence in their workplace. If a risk exists, assess and address the risk. Acquire competent professional assistance if necessary. When a duty to act is present and the employer fails to act or acts improperly, the theory of negligent security can arise!

(Modified for the purposes of this text)

James Robert JOHNSON, Personal Representative
of the Estate of Robert Wayne Johnson, Appellant,
v.
Richard E. THONI, d/b/a Thoni Oil Company of
Florida, Inc., and Thoni Service Corporation, Appellees.
No. 83-2503.
District Court of Appeal of Florida,
Third District.
July 24, 1984.

Plaintiff wrongful death action against defendant corporations which had provided management and security services to decedent's employer. The Circuit Court, Dade County, Richard S. Hickey, J., dismissed complaint, and plaintiff appealed. The District Court of Appeal, Tillman Pearson (Ret.), Associate Judge, held that wrongful death complaint which alleged that defendant corporations which provided management and security services to decedent's employer and thus had duty to all employees to provide safe workplace was sufficient to sustain claim, and was not subject to dismissal for failing to adequately allege duty or on basis of workers' compensation immunity.

Reversed and remanded.

Frances Avery Arnold, Pompano Beach, T. James Emison, Jr., Alamo, Tenn., for appellant.

Blackwell, Walker, Gray, Powers, Flick & Hoehl and James C. Blecke, Miami, for appellees.

Before HUBBART and JORGENSON, JJ., and TILLMAN PEARSON (Ret.), Associate Judge.

TILLMAN PEARSON (Ret.), Associate Judge.

The plaintiff appeals from a dismissal with prejudice of his second amended complaint for the wrongful death of his decedent. The defendants were a number of Thoni corporations, other than the one which employed the decedent. The negligence alleged is that the defendants failed in their duty to provide adequate security services which they had agreed to provide. It is further alleged that as a result of this negligence, the plaintiff's decedent was fatally shot at the Thoni filling station where he was employed.

It is unclear why the complaint was dismissed, but defendants, appellees, have suggested: (1) that the complaint fails to allege any legally recognizable duty owed by the Thoni corporations to the decedent, and (2) that the workers' compensation statute provides immunity to these corporations which had agreed to furnish the employer with inspection and advisory services related to safety of employees.

Before considering the possible bases for dismissal it should be noted that in argument counsel have referred to factual matters which they say are a result of discovery. We have not considered these arguments because we have before us a judgment of dismissal on the pleadings.

The allegations of the complaint are that each of the corporations provided management and security services to decedent's employer, employed supervisory personnel to inspect periodically the premises of the service station, and thus had a duty to all employees to provide a safe workplace or to inform decedent's employer of the steps it could take to make the workplace safer. In *Conklin v. Cohen*, 287 So.2d 56 (Fla.1973), a construction case, the supreme court stated that an owner and an architect might be held liable in a third-party suit if they had participated in the construction to the extent that they directly influenced the manner in which the work was performed. The extent of the involvement of each of the corporations in the instant case is not determinable at this stage in the proceedings. Thus, the dismissal with prejudice cannot be sustained upon the basis that the allegations of the complaint show that the duty alleged did not exist.

The argument that the defendant corporations were immune because of the provisions of the workers' compensation statute cannot be sustained under the law as set forth in *Morris v. Bryan & Fletcher, Inc.*, 373 So.2d 407 (Fla. 4th DCA 1979), and in *Greene v. Ivaco Industries, Ltd.*, 334 So.2d 347 (Fla. 1st DCA 1976). The immunity provision to which reference is made is found in Section 440.11(2), Florida Statutes (1981).[1] The complaint alleges that the corporations provided management and security services. These services do not qualify the corporations for immunity because it does not appear from the pleadings that these services were incidental to the workers' compensation or employer's liability coverage.

The facts developed may determine the existence or nonexistence of the duty and of the claimed immunity. The dismissal of the complaint with prejudice was error.

Reversed and remanded.

1. Section 440.11(2), Florida Statutes (1981), provides, as follows:

> (2) An employer's workers' compensation carrier, service agent, or safety consultant shall not be liable as a third-party tortfeasor for assisting the employer in carrying out the employer's rights and responsibilities under this chapter by furnishing any safety inspection, safety consultative service, or other safety service incidental to the workers' compensation or employers' liability coverage ...

(Modified for the purposes of this text)

418 Mass. 191
Kenneth D. FUND,[1] Administrator,[2]
v.
HOTEL LENOX OF BOSTON,
INC., & another.[3]
Supreme Judicial Court of Massachusetts,
Suffolk.
Argued April 5, 1994.
Decided July 7, 1994.

Estate of hotel guest stabbed to death in hotel room brought wrongful death action against hotel. The Superior Court, Suffolk County, Hiller B. Zobel, J., granted summary judgment for hotel on the basis that estate failed to establish necessary causal relationship between hotel's negligent security precautions and guest's death. Estate appealed. On transfer from the Appeals Court, the Supreme Judicial Court, Wilkins, J., held that estate's evidence demonstrated that jury would be warranted in concluding it was more probable than not that stabbing of guest by unidentified assailant was within reasonably foreseeable risk of harm created by hotel's negligent security precautions so as to preclude summary judgment in favor of hotel.

Summary judgment vacated.

William H. Carroll (Antonette S. Fernandez, with him), Boston, for plaintiff.

Alice Olsen Mann, Boston, for defendants.

---

[1] In the *Fund* case, there was evidence that a robbery was reasonably foreseeable and evidence of a killing of a hotel guest by a robber or robbers. Intrusions into the hotel and into guest rooms had occurred previously, and the hotel knew of them. These circumstances bearing on reasonable foreseeability, not paralleled in the case now under consideration, create a meaningful distinction between the two cases.

[2] Of the estate of Karen A. Edwards.

[3] Saunders Hotel Company, Inc. The defendants state that the Lenox Hotel is owned by Hotel Lenox of Boston, Inc., and managed by Saunders Hotel Company, Inc.

Before WILKINS, ABRAMS, NOLAN and GREANEY, JJ.

WILKINS, Justice.

Karen Edwards, a thirty-four year old attorney from Clearwater, Florida, was stabbed to death on June 1, 1989, in room 624 of the Lenox Hotel (Lenox) in Boston. She had checked into the Lenox minutes before. Her murderer was never identified. In this action the plaintiff, as administrator of Karen Edwards's estate, seeks damages for her wrongful death (G.L. c. 229,§ 2[1992 ed.]) and pain and suffering against the corporate owner and corporate manager of the hotel, respectively. We shall refer to the defendants collectively as the hotel.

The hotel moved for summary judgment. A judge of the Superior Court asked the plaintiff to assemble all the evidence on which he relied to meet his burden of proof. The plaintiff did so. In such a circumstance, the question before the motion judge was whether the evidence produced, viewed most favorably to the plaintiff, was sufficient to support the plaintiff's claim. *Kourouvacilis v. General Motors Corp.*, 410 Mass. 706, 716, 575 N.E.2d 734 (1991). The judge concluded that the plaintiff had failed to establish "the necessary causal relationship between Defendants' negligence (if any) and the decedent's death." Judgment was entered for the defendants. We transferred the plaintiff's appeal to this court on our own motion. We order that the judgment for the defendants be vacated.

[1] The hotel justifies the judge's ruling solely on the basis that the plaintiff failed to show the necessary causal connection between the hotel's alleged negligence and the stabbing. Therefore, for the purposes of the summary judgment proceedings, we assume that there was sufficient evidence that the hotel violated a duty of care owed to its guests and particularly to Karen Edwards. This case squarely puts the question whether, in circumstances in which the assailant is not identified as a person who was unlawfully on the hotel premises, causation justifying liability may be warranted on the record. In traditional tort terms, the question is whether the plaintiff has demonstrated, on the summary judgment record, that a jury would be warranted in concluding that it was more probable than not that the stabbing of

Karen Edwards was within the reasonably foreseeable risk of harm created by the hotel's negligence. See *Flood v. Southland Corp.*, 416 Mass. 62, 73, 616 N.E.2d 1068 (1993); *Mullins v. Pine Manor College*, 389 Mass. 47, 58, 449 N.E.2d 331 (1983); *id.* at 67, 449 N.E.2d 331 (O'Connor, J., dissenting); *Carey v. New Yorker of Worcester, Inc.*, 355 Mass. 450, 454, 245 N.E.2d 420 (1969). Another way of stating the issue is whether reasonable security measures probably would have prevented the type of crime that resulted in Edwards's death. See *Sharpe v. Peter Pan Bus Lines, Inc.*, 401 Mass. 788, 793, 519 N.E.2d 1341 (1988).

[2] In order to determine whether the stabbing of Karen Edwards could have been a reasonably foreseeable consequence of the hotel's negligent breach of its duty to its guest, we must first recite various facts on which the plaintiff relies in asserting that the hotel was negligent.

The hotel is in a medium to moderately high crime area. It was routine to remove transients and other trespassers from the hotel. The hotel was also aware of many incidents of larceny from hotel rooms and motor vehicles, purse-snatchings, other robberies in the alley, and break-ins, including four during the two weeks before the murder. The hotel was aware of numerous nonviolent crimes and an occasional violent crime, such as a rape and a stabbing, in the Lenox Hotel and two nearby hotels that were monitored by the same security personnel. In particular, the hotel knew its fire escape was used to gain unauthorized access to the hotel and that break-ins were occurring with some frequency in rooms connected to or close to the hotel's fire escape. The room assigned to Edwards was close to the fire escape and across a hall from an unsecured stairwell whose entry was not monitored by the hotel's security cameras. Following the murder, a knife alleged to be the murder weapon was found on the floor below Edwards's room.

The hotel had a security system that could be found to fail to meet the standard of reasonableness. At the time Edwards was stabbed, there was only one security officer on duty, assigned to cover the hotel and another hotel five blocks away. The alarm system on the fire escape repeatedly failed to detect intruders using it to gain access to rooms. A dumpster was located so that it facilitated access to and from the fire escape. No cameras monitored the door to room 624. At the time of the murder, cameras were not taping sixth floor guest areas and the quality of pictures from security cameras was "totally inadequate." There was no doorman at one hotel entrance at the time of the murder. Generally the industry agrees that there should be patrols on guest room floors. A former head of security at the hotel stated in a deposition that, only six months before the murder, he perceived the security operation at the hotel to be "in a pretty substantial state of disarray." There was, he said, no preventive patrolling. He also stated that a single security officer shared between two hotels was inadequate security.

Before her arrival at the hotel, Edwards had called to request that a refrigerator be placed in her room so that she would not have to leave the room in the evening. She brought food with her in a cooler. The refrigerator was not there when she arrived at the room shortly after checking in at 8:05 P.M. She phoned the desk, and a hotel employee told her that the refrigerator would be sent up promptly. At that time there was no security guard at the hotel. After approximately thirty minutes, a hotel employee arrived at the room. He believed the door was bolted from the inside because he could not gain access with his pass key. He attempted to call, but the line was busy. The next day Edwards's body was discovered wedged against the door to her room. The phone was off the hook. There was no evidence of a forced entry. The assailant took $200 from Edwards's wallet.

The plaintiff's evidence certainly would warrant a finding that the failure of the hotel to take reasonable precautions to protect its guests created a risk that an intruder, bent on stealing guests' property, would be able to gain undetected access to the hotel, its corridors, stairwells, fire escape, and rooms. These circumstances created the risk that a guest would encounter such an intruder who would act violently to prevent his identification and apprehension and to achieve his larcenous objective.

[3] The issue becomes more difficult, however, because the record does not show that the killer

definitely was an intruder, one against whom reasonable security precautions were intended to act as a deterrent. The hotel suggests that the assailant could just as easily have been a guest in the hotel or a guest's visitor. The likelihood of that being so is not great. The inference is not strong that a hotel guest, whose identity is known because of registration as a guest, would rob and stab another guest in her room. A visitor to a guest who might do such a thing stands in much the same position because of the risk of identification. The absence of reasonably adequate means of detecting intruders and the hotel's failure to control and monitor intruders' means of access to and egress from the hotel indicate that the risk of harm from an intruder was enhanced by the hotel's negligent omissions. That risk was sufficiently great that the less plausible possibility, that the assailant was a guest or some other person authorized to be in the hotel, should not bar the plaintiff from submitting his case to a trier of fact. On the summary judgment record a trier of fact would be warranted in concluding "that it was more probable than not that the assailant was a trespasser." *Mullins v. Pine Manor College*, 389 Mass. 47, 59, 449 N.E.2d 331 (1983).

Summary judgment for the defendants is vacated.

So ordered.

(Modified for the purposes of this text)

221 Ga.App. 1
GRIFFIN
v.
AAA AUTO CLUB SOUTH, INC.
GRIFFIN
v.
ALL SOUTH SECURITY, INC.
Nos. A95A1878. A95A2854.
Court of Appeals of Georgia.
March 15, 1996.
Reconsideration Denied March 29, 1996.

Employee brought negligence action against her employer after she was attacked by her boyfriend in employee parking lot. The Fulton State Court, Carnes, J., granted summary judgment to employer and employee appealed. The Court of Appeals, Ruffin, J., held that: (1) although employer had nondelegable duty to keep premises safe, employer was not liable for injuries to employee when she was attacked by her boyfriend in employer's parking lot, and (2) employer was not liable for injuries to employee under theory that employer provided negligent security to its employees.

Affirmed.

C. Lawrence Jewett, Jr., Mitchell G. Brogdon, Atlanta, for appellant.

Chambers, Mabry, McClelland & Brooks, Genevieve L. Frazier, Atlanta, for appellee (case no. A95A1878).

Drew, Eckl & Farnham, B. Holland Pritchard, Bruce A. Taylor, Jr., Atlanta, for appellee (case no. A95A2854).

RUFFIN, Judge.

Cynthia Griffin sued her employer, AAA Auto Club South, Inc. ("AAA") and All South Security Co. ("All South") seeking recovery for injuries sustained when her former boyfriend shot her on a Saturday night in the parking lot outside AAA's offices. She alleged that AAA breached a non-delegable duty to provide security and a safe environment for its employees, that All South negligently failed to ensure that she reached her car safely, and that AAA was responsible for All South's failure to do so on the basis of respondeat superior. In Case No. A95A1878, Griffin appeals the trial court's grant of summary judgment to AAA, and in Case No. A95A2854, she appeals the grant of summary judgment to All South. For reasons which follow, we affirm.

Case No. A95A1878

[1] Summary judgment is appropriate when the court, viewing all the facts and reasonable inferences from those facts in a light most favorable to the non-moving party, concludes that the evidence does not create a triable issue as to each essential element of the case. *Lau's Corp. v. Haskins*, 261 Ga. 491, 405 S.E.2d 474 (1991).

Viewed in a light most favorable to Griffin as the respondent below, the record shows that AAA contracted with All South for the provision of on-site, unarmed security guards at its downtown Atlanta location. On Thursday, two days prior to the shooting, Griffin notified Larry Manuel, an All South security guard, that her former boyfriend, Edward Smith, told her he was "going to buy a gun and come to AAA and shoot her." Griffin testified that she told Manuel "if you see [Smith] on the premises be on guard. If you see him on the premises or anything funny, let me know." However, there was no evidence that Smith was seen or should have been seen on Saturday by Manuel or the owner of the premises, AAA. Griffin did not request an escort to her car on Thursday night, but Manuel walked her to the door. Two days later, Griffin left the building at approximately 11:10 p.m. She did not request an escort to her car or leave with other employees, but walked to the parking lot by herself. She testified that she did not ask for an escort because she did not expect Smith to be on the premises and because she believed he had calmed down. After Griffin entered the parking lot, Smith pulled his car from around the corner of the building, jumped out of the car, and ran toward her. Griffin screamed, turned to run back toward the building, and Smith then shot her repeatedly, inflicting serious injuries.

[2] 1. Griffin contends that the trial court erred in granting summary judgment because AAA had a non-delegable duty to keep its premises and approaches safe under OCGA § 51-3-1. That statute obligates an owner or occupier of land "to exercise ordinary care in keeping the premises and approaches safe." Because the owner or occupier's duties to keep the premises and approaches safe are statutory (OCGA § 51-3-1), those duties are non-delegable even though the owner has a contract for another party to provide security. OCGA § 51-2-5(4); *Confetti Atlanta v. Gray*, 202 Ga.App. 241, 244(4), 414 S.E.2d 265 (1991). "An employer is liable for the negligence of a contractor ... [i]f the wrongful act is the violation of a duty imposed by statute." OCGA § 51-2-5(4).

[3-5] Despite this duty to keep its premises safe, "generally an intervening criminal act by a third party insulates a proprietor from liability, but even an independent criminal act could render the proprietor liable 'if the defendant (original wrongdoer) had reasonable grounds for apprehending that such criminal act would be committed' or is 'reasonably to be anticipated.'" *Clark v. Carla Gay Dress Co.*, 178 Ga.App. 157, 161, 342 S.E.2d 468 (1986). Even in such circumstances, however, the "'true ground of liability ... is the *superior* knowledge of the proprietor of the existence of a condition that may subject the invitee to an unreasonable risk of harm.' [Cit.]" *Howell v. Three Rivers Security*, 216 Ga.App. 890, 892, 456 S.E.2d 278 (1995); see also *Sailors v. Esmail Intl*, 217 Ga.App. 811, 812-813(1), 459 S.E.2d 465 (1995). Although Griffin advised the guard on Thursday of a threat from her boyfriend, it is undisputed that she had superior knowledge of his characteristics and temperament and of the nature of their relationship by Saturday. She thus had a far greater degree of foreseeability. In fact, foreseeability on the part of the security guard was solely dependant upon what Griffin said and did. Her action in going to her car without requesting an escort because she thought Smith had calmed down removed any foreseeability on AAA's part.

This case also differs significantly from those in which a jury could find a legal failure to keep the premises and approaches safe, so that its invitees in general are not subjected to the risk of foreseeable attacks by unknown assailants. This was not a random stranger attack but rather grew out of a specific private relationship which had no connection with employment whatsoever. The *place* chosen by the boyfriend for the attack just happened to be the employer's parking lot. The employer did not create or allow to exist an environment which placed Griffin at risk any more than if she had been at home or on the street.

[6] 2. Griffin also contends the trial court erred because material issues of fact remain with respect to whether AAA breached a duty to provide adequate security for its employees and whether it provided security in a negligent manner. In support of this argument, she cites Manuel's testimony that according to the express written instructions set forth in All South's post orders, security guards were responsible for making certain that employees got to their cars safely and were required both to regularly patrol the parking lot and to monitor it. However, AAA cannot be liable for performing security negligently, because "an injured person seeking to impose liability upon another for the negligent performance of a voluntary undertaking must show either detrimental reliance or an increased risk of harm." *Adler's Package Shop v. Parker*, 190 Ga.App. 68, 72(1)(b), 378 S.E.2d 323 (1989). There is no evidence that Griffin relied upon the presence of the guard and she did not ask for an escort to her car. Nor is there any contention that any action on the part of the security guard *increased* Griffin's risk; failing to take all possible actions to prevent an occurrence is not the same as increasing the risk of the occurrence. Accordingly, this enumeration of error is without merit.

3. Given our holding in Case No. A95A2854 that summary judgment to All South was proper, we need not address Griffin's remaining enumeration of error concerning whether the All South guards were AAA employees.

Case No. A95A2854

4. Griffin contends the trial court erred in granting All South summary judgment because a material issue of fact exists as to whether All South had a duty to ensure that she reached her car safely. We disagree because

the proximate cause of Griffin's injuries was not the breach of any such duty, but her superior knowledge of her boyfriend and of the risks attendant in going to the parking lot on Saturday without asking for an escort. See *Sailors*, supra at 812-813(1), 459 S.E.2d 465. Thus, the trial court did not err in granting All South summary judgment.

Judgment affirmed.

BEASLEY, C.J., and POPE, P.J., concur.

(Modified for the purposes of this text)

173 A.D.2d 906
Cornelius POLLARD, Respondent,
v.
STATE of New York, Appellant.
Supreme Court, Appellate Division,
Third Department.
May 2, 1991.

Inmate brought action against State to recover value of property stolen from locker. The Court of Claims, Orlando, J., found in favor of inmate. State appealed. The Supreme Court, Appellate Division, Mahoney, P.J., held that: (1) State lacked sovereign immunity from liability for providing allegedly negligent security for inmate's belongings, and (2) State was liable for negligence.

Affirmed.

Robert Abrams, Atty. Gen. (Denise A. Hartman, of counsel), Albany, for appellant.

Cornelius Pollard, in pro. per.

Before MAHONEY, P.J., and CASEY, WEISS, YESAWICH and HARVEY, JJ.

MAHONEY, Presiding Justice.

Appeal from a judgment in favor of claimant, entered March 16, 1990, upon a decision of the Court of Claims (Orlando, J.).

Claimant, an inmate at Green Correctional Facility in Greene County, filed a claim against the State alleging negligence on the State's part in failing to provide adequate security in the prison dormitory area, resulting in the theft of his portable cassette player and headphones from his locker. Subsequent to trial, the Court of Claims issued a decision holding that the State's lack of supervision over the dormitory area was the proximate cause of claimant's loss and awarded him the sum of $60 plus interest. This appeal ensued.

The State's first ground for reversal of the judgment in claimant's favor is that because it did not waive its sovereign immunity from liability for the exercise of discretionary judgments regarding general security measures in a prison dormitory, it cannot be held liable for claimant's loss. We disagree.

[1] While the State can, as it has done in this case, raise for the first time on appeal the defense of sovereign immunity because it "brings into question jurisdiction of the subject under the Court of Claims Act, [and] may be raised at any time" (*Heisler v. State of New York*, 78 A.D.2d 767, 768, 433 N.Y.S.2d 646), such defense, on the facts present herein, does not preclude recovery. Claimant does not challenge prison administration or allocation of staff or resources. He simply states in his claim that the State was negligent in providing security for his belongings (*see, Thomas v. State of New York*, 144 A.D.2d 882, 534 N.Y.S.2d 815 [negligent destruction of property]; *Emmi v. State of New York*, 143 A.D.2d 876, 878, 533 N.Y.S.2d 406 [negligent maintenance of State property]). As *Heisler v. State of New York*, *supra, Emmi v. State of New York*, *supra,* and *Thomas v. State of New York*, *supra,* clearly indicate, the State has a duty recognized by the law of torts; accordingly, liability may be imposed by application of general tort principles and the doctrine of sovereign immunity cannot serve as a bar to claimant's action.

[2] Turning to the issue of ordinary common-law negligence, we cannot fault the findings of the Court of Claims that claimant secured his personal belongings, including his cassette player and headphones, in an assigned metal locker with a padlock purchased from the correctional facility and that claimant did so before leaving the locker room area to report to his work assignment. Equally unassailable is the court's finding that when claimant returned, his locker was open and his personalty was missing. Further, the diagrams of the locker room area show that the station of the correction officer assigned to the dormitory area provided a clear view of claimant's locker. These evidentiary facts, juxtaposed to claimant's testimony that he observed scrape marks on his locker that were not present when he left to report to his assigned work area, satisfy claimant's burden to prove the existence of a duty by

the State, a breach of that duty and injury to claimant (*see, Akins v. Glens Falls City School Dist.*, 53 N.Y.2d 325, 333, 441 N.Y.S.2d 644, 424 N.E.2d 531; *see also, Wolfe v. Samaritan Hosp.*, 104 A.D.2d 143, 146, 484 N.Y.S.2d 168).

Regarding the State's contention of claimant's contributory negligence, we note that "this court may assess the evidence when reviewing a nonjury trial" and where, as here, we find such proof to be credible, defer to the findings of the trial court (*Town of Ulster v. Massa*, 144 A.D.2d 726, 727, 535 N.Y.S.2d 460, *lv. denied* 75 N.Y.2d 707, 554 N.Y.S.2d 476, 553 N.E.2d 1024). The record explicitly sets forth claimant's affirmative response to the State's injury, "Did you lock your locker with this combination lock when you went to the law library?" Since we find the sum awarded to be the reasonable value of the lost personalty, we affirm the judgment in favor of claimant.

Judgment affirmed, with costs.

CASEY, WEISS, YESAWICH and HARVEY, JJ., concur.

chapter eleven

Wrongful Death Actions

> *"At death, those heirs*
> *That seem the saddest,*
> *Behind their masks*
> *May be the gladdest."*
>
> Art Buck

> *"To judge the real importance of the individual,*
> *we should think of the effect his death would produce."*
>
> Francois-Gaston De Levis

At common law, any potential legal action would lie with the individual and if the individual is deceased, the action would die with the individual. Virtually all states have enacted wrongful death statutes which permit the estate of the deceased to bring an action on behalf of the deceased individual. In incidents of workplace violence where an individual may have been killed, employers should be cognizant that the surviving spouse, children, or estate of the deceased are normally empowered to bring an action against all pertinent parties. This theory of recovery is well established in the American courts, and there is substantial case law to support these types of actions.

In cases involving workplace violence, wrongful death actions are normally brought against the employer as well as the perpetrator and other related parties for the purpose of revenge, punishment, closure, financial reimbursement, or simply the money. In circumstances of workplace violence, a family member went to work just as he/she did every other day. However, on this particular day something happened, and the family member is not coming home after work. The circumstances are traumatic and sudden. After the initial shock, the normal grieving process often results in the family searching for a method of holding someone accountable. Given the fact that the individual who performed the violent act is dead or incarcerated with little or no financial resources, the family, often with prompting from television advertisements or directly from the legal profession, files a wrongful death action against the company and related parties.

For example, a disgruntled ex-employee who was recently terminated from his job returns to the workplace with a handgun and kills the personnel manager. The disgruntled employee would either commit suicide or be incarcerated for his crime. Unless the disgruntled employee has substantial financial assets, there is limited potential of financial recovery from the ex-employee for monetary damages. Additionally, if the ex-employee has any financial assets, these assets would be consumed by his defense counsel given the fact that he would not be eligible for public defender legal assistance because of the financial assets. Workers' compensation benefits could be provided to the family by the employer; however, these benefits are usually considered low in relation to the family's loss.[1]

After the initial shock and funeral services, the grieving process begins for the family. Additionally, the family has other problems facing them because the family member's salary is no longer available to the family. The benefits being provided under workers' compensation or insurance policies are often not sufficient to meet the family's financial needs. So who is going to pay for the loss of the family member and also provide for the family? The "deep pocket" is often the employer.

The employer, being a corporation or larger entity, has the potential of providing recovery to the estate of the deceased. Additionally, the other potential parties are often "judgment proof" because they lack financial assets (i.e., incarcerated perpetrator). The wrongful death action, standing alone or in combination with one or more of the negligence theories discussed throughout this text, is filed and pursued on behalf of the family or estate.

The elements required to maintain a claim for wrongful death action are normally established by state statute. In general, the family or estate of the deceased would be the proper party to bring the action, and damages can range from loss of consortium, funeral expenses, attorney fees, pain and suffering, as well as potential punitive damages. One of the primary issues in a wrongful death action is the effect of the paid workers' compensation benefits on the viability of the wrongful death action, i.e., whether workers' compensation is the exclusive remedy for the deceased family or whether a

---

[1] For example, prior to the change in workers' compensation law in Kentucky in 1997, the death benefits to a single individual without dependents was $10,000, with $2500.00 for burial costs.

wrongful death action is permitted. Defenses in a wrongful death action vary depending on the state statute and normally range from the statute of limitation defense through last clear chance and insanity.

It should be noted that the statute of limitations on wrongful death actions is usually substantially long in comparison to other causes of action (i.e., 1 to 5 years). Additionally, the claims for relief under many wrongful death statutes can be extremely broad in nature. For example, in Kentucky, the basis for a claim of relief under the general wrongful death statute can include Kentucky state constitutional issues, negligent or wrongful acts, violation of statutes, and strict liability among other claims.[2]

Another area of potential liability for employers lies in the area of wrongful death actions by the third parties. For employees, the issue of workers' compensation and the applicable death benefits would be an issue. However, for third parties who may be killed in the workplace as a result of a workplace violence incident, the potential for a wrongful death action against the employer is almost certain. For example, a disgruntled employee comes into your waiting area and kills an employee as well as a customer. In most states, the workers' compensation system would be applicable for the employee; however, the customer, being outside the workers' compensation system, may have a valid action for wrongful death against the company. Under most state statutes, the recovery of damages for death would be applicable when they are inflicted by the "negligent or wrongful act of another."[3] The negligent and wrongful acts of the company or employer are often proven through the violation of statutes or other laws, or where the individual performing the violent act was an agent or employee of the company. The issues of causation and damages are normally easily proven because of the type of workplace violence incident.

The defenses available to employers in a wrongful death case usually depend on the facts of the situation and corporate status. Several defenses to consider include the following:

- Disproof of the elements of the plaintiff's case
- Jurisdictional defense
- Statute of limitation defense
- Capacity of the party to sue
- Fraud or mistake
- The deceased contributed to his/her death
- Contributory fault of the family members or other beneficiary
- Comparative or contributory negligence standard (depending on the jurisdiction)
- Last clear chance rule
- Suicide

[2] Eades, Ronald, *Kentucky Wrongful Death Actions*, Harrison Company, Norcross, GA, 1994.
[3] *See for example, Saylor v. Hall*, 497 S.W. 2d 218 (Ky. 1973); *Ludwig v. Johnson*, 243 Ky. 533, 49 S.W. 2d 347 (1932).

- Assumption of the risk doctrine
- Self defense
- Defense of others
- Defense of property
- Insanity
- Lack of venue or proper party
- *Res judicata* or estoppel
- Workers' compensation "exclusive remedy" rule

Additionally, depending on the type of company or organization, the use of immunity protections, such as charitable immunity or governmental immunity, may be utilized if applicable.

Employers should be cognizant of the potential legal liabilities not only for their employees but for others who may be present or working in their workplace. Wrongful death actions are normally provided for under every state's laws, and wrongful death statutes normally provide for a wide variety of damages for the family or estate of the deceased, including loss of consortium, funeral expenses, mental suffering, pain and suffering, property damages, attorney fees, and punitive damages. Prudent professionals should also be aware of the "sympathy factor" when defending a wrongful death action. Due to the type of action, the jury will be sympathetic to the family because of their loss. Alienation of the jury through attack on the plaintiff during the presentation of the case can result in an undeserved verdict and escalation of the damages.

As stressed throughout this text, if the workplace violence incident is prevented in the first place, then there is no need to pursue this type of action. There is no winner in a wrongful death action. A family member is dead. The workplace violence incident has detrimentally affected everyone involved. The company is paying for the damages, in terms of monetary losses and efficacy losses. In essence, there are no winners. The only question is how much it is going to cost!

(Modified for the purposes of this text)

209 Conn. 59
Larry L. SHARP, Administrator
(ESTATE OF David C. SHARP) et al.
v.
Norbert E. MITCHELL, Sr. et al.
No. 13168.
Supreme Court of Connecticut.
Argued May 11, 1988.
Decided Sept. 6, 1988.

Estates filed wrongful death actions against the employer of three men who were asphyxiated in employer's underground fuel storage facility. The Superior Court, Danbury Judicial District, McDonald, J., granted employer's motion for summary judgment as to the counts of estates' amended complaint that alleged negligent supervision of employees, and Lavery, J., granted employer's second motion for summary judgment, on ground that estates' claim that employer had "dual capacity" was a separate cause of action and therefore was barred by the statute of limitations. Estates appealed. The Supreme Court, Arthur H. Healey, J., held that: (1) the Workers' Compensation Act, as applied to estates, was constitutional, and (2) estates' "dual capacity" claim did not relate back to estates' claim of negligent supervision of employees, for statute of limitations purposes.

Affirmed.

Joan C. Harrington, New Canaan, for appellants (plaintiffs).

Ronald D. Williams, Bridgeport, for appellees (defendants).

Before PETERS, C.J., and ARTHUR H. HEALEY, SHEA, GLASS and COVELLO, JJ.

ARTHUR H. HEALEY, Associate Justice.

This appeal arises out of a wrongful death action brought on behalf of three men who were asphyxiated in

an underground fuel storage facility while employed by
the defendant Norbert E. Mitchell Company. The plaintiffs, administrators of the estates of David C. Sharp,
Robert Vidal and Alois Entress, alleged that the defendants, Norbert E. Mitchell, Sr., and Norbert E. Mitchell,
Jr., d/b/a Norbert E. Mitchell Company, negligently
caused those deaths. The trial court, *McDonald, J.,*
granted the defendants' motion for summary judgment as
to the counts of the plaintiffs' amended complaint that
alleged negligence by the defendants, concluding that
the exclusivity provision of the Workers' Compensation
Act was constitutional. The trial court, *Lavery, J.,*
granted the defendants' second motion for summary judgment, concluding that the plaintiffs' claim that the
defendants had a "dual capacity" was a separate cause
of action and therefore was barred by the statute of
limitations.[1] We find no error.

The basic facts concerning this tragic incident are
not disputed. The plaintiffs' decedents were employed
by the defendant Norbert E. Mitchell Company, which is
in the retail petroleum business. The facilities owned
by the defendants included an underground facility that
housed seven 25,000 gallon petroleum storage tanks. This
underground storage area is approximately six feet wide,
fifty feet long and ten feet high, with access to the
tanks provided by a ladder through a thirty-six inch
manhole. On February 3, 1983, the defendant Norbert E.
Mitchell, Jr., instructed Sharp to enter the underground
area and shut off a valve on one of the tanks. When
Mitchell heard some banging noises, he sent Vidal into
the underground storage area to aid Sharp. Vidal collapsed
at the bottom of the ladder and Mitchell left the area
to call for help. Entress then descended the ladder with
a rope, but he also collapsed while on the ladder. Vidal
and Entress were pronounced dead approximately one hour
after the incident while Sharp was pronounced "brain
dead" two days later, and shortly thereafter he was
removed from life support systems.

---

[1] The plaintiffs also filed claims for loss of consortium but the
trial court, *Moraghan, J.,* granted the defendants' motion to strike
those claims. The plaintiffs are not challenging that ruling in
light of this court's decision in *Ladd v. Douglas Trucking Co.,* 203
Conn. 187, 197, 523 A.2d 1301 (1987).

Neither Sharp nor Entress was survived by dependents as defined by the Workers' Compensation Act. Their representatives were entitled to funeral expenses and their respective medical bills were paid directly to the hospital. Vidal's funeral expenses and medical bills were also paid and his dependents are currently receiving workers' compensation benefits.

The plaintiffs' first complaint was filed on August 10, 1983, and alleged that the deaths were caused by the intentional, serious and willful misconduct of the defendants. The defendants moved for summary judgment, contending that the "intentional misconduct" exception as outlined in *Jett v. Dunlap*, 179 Conn. 215, 221, 425 A.2d 1263 (1979), did not apply and therefore the action was barred by the exclusivity provision of the Workers' Compensation Act. General Statutes § 31-284(a).[2] The plaintiffs agreed, for the purposes of the summary judgment motion, that the intentional misconduct exception did not apply and subsequently amended their complaint to allege negligence by the defendants and to argue that the exclusivity provisions of the Workers' Compensation Act violated article first, § 10, of the constitution of Connecticut. On January 20, 1986, the plaintiffs further revised their complaint to allege that the defendants were negligent in designing and constructing the underground storage facility and therefore were liable under a "dual capacity" theory.[3] The defendants, who previously had filed a motion for summary judgment to the complaint that alleged negligence, planned to file another motion for summary judgment to

---

[2] General Statutes § 31-284(a) provides: "BASIC RIGHTS AND LIABILITIES. CIVIL ACTION TO ENJOIN NONCOMPLYING EMPLOYER FROM ENTERING EMPLOYMENT CONTRACTS. (a) An employer shall not be liable to any action for damages on account of personal injury sustained by an employee arising out of and in the course of his employment or on account of death resulting from personal injury so sustained, but an employer shall secure compensation for his employees as follows, except that compensation shall not be paid when the personal injury has been caused by the wilful and serious misconduct of the injured employee or by his intoxication. All rights and claims between employer and employees, or any representatives or dependents of such employees, arising out of personal injury or death sustained in the course of employment as aforesaid are abolished other than rights and claims given by this chapter, provided nothing herein shall prohibit any employee from securing, by agreement with his employer, additional benefits from his employer for such injury or from enforcing such agreement for additional benefits."

the complaint that alleged a dual capacity theory. To facilitate matters for the court, the plaintiffs amended the complaint again. In a complaint dated September 19, 1986, the plaintiffs alleged in counts one, three and five that the defendants as employers had negligently caused the deaths of their decedents. In counts two, four and six, the plaintiffs alleged that the defendants had acted negligently in a capacity other than that of an employer when they designed and constructed the underground storage area. The parties stipulated that the pending motion for summary judgment would address the claims of negligence in counts one, three and five, and that the defendants would shortly file another summary judgment motion that would address the dual capacity claims in counts two, four and six. On October 1, 1986, the defendants filed that latter motion for summary judgment.

On October 21, 1986, the trial court, *McDonald, J.*, granted the defendants' first motion for summary judgment, rejecting the argument of the plaintiffs that the Workers' Compensation Act violated article first, § 10, of the constitution of Connecticut. On February 2, 1987, the trial court, *Lavery, J.*, granted the defendants' second motion for summary judgment, dated October 1, 1986, concluding that the plaintiffs' dual capacity claims were barred by the statute of limitations and by the exclusivity provisions of the Workers' Compensation Act.

On appeal, the plaintiffs claim that the trial courts erred in: (1) holding that the Workers' Compensation Act was constitutional as applied to the plaintiffs; (2)

---

[3] The "dual capacity" doctrine describes the situation where an employer has two capacities or legal persona, such as employer and owner; see, e.g., *Ogden v. McChesney*, 41 Colo.App. 191, 193, 584 P.2d 636 (1978); or employer and manufacturer. See, e.g., *Bell v. Industrial Vangas, Inc.*, 30 Cal.3d 268, 282, 637 P.2d 266, 179 Cal.Rptr. 30 (1981). A noted authority explained the "dual capacity" as follows: "An employer may become a third person, vulnerable to tort suit by an employee if—and only if—he possesses a second persona so completely independent from and unrelated to his status as employer that by established standards the law recognizes it as a separate legal person." 2A A. Larson, Workmens' Compensation Law (1983) § 72.81, p. 14-229. In a related context, we have recently rejected the dual capacity doctrine. *Panaro v. Electrolux Corporation*, 208 Conn. 589, 545 A.2d 1086 (1988).

holding that the plaintiffs' dual capacity claims were
barred by the statute of limitations; and (3) rejecting
the dual capacity doctrine under the circumstances of
this case. We conclude that the first and second issues
raised by the plaintiffs on appeal were correctly decided
by the trial courts and therefore we do not reach the
third issue.

I.

[1] The plaintiffs argue that they should be permitted to sue the defendants for negligently causing the
deaths of their decedents because the Workers' Compensation Act, as applied to them, violates article first,
§ 10, of the constitution of Connecticut. Under the
Connecticut Workers' Compensation Act, the plaintiffs
are limited to remedies under the act. General Statutes
§ 31-274 et seq.; see, e.g., *Perille v. Raybestos-
Manhattan-Europe, Inc.*, 196 Conn. 529, 532, 494 A.2d 555
(1985); *Sullivan v. State*, 189 Conn. 550, 558, 457 A.2d
304 (1983); *Jett v. Dunlap*, supra, 179 Conn. at 217,
425 A.2d 1263. We agree with the defendants that the
Workers' Compensation Act can withstand the plaintiff's
constitutional challenge.

Article first, § 10, of the constitution of Connecticut provides: "All courts shall be open, and every
person, for an injury done to him in his person, property
or reputation, shall have remedy by due course of law,
and right and justice administered without sale, denial
or delay." The plaintiffs reason that since a common
law right to sue for wrongful death existed at the time
of the ratification of the constitution of Connecticut
in 1818 and that since such a right cannot be abolished
by the legislature without providing a reasonable alternative remedy under *Gentile v. Altermatt*, 169 Conn. 267,
283, 363 A.2d 1 (1975), appeal dismissed, 423 U.S. 1041,
96 S.Ct. 763, 46 L.Ed.2d 631 (1976), the Workers'
Compensation Act is unconstitutional as applied to them.
Although it is correct to say that article first, § 10,
protects "constitutionally incorporated" common law or
statutory rights from abolition or significant limitations, these rights include only those in existence in
1818. *Id.*; see also *Daily v. New Britain Machine Co.*,
200 Conn. 562, 585, 512 A.2d 893 (1986).

After the briefs were filed in this case, this court
announced the decision in *Ecker v. West Hartford*, 205
Conn. 219, 530 A.2d 1056 (1987), where we held that no
action for wrongful death existed at common law in
Connecticut. The *Ecker* court, after exhaustive research
and analysis of previous Connecticut cases and the
applicable law throughout the United States, concluded:
"With only a few exceptions, courts in America have
almost universally accepted, and continue to accept,
'the rule that a civil action for wrongful death was
not recognized at common law, and that no such cause of
action may be maintained except under the terms and
authority of a statute.' 61 A.L.R.3d 906, 909, and
authorities cited therein, 909-10 n. 3, and Sup. 59."
Id., 227, 530 A.2d 1056. The court also held "that no
action for wrongful death existed at common law or exists
today in Connecticut except as otherwise provided by the
legislature." *Id.*, 231, 530 A.2d 1056. It is clear that
Ecker controls this case concerning the issue of whether
a common law right to sue for wrongful death existed at
common law that would survive the adoption of the
Connecticut constitution of 1818.

The plaintiffs maintained at oral argument that *Ecker*
does not control the case before us for three reasons,
none of which is persuasive. First, the plaintiff's
summary suggest that they are suing for personal injury
and that *Ecker* involved a wrongful death claim.[4] The
decedent in *Ecker* was an employee who was struck by a
wooden canopy that fell from a building, resulting in
his death the next day. The plaintiffs in this case are
employees who were asphyxiated. Two were pronounced dead
at the scene and the other died two days later. It is
difficult to ascertain on what basis the plaintiffs can
expect this court to treat these claimants differently.
It is clear that both cases involve wrongful death and
that any cause of action instituted on that basis is
barred by *Ecker*.

In a related argument, the plaintiffs next claim that
Gentile v. Altermatt, supra, held that a negligence
action for personal injury is a constitutionally protected right and therefore their "personal injury" claim

---

[4] The plaintiffs' own brief characterizes their claim as a "wrongful
death action."

is protected by article first, § 10. This claims is also supported only by a conclusory assertion at oral argument. Since a wrongful death claim is barred by *Ecker*, the apparent logic of the plaintiffs' argument is that, nevertheless, they be permitted to sue for any personal injury that the decedents suffered before death. This court has held on numerous occasions that the exclusivity provisions of the Workers' Compensation Act bar actions by employees against their employers for job related injuries. See, e.g., *Perille v. Raybestos-Manhattan-Europe, Inc.*, supra, 196 Conn. at 532, 494 A.2d 555; *Mingachos v. CBS, Inc.*, 196 Conn. 91, 98, 491 A.2d 368 (1985); *Jett v. Dunlap*, supra, 179 Conn. at 217, 425 A.2d 1263. Although this court did not face a direct challenge to the Workers' Compensation Act as violating article first, § 10, in those cases, it is implicit in those holdings that the Workers' Compensation Act passes constitutional muster for personal injuries suffered by employees. A statute that abolishes a constitutionally incorporated right of action does not violate article first, § 10, if "the legislatively created remedy by which it is in part replaced is a reasonable alternative." *Gentile v. Altermatt*, supra, 169 Conn. at 287, 363 A.2d 1. In a related context, this court held that the benefits under the Workers' Compensation Act provided a reasonable alternative to products liability suits brought more than ten years after a defendant had parted with the product. *Daily v. New Britain Machine Co.*, supra, 200 Conn. at 585, 512 A.2d 893. It is unnecessary to itemize all the extensive alternative benefits that an employee receives in return for forfeiting a right to sue his employer for a common law tort but they include immediate payment of medical bills; General Statutes § 31-294; compensation for periods of incapacity; General Statutes § 31-307; and benefits for partial permanent disability. General Statutes § 31-308(b). To reiterate what we said in *Jett v. Dunlap*, supra, 217: "In previous decisions under the Workmen's Compensation Act we have consistently held that where a worker's personal injury is covered by the act, statutory compensation is the sole remedy and recovery in common-law tort against the employer is barred." Accord *Perille v. Raybestos-Manhattan-Europe, Inc.*, supra, 196 Conn. at 532, 494 A.2d 555; *Mancini v. Bureau of Public Works*, 167 Conn. 189, 193, 355 A.2d 32 (1974).

Lastly, the plaintiffs argue that *Ecker* does not apply because the readoption of the constitution of Connecticut in 1965 preserved all common law and statutory rights in existence in 1965. Therefore, the plaintiffs claim that since a statutory action for wrongful death[5] was extant in 1965, the readoption of the constitution of Connecticut protected that right from abrogation by the state legislature unless a reasonable alternative was provided. Although it is true that a statutory wrongful death action was well established in 1965, the plaintiffs' argument fails because the Workers' Compensation Act was also well established in 1965. Connecticut passed its initial version of the Workers' Compensation Act in 1913. Public Acts 1913, c. 138, p. 1735. When the state constitution was readopted in 1965, it was well established that an employee was prevented from instituting an action against his employer for a work related death or injury. See, e.g., *Crisanti v. Cremo Brewing Co.*, 136 Conn. 529, 531, 72 A.2d 655 (1950); *Hoard v. Sears Roebuck & Co.*, 122 Conn. 185, 188, 188 A. 269 (1936). In *Crisanti v. Cremo Brewing Co.*, supra, we held: "It is, of course, well settled that where there exists the relationship of employer and employee within the Workmen's Compensation Act the employee may recover for injuries sustained in the course of his employment only as provided by the act. The employer has no common-law liability to his employee." Since 1965, the legislature has not abolished any right that the plaintiffs seek to vindicate that existed in 1965.[6] The readoption of the state constitution in 1965 does not avail the plaintiffs.

---

[5] General Statutes (1958 Rev.) § 52-555 provides: "ACTIONS FOR INJURIES RESULTING IN DEATH. In any action surviving to or brought by an executor or administrator for injuries resulting in death, whether instantaneous or otherwise, such executor or administrator may recover from the party legally at fault for such injuries just damages together with the cost of reasonably necessary medical, hospital and nursing services, and including funeral expenses, provided no action shall be brought to recover such damages and disbursements but within one year from the date when the injury is first sustained or discovered or in the exercise of reasonable care should have been discovered, and except that no such action may be brought more than three years from the date of the act or omission complained of."

[6] The only charge in General Statutes § 52-555 Connecticut's wrongful death statute, since 1965 was an amendment in 1969 that increased the time within which to bring an action from one year to two years. Public Act 1969, No. 401 § 1.

The plaintiffs also attack *Ecker* on the ground that the intent of the framers who ratified the constitution of Connecticut of 1818 must govern what was then a constitutionally incorporated statutory or common law right. See *Gentile v. Altermatt*, supra, 169 Conn. at 284, 363 A.2d 1. The plaintiffs reason that since *Cross v. Guthery*, 2 Conn. (Root) 90 (1794), can be read to support a cause of action for wrongful death, the framers must have intended to incorporate such a right since *Cross v. Guthery*, was not expressly overruled until 1856 in *Connecticut Mutual Life Ins. Co. v. New York & New Haven R.R. Co.*, 25 Conn. 265, 273 (1856). The plaintiffs discern this purported intent merely by the fact that the four line opinion in *Cross* was not expressly discredited until after the 1818 constitution of Connecticut was ratified. Our opinion in *Ecker*, however, after careful review of the facts of *Cross*, concluded: "Although the case involved the death of the plaintiff's wife, the action appears to have sounded in contract. Therefore, we do not read the *Cross* case as having clearly established that wrongful death action existed at common law in Connecticut." *Ecker v. West Hartford*, supra, 205 Conn. at 229 n. 11, 530 A.2d 1056. Accordingly, the *Ecker* court also concluded: "Therefore, we view the case of *Connecticut Mutual Life Ins. Co. v. New York & New Haven R.R. Co.*, supra, which overruled *Cross*, not as having made a change in the Connecticut law, but rather, as having recognized that the *Cross* court was in error." *Id.*, 229, 530 A.2d 1056. The plaintiffs offer no evidence that would require us to reexamine the analysis in *Ecker*. We also note that to hold current statutes unconstitutional based on pure speculation about alleged misperceptions by those who framed our state constitution in the early 19th century is not a step taken lightly and we decline to do so in this case. The *Ecker* decision satisfactorily answers the plaintiffs' claim.

Although our interpretation of our own constitution is dispositive on this issue, we also recognize that the workers' compensation scheme has been sustained against attack on constitutional grounds, including access to the courts, in many jurisdictions throughout the country. See, e.g., *Kandt v. Evans*, 645 P.2d 1300, 1306 (Colo.1982); *Young v. O.A. Newton & Son, Co.*, 477 A.2d 1071, 1078 (Del.Super.1984); *Acton v. Fort Lauderdale Hospital*, 440 S.2d 1282, 1284 (Fla.1983); *Boyd v.*

Barton Transfer & Storage, Inc., 2 Kan.App.2d 425, 430, 580 P.2d 1366 (1978); *Schmidt v. Modern Metals Foundry, Inc.*, 424 N.W.2d 538, 542 (Minn.1988); *Linsin v. Citizens Electric Co.*, 622 S.W.2d 277, 281 (Mo.App.1981); *Roberts v. Gray's Crane & Rigging, Inc.*, 73 Or.App. 29, 35, 697 P.2d 985, review denied, 299 Or. 443, 702 P.2d 1112 (1985); *Kline v. Arden H. Verner Co.*, 503 Pa. 251, 255, 469 A.2d 158 (1983); *Edmunds v. Highrise, Inc.*, 715 S.W.2d 377, 379 (Tex.Civ.App.1986); *Messner v. Briggs & Stratton Corporation*, 120 Wis.2d 127, 134, 353 N.W.2d 363 (1984). A leading commentator has noted: "Exclusiveness clauses have consistently been held to be constitutional, under the equal protection and due process clauses of both federal and state constitutions. Attacks based on specific state constitutional provisions, such as those creating a right of action for wrongful death, have fared no better." 2A A. Larson, Workmen's Compensation Law (1988)§ 65.20, pp. 12-18-12-20.

We conclude that since an action for wrongful death was not a constitutionally incorporated right at the time of the constitution of 1818, we need not reach the issue of whether the Workers' Compensation Act provides a reasonable alternative remedy.[7]

II.

[2] The plaintiffs next argue that the trial court erred in holding that the dual capacity claims were barred by the statute of limitations. The accident in this case occurred on February 3, 1983, and the original complaint was filed on August 10, 1983. The dual capacity theory was first asserted on January 20, 1986, and then further "clarified" in subsequent complaints dated September 19, 1986, and October 20, 1986. The operative statute of limitation is General Statutes § 52-555,[8] which states in pertinent part: "[N]o action [for wrongful death] shall be brought to recover such damages and disbursements but within two years from the date when the injury is first sustained or discovered...." It is

---

[7] The trial court, *McDonald, J.*, assuming arguendo that wrongful death was a constitutionally incorporated common law right, made extensive findings that Connecticut's Workers' Compensation Act provides a reasonable alternative remedy. In view of our holding today, we do not address these findings.

clear that if the complaint on January 20, 1986, which first articulated a dual capacity theory, does not relate back to the original complaint, the plaintiffs' cause of action under that theory is barred by the statute of limitations. We must determine if the amended complaint relates back to the original complaint or states a new cause of action.

"A cause of action is that single group of facts which is claimed to have brought about an unlawful injury to the plaintiff and which entitles the plaintiff to relief. *Bridgeport Hydraulic Co. v. Pearson*, 139 Conn. 186, 197, 91 A.2d 778[1952]; *Veits v. Hartford*, 134 Conn. 428, 434, 58 A.2d 389 [1948]. 'A right of action at law arises from the existence of a primary right in the plaintiff, and an invasion of that right by some delict on the part of the defendant. The facts which establish the existence of that right and that delict constitute the cause of action.' *Pavelka v. St. Albert Society*, 82 Conn. 146, 147, 72 A. 725 [1909]. A change in, or an addition to a ground of negligence or an act of negligence arising out of the single group of facts which was originally claimed to have brought about the unlawful injury to the plaintiff does not change the cause of action. *Johnson v. Wheeler*, 108 Conn. 484, 488, 143 A. 898 [1928]; *Galvin v. Birch*, 97 Conn. 399, 401, 116 A. 908 [1922]; *O'Brien v. M & P Theatres Corporation*, 72 R.I. 289, 296, 50 A.2d 781 [1947]. It is proper to amplify or expand what has already been alleged in support of a cause of action, provided the identity of the cause of action remains substantially the same, but where an entirely new and different factual situation is presented, a new and different cause of action is stated. *Veits v. Hartford*, supra; *United States v. Memphis Cotton Oil Co.*, 288 U.S. 62, 67, 53 S.Ct. 278 [280],

---

[8] General Statutes § 52-555 provides: "ACTIONS FOR INJURIES RESULTING IN DEATH. In any action surviving to or brought by an executor or administrator for injuries resulting in death, whether instantaneous or otherwise, such executor or administrator may recover from the party legally at fault for such injuries just damages together with the cost of reasonably necessary medical, hospital and nursing services, and including funeral expenses, provided no action shall be brought to recover such damages and disbursements but within two years from the date when the injury is first sustained or discovered or in the exercise of reasonable care should have been discovered, and except that no such action may be brought more than three years from the date of the act or omission complained of.

77 L.Ed. 619 [1933]; *Seaboard Air Line Ry. v. Renn*, 241 U.S. 290, 293, [568], 36 S.Ct. 567, 60 L.Ed. 1006 [1916]." *Gallo v. G. Fox & Co.*, 148 Conn. 327, 330, 170 A.2d 724 (1961); see also *Bielaska v. Waterford*, 196 Conn. 151, 154, 491 A.2d 1071 (1985); *Giglio v. Connecticut Light & Power Co.*, 180 Conn. 230, 239, 429 A.2d 486 (1980); *Keenan v. Yale New Haven Hospital*, 167 Conn. 284, 285, 355 A.2d 253 (1974); *Baker v. Baker*, 166 Conn. 476, 486, 352 A.2d 277 (1974).

In *Giglio v. Connecticut Light and Power Co.*, supra, 180 Conn. at 239-40, 429 A.2d 486, we recognized that our relation back doctrine "is akin to rule 15(c) of the Federal Rules of Civil Procedure, which provides in pertinent part: '(c) RELATION BACK OF AMENDMENTS. Whenever the claim or defense asserted in the amended pleading arose out of the conduct, transaction or occurrence set forth or attempted to be set forth in the original pleading, the amendment relates back to the date of the original pleading.'" Further, we noted the policy behind the rule as follows: "'Rule 15(c) is based upon the concept that a party who is notified of litigation concerning a given transaction or occurrence has been given all the notice that statutes of limitation are intended to afford, 3 Moore's Federal Practice para. 15.15[3]. The objective of state statutes of limitations, to protect persons from the necessity of defending stale claims, is served under Rule 15(c), since the amendment will not relate back unless the original pleading has given fair notice to the adverse party that a claim is being asserted against him from some particular transaction or occurrence, Wright, Law of Federal Courts, p. 276 (2d ed. 1970).'" *Id.*, 240, 429 A.2d 486.

By comparing the cause of action stated in the original complaint with the cause of action stated in the complaint that first alleged the dual capacity theory, we conclude that two different causes of action were stated. In the original complaint, as amended on July 30, 1985, the plaintiffs alleged that the defendant Norbert E. Mitchell, Jr., intentionally and/or negligently caused the death of the plaintiffs' decedents by ordering them to enter an underground area, which Mitchell knew to be without adequate ventilation, contained toxic fumes, lacked oxygen, and lacked proper lighting, gauges and other safety equipment. In the later complaint that asserted the dual capacity doctrine, the plaintiffs alleged that the defendants, their agents, servants or employees acted outside the scope of their business as retailers of fuel oils, in that they negligently designed, created and constructed an underground storage area without proper ventilation, adequate warnings or adequate gauges. The actionable occurrence in the original complaint is an allegedly negligent act in supervising employees while the actionable occurrence in the dual capacity counts of the later complaints is allegedly negligent design and construction of the underground storage area. These complaints involve two different facts to prove or disprove the allegations of a different basis of liability. The fact that the same defendant is accused of the negligence in each complaint and the same injury resulted, i.e., the death of three employees by asphyxiation, does not make any and all bases of liability relate back to an original claim of negligence. The defendants did not have fair notice of the claim of negligent construction and design of the underground storage area when the original complaint merely alleged that Norbert Mitchell, Jr., was negligent in ordering the employees to enter the area. See *Giglio v. Connecticut Light & Power Co.*, supra. The evidence concerning whether the defendant Mitchell was negligent when he sent three employees into a dangerous areas is not necessarily relevant to evidence concerning who designed and constructed the underground area.

This court has faced claims that certain new amendments alleged new causes of action with varying outcomes. In *Giglio v. Connecticut Light & Power Co.*, supra, we held that adding a claim that the defendant had permitted certain defects to remain in a furnace system to a strict liability claim did not state a new cause of action. The gravamen of the cause of action was a defective furnace and the defendant was not prejudiced by the amended complaint. In *Baker v. Baker*, supra, 166 Conn. at 486, 352 A.2d 277, it was held that an amended prayer for relief asking for a divorce rather than a legal separation did not change the factual bases or series of transactions upon which the complaint was based. In contrast, we held in *Keenan v. Yale New Haven Hospital*, supra, 167 Conn. at 286, 355 A.2d 253, that a complaint alleging assault and battery based on a lack of informed consent did not relate back to a complaint that alleged medical malpractice. Likewise, in *Gallo v. G. Fox & Co.,* supra, 148 Conn. at 332, 170 A.2d 724, an amendment alleging a fall due to a foreign substance on the floor did not relate back to the original complaint alleging a fall due to a defective escalator. In *Gallo,* the defendant, the plaintiff and the accident itself (falling down) were identical in each complaint but the facts surrounding the negligent acts were different. A defective escalator and a floor made dangerous by foreign substances do not arise out of the same negligent acts. This case is similar to *Gallo* in that the defendant (Norbert Mitchell, Jr.), the plaintiffs' decedents (the three employees) and the accident itself (death by asphyxiation in an underground area) are identical in each of the different counts. It is the acts of negligence that are different. An employer's claimed negligence in sending his employees into a facility that he allegedly knows is dangerous is not the same negligence as that of a person who negligently designs and constructs a dangerous facility. See also *Patterson v. Szabo Food Service of New York, Inc.,* 14 Conn.App. 178, 183, 540 A.2d 99 (1988) (fall due to the installation of a slippery floor is separate cause of action from a fall due to a floor made slippery by spilled food).

We conclude that the amended complaint of January 20, 1986, which first alleged the dual capacity theory, stated a new cause of action and therefore was barred by the statute of limitations. In view of our conclusion that the amended complaint stated a new cause of action, we do not reach the issue of whether, under the circumstances of this case, the dual capacity doctrine would permit a cause of action by the plaintiffs against the defendants.

The trial courts properly granted the defendants' motions for summary judgment.

THERE IS NO ERROR.

In this opinion the other Justices concurred.

(Modified for the purposes of this text)

Barbara G. SANTIAGO, etc., Appellants,
v.
Edward ALLEN and Dorothy
Allen, Appellees.
No. 83-996.
District Court of Appeal of Florida,
Third District.
May 1, 1984.

Plaintiff brought a wrongful death action against owners and lessors of property on which a lounge was located. The Circuit Court, Dade County, Milton A. Friedman, J., entered summary judgment in favor of owners and lessors, and plaintiff appealed. The District Court of Appeal held that since defendants were merely owners-lessors of property, were not liquor licensees and had no authority or duty to control the premises, they could not be charged with any failure to ensure safety of lounge's patrons.

Affirmed.

Horton, Perse & Ginsberg and Edward Perse, Miami, Tew, Spittler, Berger & Bluestein, Coral Gables, for appellants.

Corlett, Killian, Hardeman, McIntosh & Levi and Richard J. Suarez and Leanne J. Frank, Miami, for appellees.

Before SCHWARTZ, C.J., and NESBITT and DANIEL S. PEARSON, JJ.

PER CURIAM.

We affirm the summary final judgment entered in favor of defendants, the Allens, on plaintiff's wrongful death claim because defendants, as mere owners and lessors of the property, had no legal duty to ensure the safety of patrons of a lounge located on that property. The plaintiff was injured during an altercation on 'the premises of the Foxxy Laidy Lounge. The building housing the lounge and the land it was on were owned by the defendants herein. The lounge was owned and operated by Astral Liquors, Inc., on the property held under a lease from the defendants. The defendants had no control over, nor any right to control, the business operations of the lounge. Further, they had no financial interest in the lounge other than their right to rent monies as the owners-lessors of the building and land where the lounge was located. Under the terms of the lease, the liquor license was to be held and renewed in the name of the lessee, the "licensee." Upon expiration of the lease, the license was to revert back to the defendants. Such arrangements have been recognized by the courts and are not contrary to public policy. *See Wright v. Cade*, 349 So.2d 833 (Fla. 1st DCA 1977), *cert. denied*, 365 S.2d 716 (Fla.1978); *Concannon v. St. John*, 384 S.2d 903 (Fla. 5th DCA 1980); *Coney v. First State Bank of Miami*, 405 So.2d 257 (Fla. 3d DCA 1981). Although the defendants are the "owners" of a reversionary interest in the liquor license, Gene Willner, sole stockholder and officer of Astral Liquors, Inc., was the "licensee" pursuant to statutory transfers of the license under Section 561.32, Florida Statutes (1981). *See Wright* (purpose of Section 561.32 is to apprise Beverage Department of the ownership and management of the *business* so it can be regulated). Since the beverage statutes are only applicable to licensees, the defendants were incapable of complying with or violating the statues. Therefore, since the defendants were merely the owners-lessors of the property, were not the licensees and had no authority or duty to control the premises, they cannot be charged with any failure to ensure the safety of the lounge's patrons. *See Robinson v. Walker*, 63 Ill.App.2d 204, 211 N.E.2d 488 (1st Dist.1965). *Cf. Barton v. Lund*, 563 P.2d 875 (Alaska 1977), *overruled in Alesna v. LeGrue*, 614 P.2d 1387 (Alaska 1980) (*Barton* held that liquor licensees cannot be civilly liable if they do not exercise, and have no power to control the business. *Alesna* held that a *licensee* may be civilly liable for business violation of state liquor laws, thus, overruling *Barton* on that point.).

The summary judgment in favor of the defendants is accordingly affirmed.

(Modified for the purposes of this text)

Jean Marie YUNKER, as trustee for the
heirs and next-of kin of Kathleen
M. Nesser, Appellant,
v.
HONEYWELL, INC., a Delaware
Corporation, Respondent.
No. C5-92-1649.
Court of Appeals of Minnesota.
March 2, 1993.
Review Denied April 20, 1993.

Wrongful death action was brought against employer after employee shot and killed co-worker. The District Court, Hennepin County, H. Peter Albrecht, J., granted summary judgment for the employer. Trustee for the co-worker's heirs and next-of-kin appealed. The Court of Appeals, Lansing, J., held that: (1) the employer could not be held liable under a negligent supervision theory where the shooting did not occur on the employer's premises and where the employer's chattels were not used to cause harm; (2) the employer was not liable for negligently rehiring the employee after his incarceration for an earlier killing; and (3) the harm to the particular co-worker was sufficiently foreseeable for the employer to owe a duty of care.

Affirmed in part, reversed in part, and remanded.

SYLLABUS BY THE COURT

In a wrongful death action predicated on a direct liability theory of negligent retention, an employer's duty of reasonable care is not limited to the scope of the employment or to actions that occur on the employer's premises.

Mark Hallberg, Mackenzie & Hallberg, P.A., Minneapolis, for appellant.

William J. Egan, Patrick J. Sauter, Eric J. Magnuson, Rider, Bennett, Egan & Arundel, 2000 Lincoln Centre, Minneapolis, for respondent.

Considered and decided by KLAPHAKE, P.J., and LANSING and CRIPPEN, JJ.

OPINION

LANSING, Judge.

On motion for summary judgment, the district court held, as a matter of law, that an employer breached no ascertainable duty of care in hiring, retaining, and supervising an employee who shot and killed a coemployee off the premises. The employee had been rehired following imprisonment for the strangulation death of another coemployee. We affirm the district court's ruling as it applies to the theories of negligent hiring and supervision, but reverse the summary judgment as it applies to negligent retention and remand that part of the action to the district court.

FACTS

Honeywell employed Randy Landin from 1977 to 1979 and from 1984 to 1988. From 1979 to 1984 Landin was imprisoned for the strangulation death of Nancy Miller, a Honeywell coemployee. On his release from prison, Landin reapplied at Honeywell. Honeywell rehired Landin as a custodian in Honeywell's General Offices facility in South Minneapolis in August 1984. Because of workplace confrontations Landin was twice transferred, first to the Golden Valley facility in August 1986, and then to the St. Louis park facility in August 1987.

Kathleen Nesser was assigned to Landin's maintenance crew in April 1988. Landin and Nesser became friends and spent time together away from work. When Landin expressed a romantic interest, Nesser stopped spending time with Landin. Landin began to harass and threaten Nesser both at work and at home. At the end of June, Landin's behavior prompted Nesser to seek help from her supervisor and to request a transfer out of the St. Louis Park facility.

On July 1, 1988, Nesser found a death threat scratched on her locker door. Landin did not come to work on or after July 1, and Honeywell accepted his formal resignation on July 11, 1988. On July 19, approximately six hours after her Honeywell shift ended, Landin killed Nesser in her driveway with a close-range shotgun blast. Landin was convicted of first degree murder and sentenced to life imprisonment.

Jean Yunker, as trustee for the heirs and next-of-kin of Kathleen Nesser, brought this wrongful death action based on theories of negligent hiring, retention, and supervision of a dangerous employee. Honeywell moved for summary judgment and, for purposes of the motion, stipulated that it failed to exercise reasonable care in the hiring and supervision of Landin. The trial court concluded that Honeywell owed no legal duty to Nesser and granted summary judgment for Honeywell.

ISSUE

Did Honeywell have a duty to Kathleen Nesser to exercise reasonable care in hiring, retaining, or supervising Randy Landin?

ANALYSIS

[1, 2] The existence of a legal duty is generally an issue for the court to decide as a matter of law. *Larson v. Larson*, 373 N.W.2d 287, 289 (Minn.1985). Whether a duty exists depends on the relationship among parties and the foreseeability of harm to others. *Erickson v. Curtis Inv. Co.*, 447 N.W.2d 165, 168-69 (Minn.1989). Public policy is a major consideration in identifying a legal duty. *Id.* at 169.

In determining that Honeywell did not have a legal duty to Kathleen Nesser arising from its employment of Randy Landin, the district court analyzed Honeywell's duty as limited by its ability to control and protect its employees while they are involved in the employer's business or at the employer's place of business. Additionally, the court concluded that Honeywell could not have reasonably foreseen Landin's killing Nesser.[9]

[3] Incorporating a "scope of employment" limitation into an employer's duty borrows from the doctrine of respondeat superior. *See Marston v. Minneapolis Clinic of Psychiatry & Neurology, Ltd.*, 329 N.W.2d 306 (Minn.1982). However, of the three theories advanced for

recovery, only negligent supervision derives from the respondeat superior doctrine, which relies on connection to the employer's premises or chattels. *See Semrad v. Edina Realty, Inc.*, 493 N.W.2d 528, 534 (Minn.1992). We agree that negligent supervision is not a viable theory of recovery because Landin was neither on Honeywell's premises nor using Honeywell's chattels when he shot Nesser.

[4] The remaining theories, negligent hiring and negligent retention, are based on direct, not vicarious, liability. *See Ponticas v. K.M.S. Inv.*, 331 N.W.2d 907, 911 n.5 (Minn.1983). Negligent hiring and negligent retention do not rely on the scope of employment but address risks created by exposing members of the public to a potentially dangerous individual. *Di Cosala v. Kay*, 91 N.J. 159, 450 A.2d 508, 515 (1982). These theories of recovery impose liability for an employee's intentional tort, an action almost invariably outside the scope of employment, when the employer knew or should have known that the employee was violent or aggressive and might engage in injurious conduct. *Id.* 450 A.2d at 515; *see also Connes v. Molalla Transp. Sys.*, 831 P.2d 1316, 1320-21 (Colo.1992); *Plains Resources, Inc. v. Gable*, 235 Kan. 580, 682 P.2d 653, 662 (1984).

I.

[5] Minnesota first explicitly recognized a cause of action based on negligent hiring in *Ponticas* in 1983.

---

[9] The parties and the district court also analyzed the issue of liability under a general theory of a duty to control third persons from committing wrongful acts. Although the "duty to control" implicates similar principles of negligence, it is not usually applied to an employment relationship, and none of the cases cited by the parties analyze a duty to control as an element of negligent hiring or retention. *See Erickson*, 447 N.W.2d 165 (whether owner of a parking ramp owed a duty to customers to use reasonable care to deter criminal activity on its premises); *Lundgren v. Fultz*, 354 N.W.2d 25 (Minn.1984) (whether a psychiatrist was negligent in allowing a patient to gain access to a handgun); *Silberstein v. Cordie*, 474 N.W.2d 850 (Minn.App.1991) (whether parents had a duty to control mentally ill child who obtained a shotgun left in an unlocked closet). *Cf. Wood v. Astleford*, 412 N.W.2d 753 (Minn.App.1987) (rejecting a duty to warn when spouse and employer had no notice of aberrant behavior or specific targets), *pet. for rev. denied* (Minn. Nov. 24, 1987).

Ponticas involved the employment of an apartment manager who sexually assaulted a tenant. The supreme court upheld a jury verdict finding the apartment operators negligent in failing to make a reasonable investigation into the resident manager's background before providing him with a passkey. The court defined negligent hiring as

> predicated on the negligence of an employer in placing a person with known propensities, or propensities which should have been discovered by reasonable investigation, in an employment position in which, *because of the circumstances of the employment*, it should have been foreseeable that the hired individual posed a threat of injury to others.

331 N.W.2d at 911 (emphasis added).

Honeywell argues that under *Ponticas* it is not liable for negligent hiring because, unlike providing a dangerous resident manager with a passkey, Landin's employment did not enable him to commit the act of violence against Nesser. This argument has merit, and we note that a number of jurisdictions have expressly defined the scope of an employer's duty of reasonable care in hiring as largely dependent on the type of responsibilities associated with the particular job. *See Connes*, 831 P.2d at 1321 (employer's duty in hiring is dependent on anticipated degree of contact between employee and other persons in performing employment duties); *Tallahassee Furniture Co. v. Harrison*, 583 So.2d 744, 750 (Fla.Dist.Ct.App.1991) (employer's responsibility to investigate an employee's background is defined by the type of work to be done by the employee), *pet. for rev. denied* 595 So.2d 588 (Fla.1992).

Ponticas rejected the view that employers are required to investigate a prospective employee's criminal background in every job in which the individual has regular contact with the public. *Ponticas*, 331 N.W.2d at 913. Instead, liability is determined by the totality of the circumstances surrounding the hiring and whether the employer exercised reasonable care. The court instructed that

> [t]he scope of the investigation is
> directly related to the severity of the
> risk third parties are subjected to by
> an incompetent employee. Although only
> slight care might suffice in the hiring
> of a yardman, a worker on a production
> line, or other types of employment where
> the employee would not constitute a high
> risk of injury to third persons, * * *
> when the prospective employee is to be
> furnished a passkey permitting admittance
> to living quarters of tenants, the em-
> ployer has the duty to use reasonable
> care to investigate his competency and
> reliability prior to employment (cita-
> tions omitted).

Id.

Applying these principles, we conclude that Honeywell
did not owe a duty to Nesser at the time of Landin's
hire. Landin was employed as a maintenance worker whose
job responsibilities entailed no exposure to the general
public and required only limited contact with coemployees.
Unlike the caretaker in *Ponticas*, Landin's duties did
not involve inherent dangers to others, and unlike the
tenant in *Ponticas*, Nesser was not a reasonably foreseeable victim at the time Landin was hired.

[6] To reverse the district court's determination
on duty as it relates to hiring would extend *Ponticas*
and essentially hold that ex-felons are inherently
dangerous and that any harmful acts they commit against
persons encountered through employment will automatically
be considered foreseeable. Such a rule would deter
employers from hiring workers with a criminal record and
"offend our civilized concept that society must make a
reasonable effort to rehabilitate those who have erred
so they can be assimilated into the community." *Id.*

Honeywell did not breach legal duty to Nesser by
hiring Landin because the specific nature of his employment did not create a foreseeable risk of harm, and
public policy supports a limitation on this cause of
action. The district court correctly determined that

Honeywell is not liable to Nesser under a theory of negligent hiring.

II.

[7] In recognizing the tort of negligent hiring, *Ponticas* extended established Minnesota case law permitting recovery under theories of negligent retention. *Ponticas*, 331 N.W.2d at 910. As early as 1889, the Minnesota Supreme Court allowed a plaintiff who had been beaten by defendant's employee to bring an action for negligence, reasoning that an employer

> had no more right * * * to knowingly and advisedly employ or allow to be employed * * * a dangerous and vicious man, than it would have to keep and harbor a dangerous and savage dog * * *.

Dean v. St. Paul Union Depot Co., 41 Minn. 360, 363, 43 N.W. 54, 55 (1889). Following *Dean*, Minnesota cases have continued to recognize the employer's duty to refrain from retaining employees with known dangerous proclivities. *See Porter v. Grenna Bakeries, Inc.*, 219 Minn. 14, 16 N.W.2d 906 (1944); *Travelers Indem. Co. v. Fawkes*, 120 Minn. 353, 139 N.W. 703 (1913); *Kresko v. Rulli*, 432 N.W.2d 764, 769 (Minn.App.1988), *pet. for rev. denied* (Minn. Jan. 31, 1989).

Although some jurisdictions apparently aggregate the theories of "negligent hiring" and "negligent retention" into a single doctrine, Minnesota case law refers to them separately, suggesting that they are related, but distinct theories of recovery. *See, e.g., Ponticas*, 331 N.W.2d at 910. The difference between negligent hiring and negligent retention focuses on when the employer was on notice that an employee posed a threat and failed to take steps to ensure the safety of third parties. The Florida appellate court has provided a useful definition:

> Negligent hiring occurs when, prior to the time the employee is actually hired, the employer knew or should have known of the employee's unfitness, and the

> issue of liability primarily focuses upon
> the adequacy of the employer's pre-
> employment investigation into the em-
> ployee's background (citations omitted).
> Negligent retention, on the other hand,
> occurs when, during the course of em-
> ployment, the employer becomes aware or
> should have become aware of problems with
> an employee that indicated his unfitness,
> and the employer fails to take further
> action such as investigating, discharge,
> or reassignment (citations omitted).

Garcia   v.   Duffy,   492   S.2d   435,   438-39
(Fla.Dist.Ct.App.1986).

[8] In analyzing whether Honeywell owed Nesser a
duty under a theory of negligent retention, we are
obligated to review the facts in a light most favorable
to Yunker. *Grondahl v. Bulluck*, 318 N.W.2d 240, 242
(Minn.1982) (appeal from summary judgment). The record
contains evidence of a number of episodes in Landin's
postimprisonment employment at Honeywell that demonstrate
a propensity for abuse and violence towards coemployees.

While at the Golden Valley facility, Landin sexually
harassed female employees and challenged a male co-worker
to fight. After his transfer to St. Louis Park, Landin
threatened to kill a co-worker during an angry confrontation following a minor car accident. In another
employment incident, Landin was hostile and abusive
toward a female co-worker after problems developed in
their friendship. Landin's specific focus on Nesser was
demonstrated by several workplace outbursts occurring at
the end of June, and on July 1 the words "one more day
and you're dead" were scratched on her locker door.

Landin's troubled work history and the escalation of
abusive behavior during the summer of 1988 relate directly
to the foreseeability prong of duty. The facts, in a
light favorable to Yunker, show that it was foreseeable
to Yunker, show that it was foreseeable that Landin
could act violently against a coemployee, and against
Nesser in particular.

[9] This foreseeability gives rise to a duty of care to Nesser that is not outweighed by policy considerations of employment opportunity. An ex-felon's "opportunity for gainful employment may spell the difference between recidivism and rehabilitation," *Haddock v. City of New York*, 75 N.Y.2d 478, 554 N.Y.S.2d 439, 444, 553 N.E.2d 987, 992 (1990), but it cannot predominate over the need to maintain a safe workplace when specific actions point to future violence.

[10] Our holding is narrow and limited only to the recognition of a legal duty owed to Nesser arising out of Honeywell's continued employment of Landin. It is important to emphasize that in reversing the summary judgment on negligent retention, we do not reach the remaining significant questions of whether Honeywell breached that duty by failing to terminate or discipline Landin, or whether such a breach was a proximate cause of Nesser's death. These are issues generally decided by a jury after a full presentation of facts.

III.

[11] Finally, we reject Honeywell's argument that the negligence action is barred by the exclusive coverage provision of the Workers' Compensation Act. Three requirements must be met in order for an injury to be exclusively compensable under the Workers' Compensation Act: the injury (1) must arise out of the employment; (2) must be in the course of the employment; and (3) must not come within the assault exception. *Foley v. Honeywell, Inc.*, 488 N.W.2d 268, 271 (Minn.1992).

To satisfy the second "in the course of" requirement, the injury must occur within the time and space boundaries of employment. *Id.* at 272. Nesser's murder occurred at her home several hours after her work shift ended; it is clear that the incident does not meet the "in the course of" test. Furthermore, a violent murder by a known assailant is almost certainly "intended to injure the employee because of personal reasons" within the assault exception of Minn.Stat. § 176.011, subd. 16 (1986). *See id.* at 273; *Bear v. Honeywell, Inc.*, 468 N.W.2d 546, 547 (Minn.1991).

DECISION

We affirm the entry of summary judgment on the theories of negligent hiring and supervision, but reverse the summary judgment on the issue of negligent retention.

Affirmed in part, reversed in part, and remanded.

(Modified for the purposes of this text)

Brigid DULLARD, as Administratrix of
the Estate of Hugh Dullard, Deceased,
Plaintiff-Appellee,

v.

The BERKELEY ASSOCIATES COMPANY and Murray Hill Proper-
ties, Inc., 35 East
Street Corp., 175 West 12th Street Company, Beatay
Holding Corporation, Bernmark Company, Skillman Avenue
Construction Corporation, Lynn Realty Corporation, Broad-
way Construction Corporation, The Cousins Company,
Stephen Perlbinder, Barton Mark Perlbinder, Bernard West,
William West, Martin Berger, Bruce Berger, Michael
Berger, Julius Perlbinder Milton West, Lillian West,
Augusta Berger Danzig, and 400 Concrete Corp. and
Perlbinder Realty Corp., individually and as a joint
venture, Defendants and Third-Party Plaintiffs-Appel-
lants-Appellees,

v.

CASTLE CONCRETE CORP., Third-Party
Defendant-Appellant-Appellee.
Nos. 731-734, Dockets 78-7457, 78-7512,
78-7560 and 78-7561.
United States Court of Appeals,
Second Circuit.
Argued April 11, 1979.
Decided Aug. 27, 1979.

Widow of construction foreman who was killed in accident at jobsite brought diversity wrongful death action against contractors and owner. The United States District Court for the Southern District of New York, Kevin T. Duffy, J., entered judgment on a verdict awarding $630,000 for wrongful death, $20,000 for conscious pain and suffering and an additional amount for funeral and hospital expenses. Appeal was taken, and the Court of Appeals, Oakes, Circuit Judge, held that: (1) there was diversity jurisdiction; (2) the district court was justified in declining to charge the jury that impossibility was a defense; (3) plaintiff adequately established a prima facie case of negligence against the company which contracted to provide a concrete super-structure; (4) when reviewed as a whole, the jury charge fairly stated the law; (5) the plain terms of the contract between a subcontractor and an intermediate contractor

established that the subcontractor was liable to indemnify only for the intermediate contractor's tort liability to third parties; (6) the record supported the
award of $20,000 for conscious pain and suffering; (7)
the award of $630,000 for wrongful death was excessive
under New York law; and (8) under all circumstances,
the judgment should not exceed $500,000 with appropriate
interest.

Judgment in accordance with opinion.

John J. Wrenn, Larkin, Wrenn & Cumisky, New York
City, for Berkeley Associates (defendants and third-party
plaintiffs-appellants-appellees other than 400 Concrete
Corporation).

Michael E. Shay, Langan & Levy, New York City, for
defendant 400 Concrete Corp.

Mark J. Aaronson, Bower & Gardner, New York City,
for third-party defendant-appellant Castle Concrete Corp.

Arthur N. Seiff, O'Dwyer & Bernstien, New York City
(Frank Durkan, O'Dwyer & Bernstien, New York City, of
counsel), for plaintiff-appellee.

Before OAKES, Circuit Judge, and PIERCE[10] and WERKER,[10]
District Judges.

OAKES, Circuit Judge:

This is an appeal in a wrongful death diversity
action from a judgment of the United States District
Court for the Southern District of New York, Kevin T.
Duffy, Judge, entered after a jury trial on October 3,
1978, and amended on October 10, 1978. The jury awarded
plaintiff $630,000 for the wrongful death of her decedent,
a construction foreman struck in the head by a falling
"4 X 4" timber; $20,000 for his conscious pain and
suffering; and $3,825.30 for funeral and hospital
expenses, for a total judgment (including prejudgment
interest) of $803,546.81. In a special verdict, the jury
fixed the proportionate liability of the parties as

---

[10] Lawrence W. Pierce and Henry F. Werker of the Southern District of
New York, sitting by designation.

follows: 39% with respect to the defendants and third-party plaintiffs. The Berkeley Associates Company and its various partners (hereinafter "Berkeley"), the owner and general contractor; 35% with respect to the defendant 400 Concrete Corporation ("400 Concrete"), which had contracted with Berkeley to perform the concrete super-structure work; and 26% with respect to the third-party defendant Castle Concrete Corporation ("Castle"), which had contracted with 400 Concrete to perform the work that 400 Concrete had agreed to perform for Berkeley. The court granted 400 Concrete's claim against Castle for indemnity of its 35% share of liability but denied both Berkeley's claim against Castle for contractual indemnity of Berkeley's 39% share and 400 Concrete's claim against Castle for indemnity of 400 Concrete's liability to indemnify Berkeley. For the reasons that follow, we are unpersuaded by the several arguments raised on appeal except for the objection that the verdict was excessive.

Decedent, a labor foreman employed by Castle, was working on East 53rd Street adjoining the construction site for a high-rise building at the time of his death. Material for the construction was stored at several locations on that street and elsewhere, and appellants used a large crane to lift the material to the upper floors of the building under construction. Decedent was struck and killed by a piece of lumber (4" X 4" X 4') that fell from an undetermined source at least ten stories above ground level. At the time of the accident Castle had stacked wood on the 29th and 31st floors, and there was some testimony that the piece of lumber fell from one of those floors. The appellants had not provided any overhead protection, or more specifically a sidewalk shed, in violation of the New York Labor law,[11] nor did they have any effective system to warn endangered workers of falling objects.

[1] Berkeley's argument that there is no diversity of jurisdiction is frivolous. When plaintiff commenced this suit, she was a citizen of Ireland. Defendants are citizens of the United States. See 28 U.S.C. § 1332(a)(2).

---

[11] N.Y.Lab.Law §§ 2(3), 200(1), & 241(6) (McKinney Cum.Supp.1978); N.Y. Board of Standards & Appeals' Industrial Code, Rules 23-1.3, -1.5(a), -1.7(a), & -1.18.

[2] Castle and Berkeley maintain that the court incorrectly charged the jury that certain provisions of the New York Labor Law, *see* note 1 *supra*, impose on the owner and general contractor the continuing duty to provide suitable overhead protection, as well as the nondelegable duty to provide reasonable safety to workers on the construction site. According to defendants, the position of the crane made it impossible to provide overhead protection and thus under New York law, *Ortiz v. Uhl*, 39 A.D.2d 143, 148, 332 N.Y.S.2d 583, 588 (1972), *aff'd*, 33 N.Y.2d 989, 353 N.Y.S.2d 962, 309 N.E.2d 425 (1974), they owed no such duty here. After reviewing the evidence, especially the photographic exhibits, we are persuaded that the defendants' proof of the impossibility of operating the crane if overhead protection were offered was so meager and dubious that the court was amply justified in declining to charge the jury that impossibility was a defense. *Ortiz* involved work at the foot of a New York State Power Authority dam, with "no practical way of providing overhead protection" over the entire surface of the gorge. 39 A.D.2d at 145, 332 N.Y.S.2d at 585. Here involved was work on a high-rise building with a limited work area where such protection could readily have been provided in the form of a sidewalk shed as set forth in the New York Board of Standards and Appeals' Industrial Code (Rule 23), without interference with the operation of the crane.

[3, 4] Castle also argues that plaintiff failed to establish a prima facie case of negligence against 400 Concrete.[12] But the jury could have found that Barton Mark Perlbinder, an officer of 400 Concrete who in that capacity had signed the contract with Castle, was on the job site almost daily, including the morning of the accident; and that another officer was also on the job site during the months of the construction. Castle points out that these officers claimed to be on the site in their capacity as officers of *Berkeley*. But the jury might have disagreed in light of Perlbinder's signing of the Castle contract. Moreover, regardless of the degree of 400 Concrete's actual control and direction of Castle's work, it would appear that 400 Concrete, as

[12] Castle makes this argument by virtue of its duty to indemnify 400 Concrete, as to which see *infra*; the latter does not dispute its liability here, its insurer apparently preferring to seek indemnity over against Castle.

a contractor, had a nondelegable duty to ensure that its construction area was so "guarded ... as to provide reasonable and adequate protection and safety to the persons employed therein." N.Y.Lab.Law § 241(6) (McKinney Cum. Supp.1978); *see Allen v. Cloutier Construction Corp.*, 44 N.Y.2d 290, 405 N.Y.S.2d 630, 376 N.E.2d 1276 (1978). The jury certainly might have found that 400 Concrete failed to provide adequate protection.

[5, 6] Appellants complain that the court erred in giving the jury a res ipsa loquitur instruction. All that the court did was to charge the jury that the law "permits but does not require" an inference of negligence against the person having exclusive control of an instrumentality that causes an accident if the accident would not ordinarily have occurred without negligence. This is a simple enough principle of circumstantial evidence and has none of the vice of creating a presumption of negligence that the mere incantation of the Latin phrase so often evokes (although in certain special relationships, e.g., carrier/passenger, it is proper to shift the burden as a matter of policy, *see* W. Prosser, *Handbook of the Law of Torts* § 40 at 231 (4th ed. 1971)). One would suppose that a timber falling on a person's head from a high building was the quintessential "thing," at least it was in Baron Pollock's mind when he dropped the phrase onto the confused heads of the legal world in the argument in *Byrne v. Boadle*, 2 H. & C. 722, 159 Eng.Rep. 299 (1863) (barrel of flour rolling out of window). After noting that Castle and 400 Concrete each contended that it was not in exclusive control of the instrumentality, the court instructed the jury that this was a question of fact. The court noted that as general contractor and owner, Berkeley was "in control of everything at the job site," and continued that the question then was whether Berkeley had exclusive control of the instrumentality.

[7, 8] Under New York law, these instructions were more than adequate, even overly generous to defendants, for New York permits a "thing to speak for itself" as a matter of inference even when plaintiff has not shown that the instrument was in the defendant's *exclusive* control; the inference may be equally applicable to several persons if "they shared a common duty and there was no indication that any one of them in particular

had actually caused the injury." *De Witt Properties, Inc. v. City of New York*, 44 N.Y.2d 417, 427, 406 N.Y.S.2d 16, 21, 377 N.E.2d 461, 466 (1978); *see also* Restatement (Second) of Torts § 328D, comment g (1965). Here there was evidence (some of which we have already recounted) that Castle 400 Concrete, and Berkeley "shared a common duty" to keep the workplace safe, and the court was justified in submitting a permissive inference charge to the jury. Although Berkeley complains of the court's language "in control of everything," the court then stated that the jury must find that Berkeley exercised *exclusive* control over the instrumentality. Reviewing the charge as a whole, we believe that the court fairly stated the law.

[9] Castle's agreement to indemnify 400 Concrete was for "any and all claims, losses, suits, damages, judgments, expenses, costs and charges of every nature and kind, both legal and otherwise, whether direct or indirect, by reason of personal injuries, death or property damage to any persons or others caused by, arising out of or occurring in connection with the work provided under the terms of this contract," even if the "claims, suits, damages and judgments for personal injuries, death or property damage ... be due to the active negligence or statutory liability" of 400 Concrete. 400 Concrete argues that this obliges Castle to indemnify 400 Concrete not only for the latter's own *tort* liability, but also for its *contractual* liability to indemnify a third party, Berkeley, for Berkeley's tort liability to plaintiff. We disagree. As the district court found, the plain terms of the contract indicate that Castle is liable to indemnify only for 400 Concrete's tort liability to third parties. If the parties had intended that this typically worded hold harmless agreement have the unusual legal effect of including 400 Concrete's *contractual* liability to indemnify a third party, one would suppose that specific language would have been inserted to manifest that intent clearly. *See Compagnie Nationale Air France v. Port of New York Authority*, 427 F.2d 951, 954 (2d Cir. 1970) (unnamed third party); *Ratigan v. New York Central Railroad*, 181 F.Supp. 228, 233-34 (N.D.N.Y.1960), *aff'd*, 291 F.2d 548, 555-56 (2d Cir.), *cert. denied*, 368 U.S. 891, 82 S.Ct. 144, 7 L.Ed.2d 89 (1961) (same).

[10] Castle and Berkeley complain that the jury award of $20,000 for conscious pain and suffering was not supportable in the record. We cannot agree. There was evidence that plaintiff's decedent was partly conscious before his death. A police officer on the scene testified that when he saw decedent a few minutes after the accident, decedent "was still alive, he was breathing, and his arms were moving to a limited degree. He was just about unconscious."

[11, 12] We do agree with appellants, however, that the $630,000 jury verdict for wrongful death was excessive under New York law. In determining whether an award is excessive under that law, a federal court may not consider the amount of the prejudgment interest, in this case an additional $148,817.90, see *Zaninovich v. American Airlines, Inc.*, 26 A.D.2d 155, 160, 271 N.Y.S.2d 866, 872 (1966). But the principal of $630,000 alone, if invested to yield a 7% annual income (a yield that may underestimate what plaintiff can obtain in the contemporary market[13]), would provide the next of kin with an annual income of $44,100, without exhausting any principal. By contrast, the income provided by decedent at the time of his death in support of his family was approximately $15,704, consisting of weekly support of $302, computed by subtracting $40 personally used by the decedent from his $300 in take-home pay plus $42 in "stamps," received under the terms of his union's contract. Given his working life expectancy of twenty-seven years, that would mean total support of approximately $424,008 which, if discounted at the same 7% rate to reflect its present value, would amount to approximately $188,250.[14]

[13, 14] This figure does not, of course, take into account decedent's overtime pay or future earning potential, or the family's loss of decedent's services, society, and parental guidance, see *Spadaccini v. Dolan*, 63 A.D.2d 110, 124, 407 N.Y.S.2d 840, 848 (1978). But

[13] The June 25, 1979, edition of the *Wall Street Journal* at 40, cols. 3 & 4, sets forth the savings yields on a variety of investments. In general, investments in the principal amount of at least $1,000 can earn over 9% interest, e.g., Treasury Notes (9%-9.4%); High Grade Corporate Bond Funds (7.9%-9.6%); Six Month Bank Certificates (8.873%-9% on minimum amount of $10,000). More speculative forms of investment, of course, may earn even higher yields.

the only testimony with respect to the first two elements was Mrs. Dullard's statement that her husband earned occasional overtime; even as to that there was no estimate as to the amount. With respect to loss of services and guidance, although we recognize that this is not only a legitimate element of damages in New York but an important one, we do not believe that it justifies more than tripling either the annual income to the survivors or the total award for loss of pecuniary support. Our survey of New York law reveals that many courts have permitted awards in excess of direct pecuniary loss to compensate for lost services and guidance, but not to the degree here. *See, e.g., Zaninovich, supra* (reducing jury verdict to $475,000, an amount yielding an annual amortized return of approximately $39,000, where direct pecuniary loss was probably no more than $20,000 per annum); *Lucivero v. Long Island Railroad*, 22 Misc.2d 674, 200 N.Y.S.2d 728 (Sup.Ct. 1960) (establishing award of $175,000 where direct pecuniary loss, assuming a working life expectancy of thirty-two years, was perhaps $166,400).[15] Moreover, it can be argued that the comparison of $44,100 (7% X $630,000) to $15,704 actual support may be unduly generous to plaintiff. First, it ignores the probability that decedent's children (aged ten, eleven, and twelve at the time of the accident) will presumably need little or no support once they complete their educations. Second, it fails to discount the $630,000 award to present value. *See Zaninovich, supra*, 26 A.D.2d at 161, 271 N.Y.S.2d at 873.

Plaintiff argues on appeal that a wrongful death award may be adjusted upward to reflect the probable effect of future inflation. New York law on this issue

[14] This amount was derived by applying the discount factor $1/(1 + i)^n$ to the annuity of $15,704, where n = the year for which present value of income is sought and i = the interest rate. The discounted annuity for the first year is therefore $15,704 × (1/1.07) or $14,677, for the second year $15,704 × (1/1.07) or $14,677, for the second year $15,704 × 1/(1.07)^2 or $13,716, and so forth. The sum of the 27 yearly amounts, so derived, is approximately $188,250, the present value of $424,008. This method if discounting allows plaintiff to recover an amount approximately equivalent to the amount of salary that decedent would have contributed to his family over the course of his work life had he lived. It is not precisely equivalent because decedent was paid in weekly and not yearly installments. It also assumes that all taxes were paid prior to the contribution to the family.

is not as clear as she supposes, however, *compare Theobald v. Grey Public Relations, Inc.*, 39 A.D.2d 902, 334 N.Y.S.2d 281 (1972) (per curiam), *and Zaninovich, supra*, 26 A.D2d at 160, 271 N.Y.S.2d at 872, *with Spadaccini, supra*, 63 A.D.2d at 124, 407 N.Y.S.2d at 848, *and Theobald, supra* (McGivern, J., dissenting). New York law is evidently as much in a "stage of uncertainty and flux," *Feldman v. Allegheny Airlines, Inc.*, 524 F.2d 384, 390 (2d Cir. 1975) (Friendly, J., concurring), as state and federal decisions are generally on the subject of inflation in the law of damages. And in the absence of any expert testimony at trial on the issue, *compare Feldman v. Allegheny Airlines, Inc.*, 382 F.Supp. 1271 (D.Conn.1974), *aff'd in part and rev'd in part*, 524 F.2d 384 (2d Cir. 1975), we are even more reluctant to make such an adjustment. Assuming, however, that some adjustment is permissible, we would still find the verdict here excessive. True, if we adopt Judge Blumenfeld's technique (in *Feldman, supra*, 524 F.2d at 387-88) of reducing the discount rate by an inflationary factor, then on one basis the net discount rate might be at this time of "double-digit" inflation zero or even a negative computation. On the other hand, this is terribly speculative and a more reasoned approach might be to follow *Feldman* and base the future discount rate upon an average of past discount rates, resulting in a net

---

[15] We note in passing that New York apparently does not require the reduction of a wrongful death award by the amount of the decedent's expected tax liability, *see Cunningham v. Rederiet Vindeggen A/S*, 333 F.2d 308 (2d Cir. 1964). (Although *Zaninovich v. American Airlines, Inc.*, 26 A.D.2d 155, 160, 271 N.Y.S.2d 866, 872 (1966), refers to income tax liability, it is unclear whether the case squarely holds that such liability is to be considered in establishing an award.) Nevertheless, the plaintiff in this case introduced evidence only as to decedent's take-home pay, not his gross income. It would be inappropriate for us to inflate the net earnings figures set forth in the text on the basis of total speculation.

Although we probably cannot consider the costs of litigation, including attorneys' fees, in passing on the excessiveness of the verdict, *Zaninovich, supra* 26 A.D.2d at 160, 271 N.Y.S.2d at 872, however unrealistic that may be, *Theobald v. Grey Public Relations, Inc.*, 39 A.D.2d 902, 903, 334 N.Y.S.2d 281, 283 (McGivern, J., dissenting), we note that in a wrongful death suit contingent fees of as much as one third of the net recovery (after deducting expenses) are presumptively permissible, N.Y. Court Rules § 603.7(e)(2) (1978); *Gair v. Peck*, 6 N.Y.2d 97, 188 N.Y.S.2d 491, 160 N.E.2d 43 (1959), *appeal dismissed, cert. denied*, 361 U.S. 374, 80 S.Ct. 401, 4 L.Ed.2d 380 (1960).

rate of approximately 1.5%. Given a total contribution
of $242,008 (computed *supra* note 4), that rate would
justify an award of approximately $346,550. The differ-
ence between the actual award of $630,000 and this amount
is too large to be justified by loss of services and
guidance, or any other legitimate element of damages.

[15] With all these considerations in mind, we
conclude that the judgment should not exceed $500,000
with appropriate interest. Although we acknowledge that
the measure of damages is a complex factual determination
involving a considerable measure of estimation which is
largely entrusted to the jury's good sense, *Bellows v.
Smith*, 50 A.D.2d 622, 375 N.Y.S.2d 43 (1975), we are
convinced that the present verdict was clearly excessive.

Accordingly, we reverse the judgment of the district
court and order a new trial solely on the issue of
damages; but we will withhold entry of judgment for
thirty days, within which time plaintiff-appellee may,
if she chooses, file with the clerk of the district
court a remittitur of all damages in excess of $500,000
(plus interest from the date of the accident). Plaintiff-
appellee shall then file in the office of the clerk of
this court a certified copy of the remittitur filed in
the district court. If plaintiff-appellee files a remittitur, the judgment of the district court, less the
amount remitted, will be affirmed; otherwise, as stated,
the judgment will be reversed and a new trial solely on
the issue of damages ordered. *See Joiner Systems, Inc.
v. AVM Corp.*, 517 F.2d 45, 49 (3d Cir. 1975); *cf. Dimick
v. Schiedt*, 293 U.S. 474, 482-83, 484-85, 55 S.Ct. 296,
79 L.Ed. 603 (1935); *Grunenthal v. Long Island Rail Road
Co.*, 393 U.S. 156, 89 S.Ct. 331, 21 L.Ed. 2d 309 (1968)
(both cases recognize practice of remittitur or conditional new trial in federal courts).

Judgment in accordance with opinion.

(Modified for the purposes of this text)

Robert Norman SIMS and Nancy
Katherine Sims, Appellants,
v.
GENERAL TELEPHONE & ELECTRONICS,
d/b/a GTE Government Systems
Corp., also d/b/a/GTE Power Systems,
also GTE, Respondent.
No. 20727.
Supreme Court of Nevada.
July 12, 1991.

Deceased janitor's parents brought wrongful death action against owner of manufacturing plant. The Eighth Judicial District Court, Clark County, Nancy A. Becker, J., entered summary judgment for owner, and parents appealed. The Supreme Court held that: (1) genuine issues of material fact, precluding summary judgment for owner, existed as to whether owner was negligent in failing to warn janitor of dangers presented by degreaser and in failing to adequately secure the degreaser area; (2) "good samaritan" law was inapplicable, since owner and its employees were under duty to take reasonable steps to rescue janitor; (30 genuine issues of material fact, precluding summary judgment for owner, existed as to whether owner's rescue attempts were negligent; and (4) genuine issues of material fact, precluding summary judgment for owner, existed as to whether owner was janitor's employer, and therefore whether exclusive remedy provisions of Nevada Industrial Insurance Act were applicable.

Reversed and remanded.

Monte J. Morris, Henderson, for appellants.

Lionel Sawyer & Collins, and David N. Frederick, Las Vegas, for respondent.

OPINION

PER CURIAM:

This is an unfortunate case. On January 22, 1987, Robert Sims (Robert), a janitor at respondent General Telephone and Electronics' (GTE) manufacturing plant, was found curled up in the bottom of a degreasing machine. At the time Sims was discovered, Robert Tate (Tate), the plant security guard, attempted to remove Sims from the degreaser, but was unsuccessful due to the heavy amount of toxic chemicals that was in the machine. Several minutes later, police arrived, and were finally able to remove Robert from the degreaser. Robert was then given emergency medical treatment and rushed to St. Rose de Lima Hospital, where he died four days later without regaining consciousness.

Robert's parents, appellants Robert Sr. and Nancy Sims, then brought this wrongful death action against GTE. In this lawsuit, appellants claim that GTE was negligent in three ways. First, appellants claim that GTE failed to provide Robert with proper warnings of the degreaser's dangers. Second, the Sims contend that GTE failed to take proper precautions against the hazards of the degreaser. Finally, appellants argue that Tate, GTE's employee, was negligent in his effort to rescue Robert.

After discovery, GTE moved for summary judgment on all issues. GTE maintained that appellants' claims were barred by the exclusive remedy provisions of the workers' compensation laws. Respondent further argued that even if GTE and Tate had behaved negligently in the manner described, appellants had produced no evidence that this negligence proximately caused Robert's death.

In October 1989, the court below first ruled on the summary judgment motion. Specifically, the district court held that genuine issues of fact existed regarding whether or not GTE was Robert's employer at the time of his death. Thus, summary judgment was denied on this issue.

Next, the court held that appellants had presented no evidence that would tend to show that any action by GTE proximately caused Robert's death. Finally, the district court held that, although appellants had failed to present any evidence that would show that Tate's rescue efforts had proximately caused Robert's death, it would allow appellants an extra month of discovery. The district court then ordered appellants to produce evidence during this period that would show that, if Tate had administered oxygen in the period before police arrived, it "would have made a difference." When appellants returned in November without any evidence in support of this causal link, summary judgment was granted.

The claims in this appeal are essentially the same as those in the court below. Namely, appellants contend that a factual dispute exists as to whether GTE or Tate's actions proximately caused Robert's death. Respondent disagrees with this contention, and also argues that, contrary to the finding of the court below, no dispute exists as to whether GTE was Robert's employer. We hold that factual disputes exist regarding GTE's negligence, Tate's negligence, and GTE's status as Robert's employer. Accordingly, we reverse the grant of summary judgment and remand for further proceedings.

I.

This case comes before us on appeal from a grant of summary judgment. The standard of review in such cases has been described many times by this court. In *Butler v. Bogdanovich*, 101 Nev. 449, 451, 705 P.2d 662, 663 (1985), we stated that "[a]n entry of summary judgment is proper only when there are no issues of fact and the moving party is entitled to an expedited judgment as a matter of law." The party opposing the motion must set forth specific facts that show that there is a genuine issue for trial. *Van Cleave v. Kietz-Mill Minit Mart*, 97 Nev. 414, 415, 633 P.2d 1220, 1221 (1981).

[1, 2] On appeal, the summary judgment standard thus requires us to determine whether a factual dispute exists with regard to each element of the cause of action. Consequently, in order to survive a summary judgment motion in negligence claim, there must be factual disputes as to: (1) duty; (2) breach; (3) actual causation; (4) legal causation; and (5) damages. *Perez v. Las Vegas Medical Center*, 107 Nev.1, 805 P.2d 589 (1991); *Beauchene v. Synanon Foundation, Inc.*, 88 Cal.App.3d 342, 346, 151 Cal.Rptr. 796 (1979).

We have, in the past, indicated our hesitance to affirm the granting of summary judgment in negligence cases, because such claims generally present jury issues. *Van Cleave*, 97 Nev. at 417, 633 P.2d at 1222. If respondent can show that one of the elements is clearly lacking as a matter of law, however, then summary judgment is proper. *Id.*

With these principles in mind, we now turn to an examination of whether a factual dispute existed with respect to each element of each of appellants' negligence claims.

II.

Appellants' first contention is that GTE's negligence prior to the accident proximately caused Robert's death. Specifically, appellants claim that GTE was negligent in the following ways: (1) it failed to warn Robert of the dangers presented by the degreaser; and (2) it failed to secure the degreaser area adequately. We will now examine these claims more closely in order to determine if a factual dispute exists.

Duty

In *Mangeris v. Gordon*, 94 Nev. 400, 402, 580 P.2d 481, 483 (1978), we stated that in order for a negligence action to succeed, the alleged wrongdoer must owe a duty of care to the person injured. We further held that in failure to warn cases, defendant's duty to warn exists only where there is a special relationship between the parties, and the danger is foreseeable. *Id.* at 403, 580 P.2d at 483.

We also discussed the duty question in *Southern Pacific Co. v. Huyck*, 61 Nev. 365, 128 P.2d 849 (1942). In *Huyck*, this court held, in the context of a suit under the Federal Employees Liability Act, that an employer has a duty to exercise due care in maintaining a safe workplace. *Id.* at 379, 128 P.2d at 855.

[3, 4] These cases clearly reveal that GTE had a duty both to warn Robert of the dangers of the degreaser machine, and to ensure that the machine did not present unreasonable dangers to those in the workplace. Under

Mangeris, where a special relationship exists, there is a duty to warn plaintiff of foreseeable dangers. Here, Robert worked in GTE's plant, and the dangers that the degreaser presented to people like Robert should have been apparent to GTE. Thus, GTE had a duty to warn Robert of the machine's dangers. Further, under *Huyck*, GTE had a duty to use reasonable care to make its workplace, including the degreaser area, safe. Thus, we now examine whether a factual dispute exists as to GTE's compliance with these duties.

Breach

The above cases show that GTE's duty was one of due care, both in ensuring that Robert was adequately warned, and in making its workplace safe. As stated in *Merluzzi v. Larson*, 96 Nev. 409, 412, 610 P.2d 739, 742 (1980), this requirement of due care "is invariably the same—one must conform to the legal standard of reasonable conduct in light of the apparent risk."

Thus, in order to *breach* this standard of care, a person must fail to exercise "that degree of care in a given situation which a reasonable man under similar circumstances would exercise." *Driscoll v. Erreguible*, 87 Nev. 97, 101, 482 P.2d 291, 294 (1971). Also, this court has noted that because reasonable people may often differ as to whether an amount of care was appropriate under the circumstances, the issue should generally be submitted to the trier of fact. *Merluzzi*, 96 Nev. at 413, 610 P.2d at 742. Consequently, in order to determine whether a factual dispute exists regarding breach, we must look at all the circumstances surrounding both the warnings given by GTE and the precautions taken with respect to the degreaser.

[5] On the warning issue, several pieces of evidence were presented. First, there were three warning signs on the degreaser tank itself. One sign had the word "DANGER" printed on it in large lettering, and then stated, in detail, various hazards and precautions to be followed by users of the machine. Another sign on the machine was entitled, "OPERATING AND CAUTION INSTRUCTIONS." This sign contained several instructions on how to operate the machine safely. Finally, the third warning sign stated in large black lettering that the

degreaser contained dangerous chemicals. This sign also described various precautions, first aid procedures, and handling and storage tips that should be followed when around the machine. On each of the three signs appears the ominous phrase: "DEATH CAN RESULT FROM CARELESSNESS."

In addition to the warning signs on the machine itself, several uncontradicted affidavits reveal that Robert was verbally warned about the dangers of the degreaser. Robert's supervisor, the plant security guard, and one of Robert's co-worker's all stated that Robert had been clearly warned never to enter the degreaser room, because of the great dangers of the machine.

Ordinarily we would agree that these verbal and written warnings were sufficient as a matter of law. However, appellants insist that the warnings lost their significance to Robert when the plant ceased operations and the degreaser room was no longer used. Appellants contend that Robert was not informed that the toxicity of the chemicals in the degreaser tank would not dissipate within a short time after its use was discontinued. Moreover, the young man, without contrary warnings, may have assumed that when degreaser room operations were shut down, the tank would have been cleaned and any dangerous chemical residues removed. Appellants conclude, therefore, that Robert should have been informed that the degreaser room would be left in a dangerous state despite the discontinuance of operations, and again told not to enter the room or approach the degreaser tank. We agree that this issue may not properly be resolved in a summary proceeding. Whether GTE's prior warnings and admonitions reasonably placed Robert on notice that the degreaser room and tank would retain its dangerous characteristics after operations were discontinued is an issue of fact to be determined by the jury. For this reason, summary judgment was improperly granted on the failure to warn issue.

[6] Appellants also contend that GTE failed in its duty to maintain a reasonably safe workplace. Specifically, appellants contend that, under the circumstances, reasonable care required GTE to take at least one of the following actions: (1) purge the tank; (2) barricade the entrance to the degreaser room (there was no door); (3) close the cover of the tank; or (4) maintain an

experienced operator on duty until the tank became harmless.

We agree with appellants that a factual dispute exists on this issue. It is quite clear that the degreaser tank was a very dangerous machine; the warnings on the tank itself reveal this, as does the fact that GTE gave specific warnings to its employees about the hazards of the tank. Furthermore, since the plant was shutting down, all of these options (with the possible exception of installing a watchman) were quite feasible. For this reason, a question of fact exists as to whether GTE's decision to rely on warnings, rather than these other precautions, was reasonable in light of the great risks posed by the tank. This is especially so given our frequent statements that questions of reasonableness are, in general, for the jury to decide.

Actual Causation

[7] Our conclusion that factual issues exist regarding the reasonableness of defendant's behavior does not, of course, end the inquiry. For unreasonable behavior is not actionable unless it *actually causes* plaintiff's injures. *See Van Cleave*, 97 Nev. at 416, 633 P.2d at 1221. Put another way, in order to satisfy this element, plaintiff must show that *but for* defendant's negligence, his or her injuries would not have occurred. *Taylor v. Silva*, 96 Nev. 738, 741, 615 P.2d 970, 971 (1980).

[8] An application of this "but for" test to the instant case reveals that, under either appellants' "failure to adequately warn [sic]" or "failure to secure" theories, a factual dispute exists on the actual causation issue. It is clear that "but for" GTE's failure to secure the degreaser area in one of the ways described above, Robert *could not*, and thus would not, have gotten into the tank. For this reason, assuming that GTE was negligent in failing to secure the area, this negligence *actually caused* Robert's death, and summary judgment should not have been granted on this issue. Moreover, if the trier of fact determines that GTE's warnings and admonitions did not reasonably forewarn Robert of the lingering hazards that would exist in the degreaser tank despite a permanent discontinuance of operations, the trier may also find that, "but for" GTE's breach of this duty,

Robert would not have entered the tank. In that regard, trial evidence may indicate that historically Robert had strictly heeded directions concerning his duties and responsibility. If so, the trier may conclude that in the face of proper warnings, Robert would have maintained his consistent attitude of compliance with instructions. Therefore, on the failure to warn issue, because the trier of fact may find a negligent breach of duty, it may also find that, as a proximate result thereof, Robert entered the tank. Summary judgment was therefore improper on this issue as well.[16]

Legal Causation

Even where it has been established that defendant's conduct has been one of the causes of plaintiff's injury, there remains the question of whether defendant will be legally responsible for the injury. This is basically a "policy issue," with the court deciding whether such considerations point in favor of making defendant responsible for the consequences that have, in fact, occurred.

[9] The main consideration in such situations is foreseeability. As stated in *Taylor v. Silva*, 96 Nev. at 741, 615 P.2d at 970, "[a] negligent defendant is responsible for all *foreseeable* consequences ... [of] his or her negligent act." (Emphasis added). This requirement means that defendant must be able to foresee that his negligent actions may result in harm of a particular variety to a certain type of plaintiff. *See Karlsen v. Jack*, 80 Nev. 201, 206, 391 P.2d 319, 321 (1964).

[10, 11] This requirement does not mean, however, that defendant must foresee the extent of the harm, or the manner in which it occurred. *Id.* Under *Karlsen*, defendant need only foresee that his or her negligent conduct could have caused a particular variety of harm to a certain type of plaintiff. Here, GTE could certainly foresee that any negligence with regard to the degreaser might result in toxic poisoning of the workers in the plant. The fact that GTE may not have been able to see the manner in which the harm occurred, i.e., it would

---

[16] We, of course, are only hypothesizing, and have no impression, or suggest none, concerning what the evidence will reflect at trial.

have been difficult to foresee that Robert would wind up in the bottom of the tank, is immaterial. Thus legal causation is present under the theories raised by appellants.

Damages

Legally recoverable damages are clearly present in this case.

III.

Appellants' next claim is that Tate's negligent rescue efforts caused Robert's death, and that GTE is liable for these actions under a theory of respondeat superior. In order to determine whether summary judgment was properly granted, we will again examine each element of appellants' theory.

Duty

At common law, there was no affirmative duty on the part of strangers to render aid to those in need of emergency assistance. *See Prosser and Keeton on Torts* § 56, at 375 (5th ed. 1984). Once someone undertook the responsibility of giving such aid, however, the common law imposed on the rescuer the duty of reasonable care. *Id.* at 378. Some observers argued that such a rule made little sense, however, because holding rescuers accountable for their ordinary negligence if they undertook a rescue, but not if they chose to ignore the situation, caused may people to choose the latter option. *Id.* Hence, a number of states passed "good samaritan" laws, which encourage the giving of emergency aid by making rescuers immune from liability for acts of ordinary negligence. *Id.*

[12] In Nevada, the "good samaritan" statute makes those who render emergency aid, gratuitously and in good faith, not liable for acts of ordinary negligence.[17] This court has never had occasion to define the meaning of the phrase "gratuitously and in good faith," as used in the Nevada statute. We now hold that this phrase limits the benefits of "good samaritan" protection to those situations in which the rescuer was not already under a duty to act. Such a construction is in line with the

obvious purpose of the statute—ensuring that those people who could ignore the situation choose instead to undertake the rescue. Further, this construction is quite consistent with the common meaning of the word "gratuitously," which is defined in one dictionary as "unnecessary or unwarranted." American Heritage Dictionary (New College Edition 1980). Thus, "good samaritan" liability applies only where rescue would otherwise be legally unnecessary or unwarranted.

[13-17] Accordingly, NRS 41.500 is not applicable in this case. For the law does sometimes require parties to act affirmatively to aid others in peril. Such a duty is generally imposed where a special relationship exists between the parties, such as innkeeper-guest, teacher-pupil, or, as here, an employer-employee.[18] *Prosser and Keeton on Torts* § 56, at 376. In these situations, the party in "control of the premises" is required to take reasonable affirmative steps to aid the party in peril. *Id.* Because GTE, and hence, its employees, were therefore already under a duty to take reasonable steps in rescuing Robert (even before any efforts were made to save him), the "good samaritan" law does not apply.[19] Consequently, GTE remained under a duty of reasonable care in performing the rescue.

[17] The Nevada "good samaritan statute" is codified as NRS 41.500. This law reads, in pertinent part, that "any person in this state [with certain exceptions not relevant to this appeal], who renders emergency care of assistance in an emergency, gratuitously and in good faith, is not liable for any civil damages as a result of any act or omission, not amounting to gross negligence, …"

Here, GTE was under a duty to rescue one of the workers who as in peril in its plant. Tate was the only person in the building that was clearly employed by GTE. Thus, the only way that GTE could perform its duty was through Tate. For this reason, it is at least arguable that Tate was involved "wholly or partly" in the performance of GTE's business when he (Tate) made the rescue attempt. Consequently, the district court correctly refused to grant summary judgment on this issue.

[18] GTE argues that even if the "good samaritan" law is not applicable, it (GTE) should still not be held liable for Tate's actions because the security guard was acting outside of the scope of employment. This argument ignores the crucial point, however, that issues pertaining to the scope of employment almost always present questions of fact. *Birkner v. Salt Lake County,* 771 P.2d 1053, 1057 (Utah 1989). The *Birkner* court went on to state that the issue must be submitted to the jury "whenever reasonable minds could differ as to whether the [employee] was involved wholly or partly in performance of the master's business."

Breach

[18] As discussed above, the question of whether reasonable care was exercised is generally one that should be decided by the jury — for it almost always involves factual inquiries. Here, appellants presented evidence that Tate could have, but did not, provide Robert with oxygen several minutes before the rescue crew arrived. A jury could have found that such behavior was unreasonable under the circumstances, and therefore the trial court properly held that a factual dispute exists regarding breach.

Actual Causation

[19] The district court's ruling granting summary judgment in favor of GTE was based on the fact that appellants failed to produce evidence that would support a finding of actual causation. Specifically, appellants did not produce medical testimony to the effect that had Robert been given oxygen ten minutes earlier, "it would have made a difference." The court therefore held that judgment against appellants on the issue of actual causation was appropriate as a matter of law.

This is a very close question. Nevertheless, as we have often stated in the past, even a slight factual dispute is sufficient to make the granting of summary judgment improper. *See, e.g., Oak Grove Investors v. Bell & Gossett Co.*, 99 Nev. 616, 623, 668 P.2d 1075, 1079 (1983). Here, Robert was still breathing (though unconscious) at the time when Tate's efforts began, but had stopped breathing when the paramedics arrived. This evidence supports an inference that the application of oxygen could have made a difference. Accordingly, we hold that a factual dispute existed on the actual causation issue.

---

[19] Although we have described this duty to rescue as arising out of the employer-employee relationship, we must note that the owner/operator of a business has an affirmative duty to attempt to rescue all workers injured on the business premises, regardless of whether these workers are his "employees" in the technical sense. Thus, the duty to rescue existed here, even though Robert may not have been an employee of GTE (see *infra*), but simply an employee of the company GTE hired (Top-Hat Maintenance) to clean the plant.

Legal Causation

[20]   Legal causation is present under this theory because it is undoubtedly *foreseeable* that, if a rescue operation is performed without proper care, the victim's condition will be made worse.

Damages

Legally recoverable damages are clearly present in this case.

IV.

Respondent's final contention is that even if material issues of fact are present under appellants' negligence theories, summary judgment was nonetheless properly granted because appellants' claims are barred by the exclusive remedy provisions of the Nevada Industrial Insurance Act (NIIA). Specifically, GTE contends that because Robert was a GTE employee, and because Robert was injured on the job, the NIIA provides Robert (and his parents) with the only available recovery for his injuries.[20]

[21]   Under NRS 616.370, the NIIA provides the exclusive remedy for workers injured on the job due to the actions of their employer. NRS 616.560 states that persons injured on the job other than by the actions of their employer (or a person employed by their employer) may bring an action at law for tort damages. Thus, the question that must be decided there is whether a factual dispute exists as to whether GTE was Robert's employer — if so, then the district court properly refused to grant summary judgment on this issue; if GTE was Robert's employer as a matter of law; however, then the lower court's grant of summary judgment may be affirmed on this ground.

[22]   This court has often held that the key to deciding such issues is whether the alleged employer exercised "control" over plaintiff. *See, e.g., Whitley v. Jake's Crane and Rigging Inc.*, 95 Nev. 819, 821, 603

---

[20] It is clear from the record that appellants received NIIA benefits as a result of this incident.

P.2d 689, 690 (1970). Whether such "control" existed, in turn, is determined by the totality of the circumstances, including: (1) the degree of supervision; (2) the existence of a right to hire and fire; (3) the right to control the worker's hours and location of employment; (4) the source of wages; (5) the extent to which the worker's activities further the alleged employer's general business concerns. *Id.* In addition, this court has noted that "in the absence of a clearly established NIIA defense, summary judgment must be denied." *Leslie v. J.A. Tiberti Construction*, 99 Nev. 494, 498, 664 P.2d 963, 965 (1983).[21]

[23] An application of these factors to the instant case reveals that the trial court properly concluded that a factual dispute exists as to whether GTE was Robert's employer. Top-Hat Maintenance hired, trained, and supervised Robert. Also, Robert was universally perceived as being employed by Top-Hat, not GTE. Top-Hat controlled the number of hours Robert worked. In addition, Top-Hat exercised the main responsibility for the decision whether to fire employees such as Robert. Finally, Top-Hat, and not GTE, was responsible for obtaining workman's compensation insurance for Robert. These pieces of evidence all point to the conclusion that whether or not Robert was employed by GTE was in dispute. Thus, the district court properly refused to grant summary judgment on this issue.

---

[21] Respondent places great reliance on several of our cases that seem to indicate that very little evidence is required to show that a defendant is an "employer" of plaintiff. Close inspection reveals that this reliance is misplaced, however, because all of the cases relied on by respondent involve an injured *construction* employee of a sub-contractor, who is attempting to sue the general contractor. *See, e.g., Hosvepian v. Hilton Hotels*, 94 Nev. 768, 770, 587 P.2d 1313, 1315 (1978).

This distinction is significant, because such cases are controlled by NRS 616.085(1), which provides that "subcontractors and their employees shall be deemed employees of the principal contractor." In non-construction situations such as this case, however, this court has taken a much more fact-specific approach to the problem. *See, e.g., Meers v. Haughthton Elevator*, 101 Nev. 283, 285 n. 3, 701 P.2d 1006, 1007 n. 3 (1985)(expressly noting that inquiry is different in non-construction situations); *Daniels v. Las Vegas Transfer and Storage*, 97 Nev. 231, 233, 627 P.2d 400, 401 (1981).

V.

Appellants allege that GTE was negligent in three ways: (1) failure to warn defendant adequately of the dangers of the degreaser; (2) failure to secure properly the degreaser; and (3) vicariously, through the actions of its employee, Robert Tate. With regard to the failure to warn claim, we have concluded that material factual issues exist concerning the adequacy of GTE's warnings and admonitions in the context of the period subsequent to the discontinuance of operations in the degreaser room. Causation issues also remain, as previously noted. Summary judgment on this issue was therefore improper.

On the failure to secure claim, however, summary judgment was also improper. It is clear that GTE had a duty to maintain a safe work place, and it is undisputed that respondent took none of the precautions suggested by appellants. Whether this inaction was reasonable under the circumstances is a question for the jury. Also, because the incident clearly would not have happened had the precautions been taken, and because the general type of injury was foreseeable, a dispute exists as to causation. Thus, factual disputes exist regarding all elements of this cause of action.

In addition, the trial court erred in granting summary judgment on the negligent rescue claim, GTE (and Tate) clearly had an affirmative duty to rescue Robert, and a duty to use reasonable care in this rescue attempt. In addition, factual disputes exist on the issue of actual and proximate causation, as well as damages.

Finally, appellants' claims are not barred by the NIIA. While the NIIA provides the exclusive remedy for injured employees against their employers, these injured employees may recover in tort against third parties. Because a factual dispute exists as to whether GTE was Robert's employer, summary judgment was properly denied on this issue.

Accordingly, the grant of summary judgment is reversed, and this matter is remanded to the district court for further proceedings consistent with the views expressed herein.

chapter twelve

Other legal considerations

"A man may as well open an oyster without a knife, as a lawyer's mouth without a fee."

Barten Holyday

"That is the beauty of the Common Law, it is a maze and not a motorway."

Lord Diplock

When designing and developing programs and procedures to prevent workplace violence, prudent professionals should not lose sight of the substantial number of protections provided to individual employees. In fact, in strictly numerical probabilities, the potential of acquiring a claim or charge for violation of an individual's rights is a greater probability than actually having a workplace violence incident. To this end, it is vitally important that employers and the professionals responsible for workplace violence prevention efforts keep in mind the individual privacy rights and related rights when establishing these programs and policies. To this end, several of the more pertinent areas of potential liability are set forth below. It should be noted that these identified areas of legal liability are not all of the potential risks with regard to privacy rights or individual rights.

Workplace privacy

A general outline for the various forms of workplace privacy issues can be found in Restatement (Second) of Torts Section 652A(2) (1977). In employment-based invasion of privacy actions, the allegations usually center upon one of the following:

a. Access to personal information in the possession of the employer
b. Unreasonable collection of information by an employer
c. Retaliation by an employer for an employee's refusal to provide personal information

    d. Unreasonable means used by an employer to collect information
    e. Personnel decisions based upon a person's off-duty activity
    f. Unwarranted disclosure of personal information about an employee by an employer
    g. Employer insults and affronts to the dignity of an individual[1]

Employers attempting to assess and identify individuals who may be prone to workplace violence usually want to acquire as much information as possible. Thus, many employers have resorted to various investigative techniques in both the pre- and post-hiring stages of employment. These include personal surveillance, security cameras, monitoring calls, as well as background investigations. A relatively new concern is the rapid increase and use of technology. E-mail and facsimile (fax) transmissions are also causing numerous problems and concerns for employers.

What legal issues may an employee maintain in tort against an employer for invasion of privacy? Basically, there are four accepted variations of the tort of invasion of privacy. These include:

    a. Intrusion upon seclusion
    b. Appropriation of name or likeness
    c. Publicity given to private life
    d. Publicity placing a person in false light[2]

A traditional invasion of privacy action probably would not be brought for the dismissal of an employee, but for other acts connected with the dismissal. The appropriation of name or likeness is unlikely to be involved in the dismissal situation, and the false light variant is difficult to distinguish from defamation. Accordingly, a dismissed employee is most likely to benefit from the intrusion and publicity given due private life variance.

Intrusion upon seclusion

The intrusion upon seclusion issue consists of an intentional interference with the plaintiff's private affairs in a manner "that would be highly offensive to a reasonable man;"[3] it does not depend on any publicity given to the information collected about the plaintiff. Most of the cases accepting the intrusion variant of the privacy tort have involved intrusion into a physical area with respect to which the plaintiff had a reasonable expectation of privacy. It's as simple as going through the desk or locker of an employee,

---

[1] *See* Report of the Committee on Employee Rights and Responsibilities, *The Labor Lawyer*, volume 10, No. 3, at p. 615 (Summer 1994).

[2] Restatement (Second) of Torts, §§ 652B, 652C, 652D, and 652E, respectively.

[3] Restatement (Second) of Tort Section 652B (1977). In *Phillips v. Smalley Maintenance Servs., Inc.*, 711 F. 2d 1524, 1532 (11th Circuit 1983), the Alabama Supreme Court, on certification, adopted the Second Restatement respecting privacy torts, and held that "acquisition of information from a plaintiff is not a requisite element of a § 652B cause of action."

who believes they have a personal right of privacy in these places. In several circumstances, the courts have agreed.

Most privacy cases involved the acquisition of information. Thus, if an employee was dismissed for reasons related to his private life, and the employee could prove that the employer had in some way unreasonably investigated his private life, a claim could be made. For example, wire tapping has been held to constitute an invasion of privacy. Polygraph examinations similarly may constitute invasion of privacy, depending upon the subjects inquired into and the laws applicable to the testing.

In the employment context, intrusion into seclusion may involve an employer's testing of employees, gathering of medical information, obtaining credit records, electronic surveillance, and obtaining background information on an employee's suitability for employment. Many other issues involving workplace investigations also address this subject. Complaints of sexual harassment require employers to investigate the allegations. Growing concerns involving workers' compensation and whether injuries are *bona fide* may require an investigation and/or surveillance. In addition, investigations may be required when employees complain of other discriminatory acts at the hands of their supervisors.

Surveillance of employees in plain view at the workplace as part of a work-related investigation is a permissible practice.[4] However, this does not permit the employer to spy on employees while they are in the bathroom or in other private settings. There is absolutely no employer protection to place surveillance cameras, one-way mirrors, or other forms of surveillance in bathrooms or other private settings.[5]

Courts have been inclined to grant employers latitude with respect to home surveillance if done as part of a claims investigation. However, there is an increased likelihood that surveillance of an employee's non-work related activities may be deemed by a court or jury to cross the line of acceptable activities. The key is the intrusiveness of the activity. As long as the surveillance is conducted on public property and does not interfere with the daily activities of the individuals being monitored, some latitude is given. A recent Kentucky Court of Appeals decision provides some insight into the extent of proof necessary for a plaintiff to get a verdict involving a surveillance claim.[6]

The underlying facts in *Kentucky Electric Steel v. May* involved an investigation of an employee who claimed a work-related injury. The employer was informed that the employee, Mr. May, was engaged in outside work while assigned to light duty. The employer hired an investigation service to determine the truth of the allegations. Videotape of Mr. May's activities was taken from a van in a public road. Neighbors were interviewed and Mrs.

---

[4] *Johnson v. Corporate View Surv. Inc.*, 602 So. 2d 385 (Ala. 1992); *Thomas v. General ELEC. Co.*, 207 F. Supp. 792 (W.D. Ky 1962).

[5] *Massey v. Victor L. Phillips Co.*, 827 F. Supp. 597 (W.D. Mo. 1993); *Brazinski v. Amoco Petroleum Additives Co.*, 6 F.3d 1176 (7th Circ. 1993).

[6] *Kentucky Electric Steel v. May*,_____S.W. 2d____, (1995 Ky. App. LEXIS 152).

May was followed on several occasions. The Mays were unaware of the surveillance until the videotape was played during a workers' compensation proceeding.

The court stated:

> "It is not uncommon for defendants and employers to investigate personal injury and workers' compensation claims. Because of the public interest in exposing fraudulent claims, a plaintiff or claimant must expect that a reasonable investigation will be made after a claim is filed. It is only when an investigation is conducted in a vicious or malicious manner not really limited and designated to obtain information needed for the defense of a legal claim or deliberately calculated to frighten or torment that the courts will not countenance it.[7]

Publicity given to private life

Under this third category, publicity given to private life, one can be subject to liability for invasion of privacy when the matter publicized is of the kind that would be highly offensive to a reasonable person and is not of legitimate concern to the public.[8] This tort of public exposure of private facts typically involves disclosure of private facts without the consent of the employee. In the context of employment, it may involve attempts to gain background information about an employer applicant, or disclosure of medical information. Unlike defamation (i.e., libel or slander), truth is not a defense. The disclosure of private facts is generally made to a wide audience. Republication of a private fact already known by the employee to fellow employees does not generally provide a cause of action. However, if an employer communicated to a larger number of people private information about the Plaintiff–Employee in connection with the employee's dismissal, a claim might be established under this theory.[9]

In *Bratt v. International Business Machines Corp.*, the court, applying Massachusetts law, held that the disclosure of information obtained when an employee used IBM's open door internal grievance policy was not an invasion of privacy because the information disclosed was not "intimate" or "highly personal."[10] The court affirmed summary judgment for the employer on this allegation. It held that disclosure of mental problems to supervisors was not an invasion of privacy because they had a legitimate need to know. It reversed summary judgment respecting disclosure of psychiatric problems

---

[7] Id.

[8] Restatement (Second) of Torts §652D.

[9] *Anderson v. Low-Rent Housing Commission* 304 N.W. 2d 239 (Iowa 1981) (False light theory, recovery permitted, public employer).

[10] *Bratt v. International Business Machines Corp.*, 785 Fed. 2d 352 (1st Cir. 1986).

by the company doctor to supervisors. It held that the expectation of privacy was much greater with respect to information disclosed in the doctor–patient setting, particularly when company policy reinforced the employee's expectation that such communication would not be divulged. The court noted that the privacy interest of the employee might be outweighed by the legitimate interest of the employer. A balancing test should be employed by the fact finder."[11]

Publicity placing a person in false light

This claim involves both inaccurate portrayals of private facts, and accurate portrayals where disclosure would be highly objectionable to the ordinary person. Such a claim generally has been difficult to maintain.[12] The key defense is whether the plaintiff has truly been placed in a "false light."

Other privacy issues

Sexual privacy is another topic encompassing a variety of employment related issues such as dating and marriage between employees, dating and sexual relationships with outsiders such as employees in competing companies or customers, extramarital relationships, sexual orientation, and even dress codes. In general, the courts have granted employers wide latitude in adopting policies addressing these issues.

Generally, employees cannot be discharged because of marital status. Few states provide statutory protection regarding marital status, and there is no specific statutory protection under federal law. Obviously, the employer cannot use this as a protected basis for discrimination or in retaliation for an employee exercising his or her rights as recognized under public policy. However, antinepotism policies have generally been upheld.

The theory behind spouses not working together is that it prevents conflict in the workplace, i.e., complaints of favoritism by co-workers, interference with workplace productivity, etc. Employers generally prevail in these types of claims.[13] In addition, many employers have policies forbidding dating among employees, especially between employees and supervisors. Discipline including termination has been upheld by the courts for violations of these policies, even in the face of invasion of privacy suits.[14]

Privacy claims have also failed where employees were fired after continuing friendships with former company officers or employees.[15] Most states, excluding California, have sided with employers when employees are discharged for dating or marrying a competitor's employee.

[11] Id. at 360.
[12] *White v. Fraternal Order of Police*, 707 F. Supp. 579 (D.D.C. 1989).
[13] *Parks v. City of Warner Robbins, GA*, 43 F.3d 609 (11th Cir. 1995), *Wright v. Metro Health Medical Ctr.*, 58 F.3d 1130 (6th Cir. 1995).
[14] *Watkins v. United Parcel Service, Inc.*, 979 F.2d 1535 (5th Cir. 1992).
[15] *Ferguson v. Freedom Forge Corp.*, 604 F.Supp. 1157 (W.D. Pa. 1985).

E-Mail

A fairly new issue that presents itself to most employers is e-mail and the potential for an invasion of privacy claim. Many employees assume their e-mail is private and cannot be accessed by anyone else. When a company then reads their e-mail — either as part of an investigation or for some other reason — the employee might sue for invasion of privacy.

As of this date, the author knows of no such successful suit, but a growing number of employees are bringing such lawsuits for invasion of privacy. Even so, they can become expensive for companies to fight even if they win, and an e-mail policy may go a long way toward preventing the suits from being brought in the first place.

Examples of some of the cases in which employees brought suit include the following:

- An employee sued after being fired for sending an e-mail in which he said he wanted to "kill the back-stabbing bastards" who managed the sales department.[16]
- Two employees at Nissan Motor Corporation were fired for sending e-mail that was critical of the manager.[17]
- A California employee sued after she discovered that employees' e-mail was being monitored.[18]

Even though these cases where brought under state law and dismissed, companies could soon face a rash of suits under a federal statute — the Electronic Communication Privacy Act of 1986.[19] This act prohibits the intentional interception or disclosure of wire, oral, or electronic communications. It does not apply if the interception is made by "the person or entity providing a wire or electronic communication service," so it would probably allow a company to read messages on its own internal e-mail system.

However, as a growing number of company e-mail systems are linked to the internet, it's not clear whether the exception would apply in such a case. The act does allow e-mail to be monitored if one of the parties to the e-mail has consented to the monitoring. Therefore, it would be important for companies who want to monitor their employees' e-mail to protect themselves by getting employees to sign-off on the monitoring in advance.

Privacy — drug testing

Another major issue regarding invasion of privacy concerns substance abuse/drug and alcohol testing. Obviously, it is undisputed that employers have a number of legitimate work-related reasons for wanting and needing

[16] *Smythe v. Pillsbury Co.*, 914 F. Supp. 97 (E.D. Pa. 1996)
[17] *Bourke v. Nissan Motor Corporation*, No. 91 Y. to C. 3979 (L.A. Cty. Superior Court).
[18] *Shoars v. Epsom America*, Nos. 90 S.W.C. 112749, 90 B.C. 7036 (L.A. Cty Superior Court).
[19] 18 USC §2510 et. seq.

to know if employees are using illegal drugs, alcohol, or other potentially harmful substances. The reasons include having a good public corporate image, reducing medical costs, lost productivity, and possible theft incidental to supporting such a habit.

Generally, U.S. Constitutional restrictions against drug testing apply only to public sector employees, as the requisite "state action" is not present for private employers. However, in the future, constitutional claims may increasingly be asserted against private-sector employers in industry subject to government-imposed drug testing requirements.[20]

Under the Fourth Amendment of the U.S. Constitution, courts have found that urinalysis infringes upon one's reasonable expectation of privacy, and thereby constitutes a search and seizure within the meaning of the Fourth Amendment. The courts then balanced the competing interest of the individual's right to privacy against the government's right to investigate misconduct. In the case of *National Treasury Employees' Union*, the U.S. Supreme Court applied a reasonableness requirement of the Fourth Amendment and approved tests performed on employees seeking promotion into highly sensitive areas of the U.S. Customs Service.[21] The courts found the reasonableness standard met because of three criteria:

1. Advance notice was provided to the employees.
2. An elaborate chain of custody and quality-control procedures was employed.
3. Individuals were given the opportunity to resubmit positive tests to a lab of their own choosing.

In another case, where railroad labor organizations filed suit to enjoin regulations promulgated by the Federal Railroad Administrations, which govern drug and alcohol testing of railroad employees, the Supreme Court found:

1. The Fourth Amendment was applicable to drug and alcohol testing.
2. However, due to the compelling government interests served by the regulations, which outweighed the employees' privacy concerns, the drug and alcohol tests mandated or authorized by the regulations were reasonable under the Fourth Amendment even though there was no requirement of a warrant or a reasonable suspicion that any particular employee might be impaired.
3. Thus, suspicionless post-accident testing of trained crews pursuant to a 1985 Federal Railroad Administration Regulation is valid.[22]

---

[20] *Schowengerdt v. General Dynamics*, 823 F.2d 1328 (9th Cir. 1987) (cause of action available under U.S. Constitution against private sector employer providing security services to U.S. Navy under "Federal Act" theory).
[21] *National Treasury Employees Union v. Von Raab*, 489 U.S. 656 (1989).
[22] *Skinner v. Railway Labor Executives' Association*, 489 U.S. 602 (1989).

Some states and at least one municipality have enacted laws that place limits on drug testing in employment. Generally, the issues include reasonable suspicion that an employee is under the influence, chain of custody issues, and guarantees of privacy. In *Wilkinson v. Times Mirror Books*, a California Appellate Court held that state constitutional right to privacy applied to private sector employees, but that the drug testing program did not violate that right because the program was reasonable and the employer had an interest in a drug- and alcohol-free workplace.[23]

Federal statutes have been enacted, such as the Omnibus Transportation Employee Testing Act of 1991[24] and the Drug Free Workplace Act of 1988.[25] Obviously, the public has the right to be secure in the knowledge that individuals employed in industries such as aviation, railroads, and trucking are not human time bombs waiting to go off as they fly an airplane, operate a train, or drive down the interstate in a heavy tractor trailer.

Drug testing must be done as quickly as possible and as accurately as possible. There are testing requirements set forth in the mandatory guidelines for Federal Drug Testing Programs, 53 Fed.Reg. 11, 1979 (April 11, 1988), and these should be followed to the letter. Employers should find a company with a well-established reputation for such testing and set up procedures with guidelines from experts in the field to avoid or minimize liability.

In a tort claim premised upon invasion of privacy for drug testing, courts have centered their inquiry into whether there has been an unreasonable intrusion into an employee's seclusion. Factors include:

1. What type of job the employee performs
2. Whether objective evidence of probable cause exists to suggest an employee is under the influence
3. The methods used to conduct the testing (i.e., does a person watch an employee provide a urine specimen or does the person wait outside the bathroom door).[26]

However, at least the 6th Circuit has ruled the right of privacy not to be implicated if the employer has a *bona fide* right to investigate.[27]

Defamation

As professionals are increasingly involved in issues such as drug testing, the potential for defamation actions are increasing. Defamation occurs when an untrue statement is communicated to a third party that tends to harm the

---

[23] *Wilkinson v. Times Mirror Books*, 215 CAL.App. 3d 1034 (1989).
[24] 49 U.S.C. §1834 (App.), 45 U.S.C. §431 (App.), 49 U.S.C. §277 (App.), for aviation, railroads, and trucking, respectively. Testing is authorized for pre-employment, random, reasonable suspicion, periodic, return to work, and post accident situations.
[25] 41 U.S.C. §5151-5160 (1990).
[26] *O'Keefe Passiac Valley Water Com'n.*, 624A. 2d 578, 582-584 (N.J. 1993).
[27] *Baggs v. Eagle-Pitcher Industrial Inc.*, 957 F. 2d 268 (6th Cir. 1992).

reputation of another so as to lower him in the estimation of the community
or to deter third persons from dealing with him. As stated by the Kentucky
Supreme Court in *McCall v. Courier-Journal & Louisville Times*, defamation is
a statement or communication to the third person which tends to:

1. Bring a person into public hatred, contempt, or ridicule
2. Cause him to be shunned or avoided
3. Injure him in his business or occupation[28]

The *prima facie* elements of defamation needed in most jurisdictions are:

a. The statement is false and defamatory
b. The statement is about the plaintiff
c. The statement is published
d. Publication is due to negligent or reckless fault of the defendant
e. Publication was not privileged
f. Publication causes injury to reputation[29]

Publication is an important element of defamation. The publication must
be shown to have been done either negligently or intentionally. Unless, the
employee's communication to the third party was privileged, no actual
malice must be proven. In another case, *Columbia Sussex Corp. v. Hay*,[30] a
hotel manager informed his entire staff that they were suspects following a
robbery in which evidence indicated the crime was an "inside job." Because
the accusation was made before the entire group, the statement was considered published. The hotel manager then subjected the entire staff to polygraph examinations.

In some circumstances, publication of the allegedly defamatory statement may encompass more than oral or written statements communicated
to a third person. Some courts recognize that "acts" can constitute publication of a defamatory statement. In a Pennsylvania case, the court refused to
grant summary judgment because an issue remained as to whether defamatory meanings could be inferred from an employer's actions in terminating
the employee, such as packing up the employee's belongings and changing
the locks on the office door.[31]

Another important aspect is the nature of the words used which have a
bearing on the damages in a defamation case. Words that are harmful by
themselves are considered defamatory *per se*. Injury may be presumed if
defamation *per se* is involved. Most causes of action based on defamation in
the employment relationship concerns statements impugning the character
of an individual or his or her abilities as an employee.

---

[28] *McCall v. Courier–Journal & Louisville Times*, 623 S.W.2d 882, 884 (Ky. 1981).
[29] *Columbia Sussex Corp. v. Hay*, 627 S.W. 2d 270, 273 (Ky. Ct. App. 1982).
[30] Id.
[31] *Doe v. Cohn Nast Ampersand Graf*, 862 F. Supp 1310 (E.D. Pa. 1994).

In *O'Brien v. Papagino's of America*, a jury found that the employer's statement that the plaintiff was terminated for drug use was not completely true. The jury also found that the employer had a retaliatory motive as well. It awarded the plaintiff damages for both defamation and wrongful termination.[32]

Truth is an absolute defense in a defamation action even where the plaintiff asserts that the alleged defamatory statements where inspired by malice and the alleged defamation is *per se* defamatory.[33]

Probably the most common affirmative defense asserted in defamation claims arising from the employment relationship is qualified privilege. The publication is qualified when circumstances exist which cast on the defendant the duty to communicate to certain other parties information concerning the plaintiff. For example, managers within the corporation may disclose to other managers rumors or comments made about employees which are defamatory. However, due to the potential for harm within the workplace setting, courts have found qualified privilege in these situations. If the publication is qualified, the presumption of malice is lost and must be proven by the plaintiff.

Americans with Disabilities Act

The Americans with Disabilities Act of 1990 (known as the ADA) has opened a huge new area of regulatory compliance which will directly or indirectly affect most companies. In a nutshell, the ADA prohibits discrimination against qualified individuals with physical or mental disabilities in all employment settings. Given the impact of the ADA on the assessment of individual employees, especially in the area of mental disability, the ADA is an important law to consider when designing and developing the workplace violence prevention program.

Structurally, the ADA is divided into five titles, and all titles possess the potential of substantially impacting the safety and health function in covered public or private sector organizations. Title I contains the employment provisions, which protect all individuals with disabilities who are in the United States, regardless of national origin and immigration status. Title II prohibits discriminating against qualified individuals with disabilities or excluding qualified individuals with disabilities from the services, programs, or activities provided by public entities. Title II also contains the transportation provisions of the Act. Title III, entitled "Public Accommodations," requires that goods, services, privileges, advantages, and facilities of any public place to be offered "in the most integrated setting appropriate to the needs of the individual."[34]

---

[32] *O'Brien v. Papagino's of America*, 780 F. 2d 1067 (1st Cir. 1986).
[33] *Bell v. Courier-Journal & Louisville Times Co.*, 402 S.W. 2d 84, 87 (Ky. 1966).
[34] ADA §305.

Title III also covers transportation offered by private entities. It addresses telecommunications. Title IV requires that telephone companies provide telecommunication relay services and that television public service announcements produced or funded with federal money include closed caption. Title V includes the miscellaneous provisions. This title noted that the ADA does not limit or invalidate other federal and state laws providing equal or greater protection for the rights of individuals with disabilities and addresses related insurance, alternate dispute, and congressional coverage issues.

Title I prohibits covered employers from discriminating against a "qualified individual with a disability" with regard to job application, hiring, advancement, discharge, compensation, training, and other terms, conditions, and privileges of employment.[35]

> Section 101 (8) defines a "qualified individual with a disability" as any person who, with or without reasonable accommodation, can perform the essential functions of the employment position that such individual holds or desires...consideration shall be given to the employer's judgment as to what functions of a job are essential, and if an employer has prepared a written description before advertising or interviewing applicants for the job, this description shall be considered evidence of the essential function of the job.[36]

The Equal Employment Opportunity Commission (EEOC) provides additional clarification as to this definition in stating "an individual with a disability who satisfies the requisite skill, experience, and educational requirements of the employment position such individual holds or desires, and who, with or without reasonable accommodation, can perform the essential functions of such position."[37]

Under the ADA, an individual has a disability if he/she has:

a. A physical or mental impairment that substantially limits one or more of the major life activities of such individual
b. A record of such an impairment
c. Is regarded as having such an impairment[38]

For an individual to be considered "disabled" under the ADA, the physical or mental impairment must limit one or more "major life activities." Under the U.S. Justice Department's regulations issued for Section 504 of the Rehabilitation Act, "major life activities" is defined as, "functions such as

---

[35] Id.
[36] ADA §101 (8).
[37] EEOC Interpretive Rules, 56 *Fed. Reg.* 35 (July 26, 1991).
[38] Subtitle A, §3(2). The ADA departed from the Rehabilitation Act of 1973 and other legislation is using the term "disability" rather than "handicap."

caring for one's self, performing manual tasks, walking, seeing, hearing, speaking, breathing, learning, and working."[39] This definition includes neither simple physical characteristics, nor limitations based on environmental, cultural, or economic disadvantages.[40]

Of particular concern in the area of workplace violence is the treatment of the disabled individual, who, as a matter of fact or due to prejudice, is believed to be a direct threat to the safety and health of others in the workplace. To address this issue, the ADA provides that any individual who poses a *direct threat* to the health and safety of others, that cannot be eliminated by reasonable accommodation, may be disqualified from a particular job.[41] The term "direct threat" to others is defined by the EEOC as meaning "a significant risk of substantial harm to the health and safety of the individual or others that cannot be eliminated by reasonable accommodation."[42] The determining factors professionals should consider in making this determination include the duration of the risk, the nature and severity of the potential harm, and the likelihood the potential harm will occur.[43]

Additionally, professionals should consider the EEOC's Interpretive Guidelines which state:

> "If an individual poses a direct threat as a result of a disability, the employer must determine whether a reasonable accommodation would either eliminate the risk or reduce it to an acceptable level. If no accommodation exists that would either eliminate the risk or reduce the risk, the employer may refuse to hire an applicant or may discharge an employee who poses a direct threat."[44]

Professionals should note that Title I additionally provides that if an employer does not make reasonable accommodation for the *known* limitations of a qualified individual with disabilities, it is considered to be discrimination. Only if the employer can prove that providing the accommodation would place an undue hardship on the operation of the employer's business can discrimination be disproved.

Section 101 (9) defines a "reasonable accommodation" as:

a. Making existing facilities used by employees readily accessible to and usable by the qualified individual with a disability" and includes:

---

[39] 28 C.F.R. §41.31. This provision is adopted by and reiterated in the Senate Report at page 22.
[40] *See Jasany v. U.S. Postal Service*, 755 F2d 1244 (6th Cir. 1985).
[41] ADA, §103(b).
[42] EEOC Interpretive Guidelines.
[43] Id.
[44] 56 *Fed. Reg.* 35,745 (July 26, 1991); *Also see Davis v. Meese*, 692 F. Supp. 505 (ED Pa. 1988)(Rehabilitation Act decision).

b. Job restriction, part-time or modified work schedules, reassignment to a vacant position, acquisition or modification of equipment or devices, appropriate adjustments or modification of examinations, training materials, or policies, the provisions of qualified readers or interpreters and other similar accommodations for . . . the QID.[45]

The EEOC further defines "reasonable accommodation" as:

1. Any modification or adjustment to a job application process that enables a qualified individual with a disability to be considered for the position such qualified individual with a disability desires, and which will not impose an undue hardship on the . . . business; or
2. Any modification or adjustment to the work environment, or to the manner or circumstances which the position held or desired is customarily performed, that enables the qualified individual with a disability to perform the essential functions of that position and which will not impose an undue hardship on the . . . business; or
3. Any modification or adjustment that enables the qualified individual with a disability to enjoy the same benefits and privileges of employment that other employees enjoy and does not impose an undue hardship on the . . . business.[46]

In essence, the company is required to make "reasonable accommodations" for any/all known physical or mental limitations of the qualified individual with a disability unless the employer can demonstrate that the accommodations would impose an "undue hardship" on the business or the particular disability directly affects the safety and health of the qualified individual with a disability or others. Included under this section is the prohibition against the use of qualification standards, employment tests, and other selection criteria that tend to screen out individuals with disabilities, unless the employer can demonstrate the procedure is directly related to the job function. In addition to the modifications to facilities, work schedules, equipment, and training programs employers should initiate an "informal interactive (communication) process" with the qualified individual to promote voluntary disclosure of specific limitations and restrictions by the qualified individual to enable the employer to make appropriate accommodations to compensate for the limitation.[47]

Companies may require medical examinations only if the medical examination is specifically job related and is consistent with business necessity. Medical examinations are permitted only after the applicant with a disability has been offered the job position. The medical examination may be given before the applicant starts the particular job, and the job offer may be con-

---

[45] ADA §101 (9).
[46] EEOC Interpretive Guidelines.
[47] Id.

ditioned on the results of the medical examination if all employees are subject to the medical examination and information obtained from the medical examination is maintained in separate confidential medical files. Employers are permitted to conduct voluntary medical examinations for current employees as part of an ongoing medical health program, but again the medical files must be maintained separately and in a confidential manner.[48]

The ADA does not prohibit companies from making inquiries or requiring medical examinations or "fit for duty" examinations when there is a need to determine whether an employee is still able to perform the essential functions of the job or where periodic physical examinations are required by medical standards or federal, state, or local law.[49]

Of particular importance to safety and health professionals is the area of controlled substance testing. Under the ADA, the employer is permitted to test job applicants for alcohol and controlled substances prior to an offer of employment under Section 104 (d). The testing procedure for alcohol and illegal drug use is not considered a medical examination as defined under the ADA. Employers may additionally prohibit the use of alcohol and illegal drugs in the workplace and may require that employees not be under the influence while on the job. Employers are permitted to test for alcohol and controlled substance use by current employees in their workplace to the limits permitted by current federal and state law. The ADA requires all employers to conform to the requirements of the Drug-Free Workplace Act of 1988. Thus, professionals should be aware that most existing pre-employment and post-employment alcohol and controlled substance programs which are not part and parcel of the pre-employment medical examination or ongoing medical screening program will be permitted in their current form.[50]

Individual employees who choose to use alcohol and illegal drugs are afforded no protection under the ADA; however, employees who have successfully completed a supervised rehabilitation program and are no longer using or addicted are offered the protection of a qualified individual with a disability under the ADA.[51]

The EEOC has also noted that it is "unlawful . . . to participate in a contractual or other arrangement or relationship that has the effect of subjecting the covered entity's own qualified applicant or employee with a disability to discrimination." This prohibition includes referral agencies, labor unions (including collective bargaining agreements), insurance companies and others providing fringe benefits, and organizations providing training and apprenticeships.[52]

---

[48] ADA §102(c)(2)(A).

[49] EEOC Interpretive Guidelines, 56 *Fed. Reg.* 35,751 (July 26, 1991). Federally mandated periodic examinations include such laws as the Rehabilitation Act, Occupational Safety and Health Act, Federal Coal Mine Health Act, and numerous transportation laws.

[50] ADA §102(c).

[51] ADA §511(b).

[52] Id.

Professionals should note that the ADA has no recordkeeping requirements, has no affirmative action requirements, and does not preclude or restrict antismoking policies. Additionally, the ADA has no retroactivity provisions.

The ADA has the same enforcement and remedy scheme as Title VII of the Civil Rights Act of 1964 as amended by the Civil Rights Act of 1991. Compensatory and punitive damages (with upper limits) have been added as remedies in cases of intentional discrimination, and there is a correlative right to a jury trial. Unlike Title VII, there is an exception where there exists a good faith effort at reasonable accommodation.[53]

Additionally, the enforcement procedures adopted by the ADA mirror those of Title VII of the Civil Rights Act. A claimant under the ADA must file a claim with the EEOC within 180 days from the alleged discriminatory event or within 300 days in states with approved enforcement agencies such as a Human Rights Commission. These are commonly called dual-agency states or Section 706 agencies. The EEOC has 180 days to investigate the allegation and to sue the employer or issue a right-to-sue notice to the employee. The employee will have 90 days from the date of this notice to file a civil action.

Compensatory and punitive damages were also made available where intentional discrimination is found. Damages may be available to compensate for actual monetary losses, for future monetary losses, and for mental anguish and inconvenience. Punitive damages are also available if an employer acted with malice or reckless indifference. The total amount of punitive damages and compensatory damages for future monetary loss and emotional injury for each individual is limited, based upon the size of the employer.

In summation, prudent professionals must exercise caution in the design and development of workplace violence prevention programs in order not to violate an individual's rights. Knowledge of the protections provided to individuals under statutory and case law are of great importance in order that discrimination of violation can be avoided, while achieving the purpose of the workplace violence program. Preplanning and preparation are key to the success of the workplace violence prevention program. However, the violation of an individual's rights can be the torpedo that sinks the program.

---

[53] Civil Rights Act of 1991, §102.

(Modified for the purposes of this text)

HILLCREST FOODS, INC.
v.
KIRITSY.
No. A97A0853.
Court of Appeals of Georgia.
July 11, 1997.
Reconsideration Denied July 24, 1997.

Restaurant customer who was injured in drive-by shooting by waitress' husband brought premises liability action against owner. The Superior Court, Clarke County, Stephens, J., denied owner's summary judgment motion, and owner appealed. The Court of Appeals, Blackburn, J., held that drive-by shooting was not foreseeable to owner so as to impose a duty to protect customer from harm that occurred.

Judgment reversed.

————————————

Chambers, Mabry, McClelland & Brooks, Genevieve L. Frazier, Jo B. Gosdeck, Atlanta, for appellant.

Gorby & Reeves, Michael J. Gorby, Mary D. Peters, Atlanta, Blakely H. Frye, Marietta, for appellee.

BLACKBURN, Judge.

[1] By interlocutory appeal, Hillcrest Foods, Inc. d/b/a Waffle House (Hillcrest) contests the denial of its motion for summary judgment as to premises liability and punitive damages in Matthew Kiritsy's action for personal injury damages. This case arises out of a drive-by shooting at a Waffle House where Kiritsy was a patron. The trial court granted Hillcrest's motion for summary judgment on Kiritsy's claim for negligent retention of its employee Letitia Johnson, the intended victim, who was the wife of the shooter, Nathaniel Johnson. The trial court did not rule on Hillcrest's motion for summary judgment as to punitive damages, which is therefore deemed denied. See *Kim v. Tex Financial Corp.*, 223 Ga.App. 528, 529(2), 479 S.E.2d 375 (1996).

The standard of review of the denial of a defendant's motion for summary judgment is a *de novo* review of the evidence of record with all reasonable inferences therefrom viewed in the light most favorable to the nonmoving party. The purpose of the review is to determine whether there remains a question for jury determination as to at least one material fact upon which plaintiff's case rests. If defendant points out the failure of plaintiff to establish any single element essential to the claimed cause of action, then defendant is entitled to summary judgment as to such claim.

"A *defendant* may [establish entitlement to summary judgment] by showing the court that the documents, affidavits, depositions and other evidence in the record reveal that there is no evidence sufficient to create a jury issue on at least one essential element of plaintiff's case. *If there is no evidence sufficient to create a genuine issue as to any essential element of plaintiff's claim, that claim tumbles like a house of cards. All of the other disputes of fact are rendered immaterial.* A defendant who will not bear the burden of proof at trial need not affirmatively disprove the nonmoving party's case; instead, the burden on the moving party may be discharged by pointing out by reference to the affidavits, depositions and other documents in the record that there is an absence of evidence to support the nonmoving party's case. If the moving party discharges this burden, the nonmoving party cannot rest on its pleadings, but rather must point to specific evidence giving rise to a triable issue. OCGA § 9-11-56(e)." (Citation omitted; emphasis in original and supplied.) *Lau's Corp. v. Haskins*, 261 Ga. 491, 405 S.E.2d 474 (1991); see *Artlip v. Queler*, 229 Ga.App. 775, 776-777, 470 S.E.2d 260 (1996).

It is undisputed that while Kiritsy was a patron at the Waffle House on West Broad Street in Athens at approximately 1:00 a.m. on September 16, 1994, three or four shots were fired from a gun in a vehicle traveling on West Broad Street. Kiritsy was struck by two bullets and is paralyzed from the waist down. Although no arrests have been made, a witness to the shooting, who had been a passenger in the vehicle from which the shots were fired, testified in his deposition that Nathaniel Johnson fired the shots in an attempt to kill his wife, Letitia Johnson, a waitress at the Waffle House.

1. Hillcrest maintains that the trial court erred in failing to grant its motion for summary judgment with regard to Kiritsy's claims for damages based upon premises liability. Hillcrest contends that: there is no causal relationship between its conduct and the criminal act of a drive-by shooting; there is no evidence of prior, substantially similar crimes on its property or the approaches thereto and therefore the shooting was not foreseeable; and that there was no reasonable action which it could have taken which would have prevented the incident in any event.

Hillcrest filed its motion for summary judgment on all counts and outlined the record upon which it relied. The trial court granted summary judgment to Hillcrest on plaintiff's claim that Hillcrest was liable for his injuries based upon its negligent retention of Letitia Johnson. Kiritsy contends that Hillcrest negligently retained Johnson after it became aware of prior disputes between Johnson and her husband.

In its order, the trial court found that "plaintiff alleges that the direct cause of his injuries was the intervening criminal act of a third party, Nathaniel Johnson" and that "it violates public policy and is patently absurd to suggest that, because Letitia Johnson was a victim of domestic violence, she should have been terminated." The trial court held that a question of fact regarding foreseeability remained as to the premises liability claim in that prior similar criminal acts are not the only way of proving foreseeability, relying on *Doe v. Prudential-Bache/A.G. Spanos Realty, etc.*, 222 Ga.App. 169, 474 S.E.2d 31 (1996) and *Wallace v. Boys Club of Albany, etc.*, 211 Ga.App. 534, 439 S.E.2d 746 (1993).

The trial court stated in its denial order that "[t]he shooting of the Plaintiff was a crime against the person. Plaintiff presented evidence of prior crimes against persons committed on the premises as follows: 1) December 8, 1991, Tamra White was punched in the face; 2) July 21, 1992, an employee of the Waffle House was the victim of an aggravated assault and attempted rape; 3) July 20, 1993, an unidentified person attempted to run over John Batch with an automobile; 4) January 1, 1994, an unidentified male robbed a patron at gunpoint on the premises; 5) February 6, 1994, Keith Hodapp was attacked on Defendant's premises by Trey Ford and struck with fists; 6) March 20, 1994, a fight broke out at the Waffle House wherein the victim was punched in the face resulting in the offender being charged with simple battery, impersonating a law enforcement officer and giving false information. There was also evidence from employee Sarah Collins that fist fights and arguments on the premises were quite common." For purposes of summary judgment analysis, we accept all these findings as true, such view being most favorable to the nonmovant.

[2, 3] In order for plaintiff to establish the negligent tort claim, it must show that Hillcrest violated its duty to plaintiff, and that such violation was the proximate cause of plaintiff's injuries. See OCGA § 51-1-1. Hillcrest's customers, having been induced to come onto its property by implied invitation, are invitees. The duty of a proprietor to an invitee is to "exercise ordinary care in keeping the premises and approaches safe." OCGA § 51-3-1. "The proprietor is not the insurer of the invitee's safety, but is bound to exercise ordinary care to protect the invitee from unreasonable risks of which he or she has superior knowledge." (Citation omitted.) *Lau's Corp.*, supra at 492(1).

[4] In order to establish liability of a proprietor for the independent criminal acts of another which result in injury to one of its customers, it must generally be shown that the subject criminal act occurred on the premises or approaches of the proprietor's property or on adjacent property over which the proprietor has exercised some degree of control, that the proprietor had reasonable notice of prior, similar criminal acts of such a nature that the proprietor was put on notice of the likelihood of the subject criminal act being committed, and that given such notice, the proprietor could reasonably have acted thereon to protect its customers. See OCGA § 51-3-1; *Sturbridge Partners, Ltd. v. Walker*, 267 Ga. 785, 482 S.E.2d 339 (1997); and *Lau's Corp.*, supra.

[5]   We first address the issue of whether Hillcrest's duty under OCGA § 51-3-1 to protect its customers while on its premises or approaches extends to acts which are initiated from a public thoroughfare over which Hillcrest has no authority and has not otherwise exercised control. We have held that a proprietor "has a duty, when [it] can reasonably apprehend danger to a customer from the misconduct of other ... *persons on the premises*, to exercise ordinary care to protect the customer from injury caused by such misconduct." (Citation and punctuation omitted; emphasis supplied.) *Adler's Package Shop v. Parker*, 190 Ga.App. 68, 69(1)(a), 378 S.E.2d 323 (1989).

Although the premises owner has a duty to keep the premises and *approaches* safe, see OCGA § 51-3-1, "[i]f the approach is a public way, the occupier's duty ... is to exercise due care within the confines of his right in the public way." (Punctuation omitted.) *Reed v. Ed Taylor Constr. Co.*, 198 Ga.App. 595, 597, 402 S.E.2d 346 (1991).

[6]   No evidence has been introduced showing that Hillcrest used the public road in front of the Waffle House in any manner other than that consistent with the general motoring public. The premises owner's duty to keep the *approaches* safe does not generally extend to the busy public thoroughfare on which the premises is located, where the owner has no legal right to control such thoroughfare. OCGA § 51-3-1 generally contemplates approaches over which the proprietor has a right of control. Hillcrest had no right to control the use of the public road adjacent to the Waffle House and thus no duty arose to it based upon its use or control thereof. See *Rischack v. City of Perry*, 223 Ga.App. 856, 479 S.E.2d 163 (1996).

[7]   The evidence indicates that several prior crimes against persons, as outlined above, had occurred on Hillcrest's property during the three years prior to the drive-by shooting. None of the prior crimes involved a drive-by shooting, and there is nothing in the record to put Hillcrest on notice of the likelihood of a future drive-by shooting.

[8]   Kiritsy contends that Hillcrest should have anticipated Johnson's criminal act because it had knowledge of an incident at the Waffle House in which Johnson

had threatened to shoot a male employee who had been
harassing Johnson's wife. That incident occurred over
five months prior to the drive-by shooting at issue.
Johnson did not display a gun during that verbal incident,
and the male employee was fired for harassment, thereby
removing any further threat of such an incident occurring.
While this might show Johnson's violent verbal tendencies
in defense of his wife, his threats to the male employee
did not put Hillcrest on notice that Johnson might
attempt to kill his wife, or that he was physically
violent toward his wife or Hillcrest's customers. Plaintiff contends that the tumultuous relationship between
the Johnsons was known to Hillcrest. Even if this were
so, there is nothing in such relationship which would
have put defendant on notice of the likelihood of a
drive-by shooting.

A drive-by shooting is a totally different crime from
those in which all actions constituting the crime occur
on the defendant's property. In this drive-by shooting,
the perpetrator was not even on the defendant's property,
but rather was traveling on a busy public thoroughfare.
It is difficult to imagine what effective action Hillcrest
could reasonably have taken which could have prevented a
drive-by shooting even had there been a prior such event.

Georgia courts have not dealt previously with a
proprietor's duty in a drive-by shooting case; therefore,
we look to other jurisdictions who have addressed this
situation. In *Thai v. Stang*, 214 Cal.App.3d 1264, 263
Cal.Rptr. 202, 206 (1989), the California Court of Appeals
determined that a drive-by shooting in a roller skating
rink parking lot was an unforeseeable criminal attack
by a third party. The Court determined that despite
evidence of prior crimes against persons on the owner's
property, "the random nature of drive-by shootings ...
makes them difficult to police against." *Id.*, 263
Cal.Rptr. at 207. The court questioned the property
owner's ability to have prevented the injuries sustained
by the plaintiff under such circumstances, and it noted,
"[w]hile we do not minimize the policy of preventing
future harm, imposing a duty to protect against drive-
by shootings would place an onerous burden on business
owners. Moreover, when the burden of preventing future
harm is great, a high degree of foreseeability is
required." *Id.*

The trial court's reliance on *Wilks v. Piggly Wiggly Southern*, 207 Ga.App. 842, 429 S.E.2d 322 (1993) as authority for the proposition that the law does not require that the criminal act occur on the premises or approaches of the proprietor's premise in order to establish foreseeability is misplaced. In *Wilks*, the victim was mugged around the corner from the defendant's property. *Id.* We reversed the trial court's grant of the property owner's motion for summary judgment, however, because the property owner had allowed the attackers to loiter on his premises waiting for victims. *Id.* at 844, 429 S.E.2d 322. *Wilks* is factually distinguishable from the present case and does not require a different result.

The trial court's reliance on *Doe* (a case involving a rape which occurred in a parking garage where the court acknowledged that if the defendant admitted the existence of a dangerous condition, foreseeability may be established even in the absence of a prior substantially similar criminal act) and on *Wallace* (A case involving the failure to properly supervise a child who was injured after wandering off from the premises of the defendant) is also misplaced as the cases are factually distinguishable. There has been no admission by Hillcrest in the instant case and the criminal conduct was not initiated on its property. Therefore, *Doe* is inapplicable to the present analysis. *Wallace* was not a true premises liability case, as it involved a failure to supervise a child, and has no application to the present case.

[9] Without foreseeability that a criminal act will occur, no duty on the part of the proprietor to exercise ordinary care to prevent that act arises. Our Supreme Court has reaffirmed the requirement of substantially similar criminal acts to establish the foreseeability of the proprietor in *Sturbridge Partners, Ltd. v. Walker*, supra. There, the Supreme Court affirmed this Court's holding that notice of prior burglaries of vacant apartments was sufficient to make foreseeable to the landlord the potential for violence in the event of the burglary of an occupied unit. Off-premises conduct or injury was not involved in *Walker*, and it does not change the general law in such cases.

Under the facts of this case, the plaintiff has failed to establish that the drive-by shooting was foreseeable by the defendant. While there may be situations in which off-premises criminal conduct will not preclude a proprietor's liability, any such case would necessarily involve a higher degree of foreseeability, not here present, than would those cases involving on-premises conduct. Although a premises owner must exercise ordinary care to protect its customers from the unreasonable risks of which the proprietor is aware, the owner is the insurer of its customers' safety. See *Lau's Corp.*, supra. The law must be applied uniformly to the parties even when egregious injuries have occurred. Hillcrest did not ensure Kiritsy's safety, and under the facts herein involved, it cannot be responsible for Kiritsy's injuries because the drive-by shooting was not a foreseeable act. It was an act of terrorism that could have occurred anywhere that the intended victim happened to be. Hillcrest had no basis to foresee such event, and there was no effective action which it could reasonably have taken to prevent said act under the circumstances of this case. Hillcrest and Kiritsy are both victims of the despicable shooting for the same reason. That being the perpetrator's decision to shoot Letitia Johnson at the subject time and location. The shooting was a transitory act that could have been carried out at any time and place that the intended victim happened to be.

Although Kiritsy's security expert averred that Hillcrest was negligent in failing to hire security guards, the expert did not indicate how a security guard would have been a deterrent for a drive-by shooting. Furthermore, the expert did not recommend the use of bullet-proof glass.

The trial court erred in denying Hillcrest's motion for summary judgment on Kiritsy's claims based on premises liability. Even assuming Hillcrest had knowledge that the Johnsons had a tumultuous martial relationship, we cannot say that such information translates into notice of potential danger to customers. For to do so would jeopardize the job of every employed victim of domestic violence. Nor can we say that verbal confrontations are indicative of a future drive-by shooting. To so hold would threaten the job security of every person in an abusive relationship.

Our determination regarding Hillcrest's duty makes it unnecessary to address Kiritsy's additional arguments.

2. Because of our holding in Division 1, Hillcrest is not liable for punitive damages and the trial court erred in denying Hillcrest's motion for summary judgment on this issue.

Judgment reversed.

POPE, P.J., and JOHNSON, J., concur.

(Modified for the purposes of this text)

Robin K. WHITE
v.
RANSMEIER & SPELLMAN.
Civil No. 95-626-JD.
United States District Court.
D. New Hampshire.
Oct. 10, 1996.

Employee brought action against her employer, alleging that she was terminated in violation of the Americans with Disabilities Act (ADA) and Title VII, and employer counterclaimed for damages. On employee's motion to dismiss counterclaims, the District Court, DiClerico, Chief Judge, held that: (1) employer failed to state claim against employee for intentional interference with contractual relations under New Hampshire law; (2) no claim would be recognized under New Hampshire law for prima facie tort; and (3) employer's claim against at-will nonmanagerial employee was outside scope of any cause of action for breach of duty of loyalty recognized under New Hampshire law.

Motion to dismiss granted.

Andrea P. Thorn, Andover, MA.

Robert J. Gilbert, Andover, MA.

Garry R. Lane, Concord, NH.

ORDER

DiCLERICO, Chief Judge.

The plaintiff, Robin White, brought this action seeking damages related to her termination from her position as a legal secretary with the defendant law firm, Ransmeier & Spellman. The defendant filed a five-count counterclaim against the plaintiff, seeking damages from the events that precipitated the plaintiff's termination. Before the court is the plaintiff's consolidated motion to dismiss the defendant's counterclaims and motion to strike certain portions of the defendant's prayer for relief (document no. 9).

Background[1]

The plaintiff was employed by the defendant as a legal secretary from August 1989 until her termination on November 1, 1994. It is not disputed that she performed her job duties adequately throughout the duration of her employment.

In March 1994, the plaintiff informed one of the lawyers at the firm that she wanted to leave her job because she had had an affair with Richard Meaney, the firm's legal administrator. The matter was brought to the attention of at least one member of the firm's executive committee, and the situation apparently was diffused when Meaney submitted his resignation on or about June 1, 1994.

However, in the weeks that followed Meaney's resignation, the plaintiff began to feel that other employees of the firm were blaming her for Meaney's departure. The plaintiff harassed one employee who had been friendly with Meaney by making phone calls to the employee's home late at night and then hanging up, by sneaking up behind the employee, and by staring at her without saying anything. The plaintiff also stared threateningly at other female employees, and on one occasion, bumped an employee as she was walking down a flight of stairs.

Meanwhile, in August 1994, Meaney secured a position as a legal administrator with Greeley, Walker & Kowan, a Honolulu, Hawaii law firm. On September 2, 1994, an unidentified female caller, believed to be the plaintiff, placed a telephone call to inform the lawyers at Greeley, Walker that a bomb had been planted at their offices. In the next several weeks, a caller also believed to be the plaintiff made phone calls threatening employees at Greeley, Walker; placed "hang-up" phone calls to Meaney's wife, who had not yet left for Hawaii; made harassing calls to Meaney in Hawaii; and called Meaney's daughters' schools, informing school administrators that "Dick Meaney is going to have his house burned down with this kids inside it," and that one of Meaney's daughters was

---

[1] The facts relevant to the instant motion either are not in dispute or have been alleged by the defendant.

about to be murdered because of a grudge against her father. Ransmeier & Spellman was informed of these actions.

The plaintiff's conduct continued into October 1994, during which the plaintiff made death threats to two Ransmeier & Spellman employees and continued to harass and threaten Greeley, Walker and its employees. After consulting with the police, the New Hampshire Attorney General, the Greeley, Walker firm, and an unspecified number of workplace violence experts, the firm decided to terminate the plaintiff.

On December 29, 1995, the plaintiff commenced the instant action against Ransmeier & Spellman, alleging that she had been terminated in violation of the Americans with Disabilities Act, Title VII of the Civil Rights Act of 1964. N.H.Rev.Stat. Ann. ("RSA")§ 354-A, and, in addition, asserting a variety of common-law theories. On March 20, 1996, the defendants filed an answer and five-count counterclaim, seeking, *inter alia*, damages, enhanced compensatory damages, and attorney's fees, and alleging (1) interference with business relations; (2) prima facie tort; (3) intentional tort to cause harm; (4) breach of the implied covenant of good faith and fair dealing; and (5) breach of duty of loyalty.

DISCUSSION

I. *Motion to Dismiss*

A motion to dismiss under Fed.R.Civ.P. 12(b)(6) is one of limited inquiry, focusing not on "whether a [claimant] will ultimately prevail but whether the claimant is entitled to offer evidence to support the claims," *Scheuer v. Rhodes*, 416 U.S. 232, 236, 94 S.Ct. 1683, 1686, 40 L.Ed.23 90 (1974). Accordingly, the court must take the factual averments contained in the defendant's counterclaim as true, "indulging every reasonable inference helpful to the [defendant's] cause." *Garita Hotel Ltd. Partnership v. Ponce Fed. Bank*, 958 F.2d 15, 17 (1st Cir.1992); *see also Dartmouth Review v. Dartmouth College*, 889 F.2d 13, 16 (1st Cir.1989). In the end, the court may grant a motion to dismiss under Rule 12(b)(6) "'only if it clearly appears, according to the facts alleged, that the [defendant] cannot recover on any viable theory.'" *Garita*, 958 F.2d at 17 (quoting *Correa-Martinez v. Arrillaga-Belendez*, 903 F.2d 49, 52 (1st Cir.1990)).

A. Interference with Contractual Relations

[1] Although the defendant has styled count I of its counterclaim as interference with business relations, its own citations to *Demetracopoulos v. Wilson*, 138 N.H. 371, 640 A.2d 279 (1994), and *Jay Edwards, Inc. v. Baker*, 130 N.H. 41, 534 A.2d 706 (1987) indicate that the claim is properly brought under the rubric of intentional interference with contractual relations. To succeed on such a theory, Ransmeier & Spellman must show that the plaintiff improperly and intentionally interfered with an existing contractual relationship between Ransmeier & Spellman and a third party. *See Demetracopoulos*, 138 N.H. at 373-74, 640 A.2d at 281; *Montrone v. Maxfield*, 122 N.H. 724, 726, 449 A.2d 1216, 1217 (1982); *Restatement (Second) of Torts* § 766 (1979).

[2] Ransmeier & Spellman's claims hinge on its assertions that the plaintiff improperly interfered with the firm's relationships with its clients and its employees by engaging in conduct that she knew was substantially certain to cause a loss of billable hours and to hinder the firm's employees from performing their duties. *See* Memorandum in Support of Objection to Motion to Dismiss and to Strike ("Defendant's Memorandum") at 9-10. However, it has failed to allege any facts demonstrating that the plaintiff's conduct caused Ransmeier & Spellman not to perform its contractual obligations to third parties.[2] As such, Ransmeier & Spellman has failed to state a claim for interference with contractual relations.

The motion to dismiss count I of the defendant's counterclaim is granted.

B. Prima Facie Tort & Intentional Tort to Cause Harm

[3] Counts II and III of the defendant's counterclaim seek relief for conduct that was "outrageous[,]

intentional, ... and caused harm to the firm." Defendant's Memorandum at 11. Although some jurisdictions have recognized various versions of "umbrella" liability for intentional torts, *see, e.g., Gray v. Bicknell*, 86 F.3d 1472, 1481 (8th Cir.1996) (elements of prima facie tort under Missouri law); *Twin Laboratories, Inc. v. Weider Health & Fitness*, 900 F.2d 566, 571 (2d Cir.1990) (elements of prima facie tort under New York law); *see Restatement (Second) of Torts* § 870 (1979), New Hampshire has not recognized such a cause of action. In light of the New Hampshire Supreme Court's silence on this issue and the potential breadth of the theories the defendant has presented, the court declines to recognize these causes of action.

The plaintiff's motion to dismiss counts II and III of the counterclaim is granted.

C. Breach of the Implied Covenant of Good Faith and Fair Dealing

[4, 5]   In count IV of its counterclaim, the defendant alleges that the plaintiff violated the implied covenant of good faith and fair dealing in her at-will employment contract by engaging in behavior that was "inconsistent with common standards of decency, fairness, and reasonableness and with the parties' agreed-upon common purposes and justified expectations." Defendant's Memorandum

---

[2] The court notes that the New Hampshire Supreme Court has not adopted § 766A of the Restatement (Second) of Torts, which creates a cause of action for a party whose performance of a contract with a third party is made more burdensome by the improper and intentional acts of a defendant, and appears not to have extended the tort of interference with contractual relations to cover such circumstances. In the absence of guidance from the New Hampshire Supreme Court, the court declines to recognize such a cause of action. *Accord Gemini Physical Therapy & Rehab., Inc. v. State Farm Mut. Auto. Ins. Co.*, 40 F.3d 63, 66 (3d Cir.1994)(declining to recognize cause of action based on § 766A under Pennsylvania law); *see also Price v. Sorrell*, 784 P.2d 614, 616 (Wyo.1989)(causing contract to be more costly to perform "too speculative and subject to abuse to provide a meaningful basis for a cause of action"). *See generally Windsor Sec., Inc. v. Hartford Life Ins. Co.*, 986 F.2d 655, 659-63 (3d Cir.1993)(distinguishing between inducing third party not to perform contract with plaintiff and hindering plaintiff from performing contract with third party).

at 12 (quoting *Centronics Corp. v. Genicom Corp.*, 132 N.H. 133, 140, 562 A.2d 187, 191 (1989)). However, the theory upon which the defendant relies does not convert every potentially tortious act between parties to a contract into a breach of the implied covenant. Rather, it only permits recovery "under an agreement that appears by word or silence to invest one party with a degree of discretion in performance sufficient to deprive another party of a substantial proportion of the agreement's value." *Id.* at 143, 562 A.2d at 193; *see, e.g., Griswold v. Heat Inc.*, 108 N.H. 119, 124, 229 A.2d 183, 187 (1967) (contract under which party was to provide "such services, as he, in his sole discretion, may render" obligated party to provide a level of services consistent with good faith); *Howtek v. Relisys*, 94-297-JD, slip op. at 6-7 (D.N.H. Feb. 1, 1996) (express agreement between designer and manufacturer to negotiate in good faith for manufacture of additional products requires designer to inform manufacturer of its intention to market new items in related field and to possess genuine willingness to entertain reasonable offers to manufacture such products).

Unlike *Griswold* or *Howtek*, the at-will employment contract between the plaintiff and defendant neither expressly nor implied granted the plaintiff the discretion to engage in conduct that could have frustrated the defendant's purpose in contracting. Indeed, the conduct in which the plaintiff is alleged to have engaged was wholly independent of her obligations under her employment contract. Thus, regardless of whether on balance the plaintiff's conduct outweighed her contributions to the firm, her conduct could not have deprived the defendant of the benefit of the bargain of its employment contract with the plaintiff. Although the alleged conduct is, by all accounts, inconsistent with common standards of decency and may give rise to some form of tort or criminal liability, the plaintiff's allegations are insufficient to justify a damage award for breach of the implied covenant of good faith and fair dealing.

The plaintiff's motion to dismiss count IV of the counterclaim is granted.

D. Breach of Loyalty

[6] In count V the defendant seeks relief for the plaintiff's breach of her duty of loyalty. This court has recognized that under New Hampshire law:

> an employee holding a position of trust and confidence, such as a supervisor, manager, director, or officer, owes a fiduciary duty of loyalty to her employer. The duty demands that the employee act solely for the benefit of the employer, never to the employer's detriment. Detrimental behavior could include misappropriating a business opportunity of the employer, use of confidential information, or soliciting clients of the company for the employee's competing business.

Liberty Mutual Ins. v. Ward, 93-610-L, 1994 WL 269283, slip op. at 9 (D.N.H. July 11, 1994). The defendant's claim, which seeks recovery for acts of an at-will, nonmanagerial employee, falls outside the scope of *Ward's* conception of a breach of duty of loyalty, and beyond the scope of any cause of action for breach of the duty of loyalty recognized by the New Hampshire Supreme Court.

The plaintiff's motion to dismiss count V of the counterclaim is granted.

II. *Motion to Strike*

The plaintiff has moved to stroke paragraph C of the defendant's prayer for relief, which seeks an award of "damages, enhanced compensatory damages, and attorney's fees on Ransmeier & Spellman's counterclaims." As the court has already dismissed the counterclaims, the motion to strike paragraph C is moot.

The plaintiff also has moved to strike paragraph E of the defendant's prayer for relief, which asks the court to "order [the plaintiff's] attorney's to pay the legal costs and expenses of this action due to their lack of a thorough investigation before bringing this action." The defendant has withdrawn its request, acknowledging its failure to comply with Fed.R.Civ.P. 11(c)(1)(A). Accordingly, the motion to strike paragraph E is moot.

CONCLUSION

The plaintiff's motion to dismiss the defendant's counterclaims (document no. 9) is granted. The plaintiff's motion to strike portions of the defendant's prayer for relief (document no. 9) is moot.

SO ORDERED.

(Modified for the purposes of this text)

89 N.Y.2d 172
Ricky BROWN et al., Appellants,
v.
STATE of New York, Respondent.
Court of Appeals of New York.
Nov. 19, 1996.

Nonwhite citizens who were allegedly subjected to unconstitutional searches brought action against state, alleging various state and federal constitutional and civil rights violations. The Court of Claims, Hanifin, J., dismissed action, and plaintiffs appealed. The Supreme Court, Appellate Division, 221 A.D.2d 681, 633 N.Y.S.2d 409, affirmed, and appeal was taken. The Court of Appeals, Simons, J., held that: (1) Court of Claims' jurisdiction is not limited to common-law tort causes of action and damage claims against state based upon violations of the State Constitution come within jurisdiction of the Court of Claims; (2) state was not a "person" within meaning of statute providing damage remedy for deprivation of federal constitutional right; (3) cause of action to recovery damages may be asserted against state for violation of equal protection and search and seizure clauses of the State Constitution; and (4) Court of Claims had jurisdiction over claims for negligent training and supervision.

Affirmed as modified.

Bellacosa, J., filed dissenting opinion.

────────────

Whiteman, Osterman & Hanna, Albany (Scott N. Fein, Paul C. Rapp, Philip H. Dixon, D. Scott Bassinson and Lisa M. Codispoti, of counsel), for appellants.

Dennis C. Vacco, Attorney General, Albany (Denise A. Hartman and Peter H. Schiff, of counsel), for respondent.

James A. Gardner, Stephentown, Christopher Dunn and Norman Siegel, New York City, for New York Civil Liberties Union, amicus curiae.

Stenger and Finnerty, Buffalo (Michael L. Jackson of counsel), and Michael Deutsch, New York City, for Center for Constitutional Rights, amicus curiae.

OPINION OF THE COURT

SIMMONS, Judge.

This is a class action brought on behalf of nonwhite males who were stopped and examined by police officers between September 4, 1992 and September 9, 1992 while the police were investigating a crime in the City of Oneonta. The claimants seek monetary damages from the defendant alleging illegal and unconstitutional acts by the State of New York, the New York State Police, the State University of New York and the State University of New York, College at Oneonta (SUCO) and various officers and employees of those entities.

Before answering, the State moved to dismiss alleging that the claim was facially defective because the court lacked subject-matter jurisdiction and the claim failed to state any cause of action. The Court of Claims granted the motion, holding (a) that constitutional torts are not cognizable in the Court of Claims; (b) that direct actions for violations of the New York State Constitution's Bill of Rights, specifically the right to be free from unreasonable searches and seizures and the right to equal protection under the law, are not cognizable claims in any court in the State absent some link to a common-law "traditional" tort; (c) that actions for negligent training and supervision are not cognizable claims in the Court of Claims where the underlying harm — in this case, constitutional violations — are themselves not matters within the court's jurisdiction; and (d) that actions based on 42 U.S.C. § 1981 do not lie against States. The Appellate Division affirmed.

The primary issues presented to this Court are whether, absent either a statute expressly authorizing such claims or a traditional common-law tort theory supporting money damages, the Court of Claims has subject-matter jurisdiction of these tort claims against the State and whether claimants state causes of action against defendant based upon rights secured to them by the State and Federal Constitutions and various State statutes.

The order should be modified. The Court of Claims has subject-matter jurisdiction of the claim. Notwithstanding jurisdiction to hear claims based on 42 U.S.C. § 1981, however, claimants do not state causes of action under that statute. The causes of action seeking damages based upon provisions of the New York Constitution are facially sufficient and should be reinstated.

I.

The claims arise out of an incident occurring in the early morning of September 4, 1992 when a 77-year-old white woman was reportedly attacked at knifepoint in a house outside Oneonta city limits, near the State University campus. The victim described her assailant as a black male and police determined that he may have cut his hand during the alleged attack.

Having failed to identify a suspect during the morning following the attack, the New York State Police and SUCO security personnel prevailed upon SUCO officials to prepare a computer generated list from the University computer system containing the name and address of every African-American male attending the University. Using this list, State Police, SUCO security personnel and local law enforcement officers sought to question each student named on it. African-American students were interrogated in their dormitories, on the SUCO campus, in off-campus apartments and on the streets in and around the City of Oneonta. The interrogations were systematic, consisting of a "stop" followed by questions regarding potential involvement in the incident, requests for alibis, and an inspection of the students' hands and forearms.

When these efforts failed to yield any suspects, the State Policy and local law enforcement officials embarked on a five-day "street sweep" in which every nonwhite male found in and around the City of Oneonta was stopped and similarly interrogated. In the nearly four years since the incident, no one has been arrested for the crime.

Claimants instituted this action asserting that the conduct of the defendants was racially motivated and denied them rights guaranteed by the State and Federal Constitutions.[3]

II.

These claims sound in constitutional tort.[4] Analysis starts by defining what is meant by that term.

[1] A constitutional tort is any action for damages for violation of a constitutional right against a government or individual defendants. Constitutional tort claims were first recognized after the Civil War when Congress authorized civil damage actions against those "who, under color of" State law, or custom have deprived others of constitutional rights (Act of Apr. 20, 1871, ch. 22, § 1, 17 U.S. Stat. 13). Those statutes, now codified in 42 U.S.C. § 1981 *et seq.* remained relatively obscure until the 1961 decision of the Supreme Court in *Monroe v. Pape*, 365 U.S. 167, 81 S.Ct. 473, 5 L.Ed.2d 492. In *Monroe*, the Court held that a plaintiff whose constitutional rights have been infringed by one acting under color of State law can bring a Federal action under Section 1983 even where the State provides an adequate remedy at common law (*but see*, Whitman, *Constitutional Torts*, 79 Mich. L. Rev. 5, 8). The statute was intended to create "a species of tort liability" in favor of persons deprived of their constitutional rights (*see, Carey v. Piphus*, 435 U.S. 247, 253, 98 S.Ct. 1042, 1047, 55 L.Ed.2d 252 [quoting *Imbler v. Pachtman*, 424 U.S. 409, 417, 96 S.Ct. 984, 988, 47 L.Ed.2d 128]).

In addition, in 1971, the Supreme Court recognized a cause of action for damages based upon duties defined in the Federal Constitution (*see. Bivens v. Six Unknown Fed. Narcotics Agents*, 403 U.S. 388, 91 S.Ct. 1999, 29 L.Ed.2d 619). The Court did not predicate recovery on

---

[3] The claim was brought as a class action on behalf of two distinct aggrieved classes. Class I, those persons whose names appeared on the computer generated list wrongfully generated by SUCO officials, and Class II, those persons wrongfully stopped, questioned and examined by law enforcement officials in the absence of any articulable suspicion. Only claims involving Class II are being appealed.

[4] The term "constitutional tort" has been attributed to Professor Marshall Shapo who used it in an article on the subject in the Northwestern University Law Review 35 years ago (*see*, Burnham, *Separating Constitutional and Common Law Torts: A critique and a Proposed Constitutional Theory of Duty*, 73 Minn. L. Rev. 515, n. 2). It is now used generally by courts and commentators.

the civil rights statutes but implied a cause of action for damages based on the guarantees against unlawful searches and seizures contained in the Fourth Amendment. A number of States have similarly recognized causes of action against individuals and governments for constitutional torts based upon local law (*see, e.g., Widgeon v. Eastern Shore Hosp. Ctr.,* 300 Md. 520, 479 A.2d 921; *Gay Law Students Assn. v. Pacific Tel. & Tel. Co.,* 24 Cal.3d 458, 156 Cal.Rptr. 14, 595 P.2d 592; *Phillips v. Youth Dev. Program,* 390 Mass. 652, 459 N.W.2d 453; *Newell v. City of Elgin,* 34 Ill.App.3d 719, 340 N.E.2d 344; *see generally,* Friesen, State Contitutional Law ¶ 7.05[2]; *and see,* ¶ 7.07[1]] [for a list of States viewing favorably damage remedies for violation of State constitutional provisions]).

[2] Although the Supreme Court has drawn on common-law principles to define the scope of liability in these actions, and constitutional and common-law torts frequently protect similar interests, the causes of action are not coextensive (*see generally,* Whitman. *op. cit.,* at 14; Wells and Eaton *Substantive Due Process and the Scope of Constitutional Torts,* 18 Ga. L. Rev. 201, 233; *and see, Carey v. Piphus,* 435 U.S., at 258, 98 S.Ct., at 1049, *supra*). The common law of tort deals with the relation between individuals by imposing on one a legal obligation for the benefit of the other and assessing damages for harm occasioned by to failure to fulfill that obligation (Prosser and Keeton, Torts § 53, at 356 [5th ed.]). Common-law duties arise in virtually all relationships and protect against most risks of harm. Constitutional duties, by contrast, address a limited number of concerns and a limited set of relationships. Constitutions assign rights to individuals and impose duties on the government to regulate the government's actions to protect them. It is the failure to fulfill a stated constitutional duty which may support a claim for damages in a constitutional tort action.

Claimants ask that we recognize a damage remedy against the State based on the New York Constitution as Congress, the Supreme Court and several State courts have done before us based on the Federal and State Constitutions.

III.

Jurisdiction

The first question presented is the jurisdiction of the Court of Claims to entertain constitutional tort claims.

Under the common law, a State is immune from suit unless it waives its sovereign immunity. The provisions applicable here are contained in article VI, § 9 of the State Constitution, which continues the Court of Claims and authorizes the Legislature to determine its jurisdiction, and the Court of Claims Act, which contains the waiver of immunity and the jurisdictional and procedural provisions necessary to implement the constitutional section. The Court of Claims declined to exercise jurisdiction in this case because it believed the statues were not sufficiently broad to waive the State's immunity suit for constitutional torts.

Sovereign immunity has been described as an "outmoded" holdover of the notion that the King can do no wrong (Breuer, The New York State Court of Claims: Its History, Jurisdiction and Reports, at 13). While the State and its agencies must pay for property taken for public purposes, in the absence of consent, immunity is otherwise a complete protection under the common law (*see generally*, Restatement [Second] of Torts § 895B). In the past, New York waived immunity and compensated aggrieved parties for very few claims and they were adjusted by a variety of tribunals with limited jurisdiction. Any others were satisfied, if at all, by private bills addressed to the Legislature's sense of justice. The inequity and inefficiency of such a system became apparent over time and the method for handling claims against the State has gradually evolved to the present system in which jurisdiction over such matter is vested in the Court of Claims (*see*, Breuer, *op. cit.*, at 13 *et seq.* for a history of the subject).

The present Court of Claims Act was adopted in 1939. One commentator observed, it confers jurisdiction on the court to hear and determine "almost every conceivable kind of action against the State" (*see*, Breuer, *op. cit.*, at 23). Subdivision (2) of Section 9 of the present Act confers jurisdiction on the court "[t]o hear and

determine a claim of any person, corporation or municipality against the state for the appropriation of any
real or personal property or any interest therein, for
the breach of contract, express or implied, or for the
torts of its officers or employees while acting as such
officers or employees."

In *Smith v. State of New York*, 227 N.Y. 405, 409-410,
125 N.E. 841, *rearg. denied* 229 N.Y. 571, 129 N.E. 918,
we stated as a general rule that the jurisdiction of
the Court of Claims is to be construed broadly and waiver
of immunity narrowly. Claimant in *Smith* sought damages
from the State for personal injuries allegedly sustained
as the result of the State's negligence. We construed
Section 264 of the Code of Civil Procedure, a predecessor
to Section 9, as granting the Court of Claims jurisdiction
of the matter, stating that its jurisdiction was of the
"broadest character" (at 409, 125 N.E. 841). We denied
liability, however, concluding that although the State
had waived its immunity from suit, it had not waived
its immunity from liability: the Court had jurisdiction
to hear the claim, but the claim failed because the
State had not waived its substantive liability (*id.*, at
409-410, 125 N.E. 841).

The jurisdiction of the Court of Claims is today,
as it was characterized in *Smith*, of the "broadest
character," but the *Smith* Court's interpretation of the
waiver provision of Section 264 was at odds with the
public policy which seeks to reduce rather than increase
the obstacles to recovery of damages, whether defendant
is a private person or a public body (*see, Abbott v.
Page Airways*, 23 N.Y.2d 502, 507, 297 N.Y.S.2d 713, 245
N.E.2d 388; *see also, Bing v. Thunig*, 2 N.Y.2d 656, 666,
163 N.Y.S.2d 3, 143 N.E.2d 3 ["(1)iability is the rule,
immunity the exception"], quoted with approval in *Abbott,
supra*, at 507, n. 2, 297 N.Y.S.2d 713, 245 N.E.2d 388).
Thus, the Legislature subsequently enacted a new statute
to overcome the ruling in *Smith*. That revision, the
substance of which was incorporated into the statute now
before us, "extended, supplemented and enlarged" the
waiver to remove the defense of sovereign immunity for
tort actions (*Jackson v. State of New York*, 261 N.Y.
134, 138, 184 N.E. 735, *rearg. denied* 261 N.Y. 637, 185
N.E. 771; *see*, Breuer, *op. cit.*, at 27). The present
statute provides:

> "[t]he state hereby *waives its immunity
> from liability and action* and *hereby
> assumes liability* and consents to have
> the same determined in accordance with
> the same rules of law as applied to
> actions in the supreme court against
> individuals or corporations" (Court of
> Claims Act § 8 [emphasis added]).

The waiver includes all claims over which the Court of Claims has jurisdiction — appropriation, breach of contract and torts — and applies the rule of respondeat superior to the State (*see*, Court of Claims Act § 9[2]; *Jackson v. State of New York, supra*, at 138, 184 N.E. 735).[5]

The State contends that the waiver is limited to traditional common-law torts. It notes that damage actions under the Federal civil rights statutes, although authorized following the Civil War, were virtually unknown until well after the Court of Claims Act was enacted and damage claims brought directly under the Federal Constitution against Federal officials were not formally recognized until *Bivens* was decided in 1971. Thus, the State reasons, it cannot be said that the Legislature intended to confer jurisdiction upon the Court of Claims to redress constitutional torts when it enacted the present statute.

In attempting to discover the legislative intention, it is well to recognize that the word tort has no established meaning in the law. Broadly speaking, a tort is a civil wrong other than a breach of contract (*see*,

---

[5] The dissent maintains that the Legislature did not intend to broaden the State's existing liability when enacting Section 8 of the Court of Claims Act in 1939 (dissenting opn., at 200, at 240 of 652 N.Y.S.2d, at 1146 of 674 N.E.2d, citing as authority the Memoranda of Judge Barrett and Senator Feinberg). As a general statement, that is true but what the dissent fails to explain is that Section 8 was derived from former Section 12-a, the section which was enacted in 1929 to overcome the deficiencies of the statute interpreted in *Smith v. State, supra*. It is essentially the same statute interpreted by this Court in *Jackson v. State, supra* in 1933. Thus, the 1939 Act did not as Judge Barrett noted, change the substance of the existing law but the law in effect at the time the Court of Claims Act was enacted in 1939 was certainly not the law interpreted in *Smith v. State of New York, supra*, as the dissent would have us believe.

Prosser and Keeton, *op. cit.*, § 1). There are no fixed categories of torts, however, and no restrictive definitions of the term (*see, Advance Music Corp. v. American Tobacco Co.*, 296 N.Y. 79, 70 N.E.2d 401; *see also*, Prosser and Keeton, *op. cit.*) Indeed, there is no necessity that a tort have a name; new torts are constantly being recognized (*see*, the extensive analysis by Justice Breitel, as he then was, in *Morrison v. National Broadcasting Co.*, 24 A.D.2d 284, 266 N.Y.S.2d 406, *revd. on other grounds* 19 N.Y.2d 453, 280 N.Y.S.2d 641, 227 N.E.2d 572; *see also*, 24 A.D.2d 284, 266 N.Y.S.2d 406, 16 A.L.R.3d 1175). Tort law is best defined as a set of general principles which, according to Prosser and Keeton, occupies a "large residuary field" of law remaining after other moore clearly defined branches of the law are eliminated. (Prosser and Keeton, *op cit.*, § 1, at 2.)

Inasmuch as there is no clear definition by which wrongs are classified as torts, the Legislature could not have used the term when enacting Section 9(2) in 1939 with a precision that would limit the jurisdiction of the Court of Claims solely to common-law torts or those recognized at the time. It is more likely that the term was used generally to indicate a branch of the law broader than the then-existing categories and subject to expansion as new wrongs supporting liability were recognized.

Indeed, there is evidence that the Court of Claims accepts this view for it has entertained jurisdiction over new torts recognized after the Act was adopted (*see, e.g., Doe v. State of New York*, 155 Misc.2d 286, 297-298, 588 N.Y.S.2d 698, *mod.* 189 A.D2d 199, 595 N.Y.S.2d 592 [applying the rule in *Bovsun v. Sanperi*, 61 N.Y.2d 219, 473 N.Y.S.2d 357, 461 N.E.2d 843 (1984)]) and it has frequently retained jurisdiction of claims seeking damages for constitutional torts in the past, albeit without discussion (*see, Vaughan v. State of New York*, 272 N.Y. 102, 5 N.E.2d 53, *appeal dismissed* 300 U.S. 638, 57 S.Ct. 510, 81 L.Ed. 855; *Brenon v. State of New York*, 31 A.D.2d 776, 297 N.Y.S.2d 88; *Frady v. State of New York*, 19 A.D.2d 783, 242 N.Y.S.2d 95; *Periconi v. State of New York*, 91 Misc.2d 823, 398 N.Y.S.2d 959; *Dean v. State of New York*, 111 Misc.2d 97, 443 N.Y.S.2d 581, *affd.* 91 A.D.2d 805, 458 N.Y.S.2d 899; *Herman v. State of New York*, 78 Misc.2d 1025, 357

N.Y.S.2d 811; *Hook v. State of New York*, 15 Misc.2d 672,
181 N.Y.S.2d 621). To be sure, there also have been
Court of Claims decisions, most unpublished, denying
jurisdiction to litigate constitutional wrongs, and the
State would distinguish the cited cases as actions
involving constitutional torts joined with common-law
torts. That has not been uniformly true, however; some
of the cited claims involved no common-law cause of
action and others asserted separate causes of action
involving only the violation of a constitutional duty
and those constitutional tort claims were sustained (*see,
e.g., Dean, supra; Periconi, supra, Herman, supra*).

The State also contends that the waiver contained in
Section 8 does not reach this claim because it is limited
to liability actions similar to those which may be
brought in Supreme Court against individuals and corporations (*see*, Court of Claims Act §§ 8, 12[1]). Individuals and corporations, it claims, cannot be sued for
constitutional violations.

Admittedly, there are few constitutional tort actions
against individuals and corporations in Supreme Court
because the Constitutions do not generally restrict the
actions of private parties (*see, e.g., SHAD Alliance v.
Smith Haven Mall*, 66 N.Y.2d 496, 498 N.Y.S.2d 99, 488
N.E.2d 1211 [holding that article I, § 8 of the State
Constitution, which guarantees the right of free speech,
does not apply to individuals or corporations]). There
are, however, some constitutional provisions that explicitly regulate private conduct and the prohibition against
discrimination contained in Section 11 is one of them.
Article I, § 11 prohibits discrimination by "any * * *
person or by any firm, corporation, or institution, or
by the state." Thus, the rights guaranteed by that
constitutional provision may be enforced in Supreme Court
to recover damages for private acts of discrimination
although enabling legislation was required before the
action could be maintained because the provision was not
self-executing (*see*, Executive Law § 297[9]; Civil Rights
Law § 40-d).

Furthermore, the State and Federal courts have concurrent jurisdiction over constitutional tort claims
asserted under the procedures authorized by the Federal
civil rights statutes (*see, Maine v. Thiboutot*, 448 U.S.

1, 3, n. 1, 100 S.Ct. 2502, 2503 n. 1, 65 L.Ed.2d 555;
Martinez v. California, 444 U.S. 277, 283-284, n. 7,
100 S.Ct. 553, 558, n. 7, 62 L.Ed.2d 481, 1 Friesen,
op. cit., ¶ 7.03[2]) and New York courts have consistently
accepted jurisdiction of such claims against "individuals
or corporations" (see, e.g., *Town of Orangetown v. Magee*,
88 N.Y.2d 41, 643 N.Y.S.2d 21, 665 N.E.2d 1061; *Cox v.
City of New York*, 40 N.Y.2d 966, 390 N.Y.S.2d 819, 359
N.E.2d 329; *DiPalma v. Phelan*, 179 A.D.2d 1009, 578
N.Y.S.2d 948, affd. 81 N.Y.2d 754, 593 N.Y.S.2d 778,
609 N.E.2d 131; *Manti v. New York City Tr. Auth.*, 165
A.D.2d 373, 568 N.Y.S.2d 16; *Clark v. Bond Stores*, 41
A.D.2d 620, 340 N.Y.S.2d 847; see also, 1 Civil Actions
Against State and Local Government, Its Divisions,
Agencies and Officers §§ 7.90-7.97 [Shepards/McGraw-Hill,
2d ed.]).

Thus, while the analogy between a government and an
individual or corporation contained in sections 8 and
12 of the Act has some inherent limitations because
individuals "do not do the same things in the same way
as does the State" (Davison, Claims Against the State
of New York ¶ 11.03, at 76-77; *and see, Newiadony v.
State of New York*, 276 A.D. 59, 93 N.Y.S.2d 24), the
causes of action asserted by claimants are sufficiently
similar to claims which may be asserted by individuals
and corporations in Supreme Court to satisfy the statutory
requirement.

[3] Accordingly, we conclude that the Court's jurisdiction is not limited to common-law tort causes of
action and that damage claims against the State based
upon violations of the State Constitution come within
the jurisdiction of the Court of Claims.

IV.

The Causes of Action

A.

[4] Claimants' first five causes of action[6] are based
on violations of Section 1981, the enabling act which
provides a damage remedy for the deprivation of Federal

constitutional rights.[7] We conclude they must be dismissed for failure to state causes of action.

In *Monell v. New York City Dept. of Social Servs.*, 436 U.S. 658, 98 S.Ct. 2018, 56 L.Ed.2d 611, the Supreme Court held that no suit would lie against the State under Section 1983 because the State was not a "person" within the meaning of the statute.[8] It also held that the doctrine of respondeat superior had no application in actions based on the statute. The Court reasoned that the State is not a "person" because a waiver of its immunity (whether based on the Eleventh Amendment or on

---

[6] Claimants alleged the following claims:

Claim 1—Racially motivated violation of Fourth Amendment of United States Constitution, thereby violating 42 U.S.C.§ 1981;

Claim 2—Racially motivated violation of the Equal Protection Clause of the Fourteenth Amendment of the United States Constitution based on Fourth Amendment violations, thereby violating 42 U.S.C.§ 1981;

Claim 3—Racially motivated violation of the New York State Constitution, article I,§ 12 and New York Civil Rights Law § 8, thereby violating 42 U.S.C.§ 1981;

Claim 4—Racially motivated violation of the Equal Protection Clause of the Fourteenth Amendment of the United States Constitution based upon violations of article 1,§ 12 of the New York State Constitution, thereby violating 42 U.S.C.§ 1981;

Claim 5—Racially motivated violation of article I,§ 11 of the New York Constitution and New York Civil Rights Law § 40-c, thereby violating 42 U.S.C.§ 1981;

Claim 6—Conspiracy to violate civil rights based upon 42 U.S.C.§ 19

Claim 7—Racially motivated violation of article I,§ 12 of the New York Constitution and New York Civil Rights Law § 8;

Claim 8—Racially motivated violation of article I,§ 11 of the New York Constitution and New York Civil Rights Law § 40-c;

Claim 9—Violation of New York State Civil Rights Law § 40-c based upon harassment and discrimination as defined in Penal Law § 240.

Claim 10—Violation of Personal Privacy Protection Law, New York Public Officers Law, article 6-A,§§ 91-99 (abandoned on this appeal);

Claim 11—Negligent training and/or supervision of officers and investigators.

[7] 42 U.S.C.§ 1981 provides:

"(a) All persons within the jurisdiction of the United States shall have the same right in every State and Territory to make and enforce contracts, to sue, be parties, give evidence, and to the full and equal benefit of all laws and proceedings for the security of persons and property as is enjoyed by white citizens, and shall be subject to like punishment, pains, penalties, taxes, licenses, and exactions of every kind, and to no other. * * *

"(c) The rights protected by this section are protected against impairment by nongovernmental discrimination and impairment under color of State law."

historic common-law principles), must be expressly stated before a State may be sued in Federal courts (*see, Will v. Michigan Dept. of State Police*, 491 U.S. 58, 63-65, 109 S.Ct. 2304, 2308-2309, 105 L.Ed.2d 45).

Claimants have based their claims on Section 1981. They maintain that section provides a basis of liability independent of Section 1983. In *Jett v. Dallas Ind. School Dist.*, 491 U.S. 701, 109 S.Ct. 2702, 105 L.Ed.2d 598, however, the Supreme Court held that Section 1983 provides the exclusive Federal damages remedy for violation of the rights guaranteed by Section 1981. The State is not a "person" within the statute and it cannot be liable in an action based on Section 1981 (*Jett, supra*, 491 U.S., at 731, 109 S.Ct., at 2720; *Dennis v. County of Fairfax*, 55 F.3d 151, 156, n. 1[45th Cir.1995]; *Tarpley v. Greene*, 684 F.2d 1, 11, n. 25 [D.C.App.1982], cited with approval in *Jett, supra*, 491 U.S., at 735, 109 S.Ct., at 2722-2723). Inasmuch as the *Jett* ruling controls claimants' first five causes of action they fail.

Claimants contend, that the Civil Rights Act of 1991, passed after the *Jett* decision, added subdivision (c) to Section 1981 for the purpose of overruling the holding in *Jett* and providing a broader avenue of relief to claimants. The legislative history accompanying the Act does not address the Supreme Court's holding in *Jett*. It mentions subdivision (c) only briefly and states that it was added to reaffirm (*Runyon v. McCrary*, 427 U.S. 160, 96 S.Ct. 2586, 49 L.Ed.2d 415 [holding private parties liable under 42 U.S.C.§ 1981]; *see,* H.R.Rep. No. 40[I], 102d Cong., 1st Sess. 92, 141, reprinted in 1991 U.S.Code Cong. & Admin. News 630, 670). This rationale was not discussed in the Congressional Debate, however (137 Cong. Rec. S15473, S15483). Federal courts, with little legislative history to guide them, have held

---

[8] 42 U.S.C.§ 1983—"Civil action for deprivation of rights"
"Every person who, under color of any statute, ordinance, regulation, custom, or usage, of any State or Territory or the District of Columbia, subjects, or causes to be subjected, any citizen of the United States or other person within the jurisdiction thereof to the deprivation of any rights, privileges, or immunities secured by the Constitution and laws, shall be liable to the party injured in an action at law, suit in equity, or other proper proceeding for redress. For the purposes of this section, any Act of Congress applicable exclusively to the District of Columbia shall be considered to be a statute of the District of Columbia."

conflicting views as to whether the amendment was meant to overrule *Jett* (compare *Ebrahimi v. City of Huntsville Bd. of Educ.*, 905 F. Supp. 993, 995, n. 2 [N.D.Ala.]; *Johnson v. City of Fort Lauderdale*, 903 F.Supp. 1520, 1523 [S.D.Fla.]; with *Federation of African Am. Contrs. v. City of Oakland*, 96 F.3d 1204 [9th Cir.]; *Robinson v. Town of Colonie*, 878 F.Supp. 387, 405, n. 13 [N.D.N.Y.]; *La Compania Ocho v. United States Forest Serv.*, 874 F.Supp. 1242, 1251 [D.N.M.]). We think the likeliest explanation for the amendment is the one adopted without discussion by the Court of Appeals in (*Dennis*, 55 F.3d at 156, n. 1, supra): that the amendment was intended to codify the rule in *Runyon v. McCrary*, supra.

Accordingly, Claims 1-5, based on 42 U.S.C.§ 1981, were properly dismissed.

B.

New York has no enabling statute similar to those contained in the Federal civil rights statutes permitting damage actions for the deprivation of constitutional rights. Thus, if we are to recognize a damage remedy it must be implied from the Constitution itself. The analysis is similar to that used by the Supreme Court when it recognized causes of action based on the Federal Constitution in *Bivens* (supra [search and seizure]) and (*Davis v. Passman*, 442 U.S. 228, 99 S.Ct. 2264, 60 L.Ed.2d 846 [equal protection]).

[5, 6] A civil damage remedy cannot be implied for a violation of the State constitutional provision unless the provision is self-executing, that is, it takes effect immediately, without the necessity for supplementary or enabling legislation (*see generally*, Friesen, State Constitutional Law ¶ 7.05[1], quoting from Cooley, Constitutional Limitations [7th ed.]; 16 C.J.S., Constitutional Law,§ 46). In New York, constitutional provisions are presumptively self-executing (*see, People v. Carroll*, 3 N.Y.2d 686, 691, 171 N.Y.S.2d 812, 148 N.E.2d 875).

[7, 8] Manifestly, article I, § 12 of the State Constitution and that part of section 11 relating to equal protection are self-executing. They define judicially enforceable rights and provide citizens with a basis for judicial relief against the State if those

rights are violated. Actions of State or local officials which violate these constitutional guarantees are void (*see, e.g., Foss v. City of Rochester,* 65 N.Y.2d 247, 491 N.Y.S.2d 128, 480 N.E.2d 717 [equal protection]; *People v. Griminger,* 71 N.Y.2d 635, 529 N.Y.S.2d 55, 524 N.E.2d 409 [search and seizure]).

[9] The violation of a self-executing provision in the Constitution will not always support a claim for damages, however (*see, Shields v. Gerhart,* 163 Vt. 219, 658 A.2d 924; *Figueroa v. State of Hawaii,* 61 Haw. 369, 604 P.2d 1198; *and see generally,* Friesen, *op.cit,* ¶ 7.05). The substantive right may be firmly established, as in the case of sections 11 and 12, but it remains to determine whether the remedy of damages for the invasion of those rights will be recognized.

C.

The State courts that have implied damage causes of action have traditionally rested their decisions on (1) the reasoning contained in the Restatement (Second) of Torts § 874A, (2) analogy to a *Bivens* action, (3) common-law antecedents of the constitutional provision at issue, or a combination of all three (*see generally,* Friesen, *op. cit.;* Baker, *The Minnesota Constitution as a Sword; The Evolving Private Cause of Action,* 20 Wm. Mitchell L.Rev. 313; Bandes, *Reinventing Bivens: The Self-Executing Constitution,* 68 S. Cal. L. Rev. 289).

Section 874A of the Restatement (Second) of Torts states that a court may imply a civil remedy from legislative or constitutional provisions, even though one is not expressly provided, if it determines that the remedy is appropriate in furtherance of the purpose of the provision and needed to assure its effectiveness (*see also,* comment *d; People v. Carroll, supra,* at 690-691, 171 N.Y.S.2d 812, 148 N.E.2d 875).

Some courts have relied on the reasoning of *Bivens,* 403 U.S. 388, 91 S.Ct. 1999, 29 L.Ed.2d 619, *supra.* In *Bivens,* the Supreme Court implied a cause of action for damages against Federal officials who violated the search and seizure provisions of the Fourth Amendment. The underlying rationale for the decision, in simplest terms, is that constitutional guarantees are worthy of protection

on their own terms without being linked to some common-
law or statutory tort, and that the courts have the
obligation to enforce these rights by ensuring that each
individual receives an adequate remedy for violation of
a constitutional duty. If the remedy is not forthcoming
from the political branches of government, then the
courts must provide it by recognizing a damage remedy
against the violators much the same as the courts earlier
recognized and developed equitable remedies to enjoin
unconstitutional actions. Implicit in this reasoning is
the premise that the Constitution is a source of positive
law, not merely a set of limitations on government.

The *Bivens* analysis illustrates the Restatement principle. Although its use in the Federal courts has been
narrowed somewhat by the Supreme Court (*see, Federal
Deposit Ins. Corp. v. Meyer*, 510 U.S. 471, 114 S.Ct.
996, 127 L.Ed.2d 308; *Schweiker v. Chilicky*, 487 U.S.
412, 108 S.Ct. 2460, 101 L.Ed.2d 370; *United States v.
Stanley*, 483 U.S. 669, 107 S.Ct. 3054, 97 L.Ed.2d 550;
see generally, Bandes, *Reinventing Bivens: The Self-
Executing Constitution, op. cit.*), it is well recognized
and has been applied to support a number of State
decisions (*see e.g., Widgeon v. Eastern Shore Hosp. Ctr.*,
300 Md. 520, 479 A.2d 921, *supra; Gay Law Students Assn.
v. Pacific Tel. & Tel. Co.*, 24 Cal.3d 458, 156 Cal.Rptr.
14, 595 P.2d 592, *supra; Phillips v. Youth Dev. Program*,
390 Mass. 652, 459 N.E.2d 453, *supra; Newell v. City of
Elgin*, 34 Ill.App.3d 719, 340 N.E.2d 344, *supra;* and
see generally, Friesen, *op. cit.*, ¶ 7.05[2]; ¶ 7.07[1]).

Finally, the courts have looked to the common-law
antecedents of the constitutional provision to discover
whether a damage remedy may be implied. New York's first
Constitution in 1777 recognized and adopted the existing
common law of England and each succeeding Constitution
has continued that practice. Thus, in some cases, there
exist grounds for implying a damage remedy based upon
preexisting common-law duties and rights.

V.

[10] Using these analytical tools, we conclude that
a cause of action to recover damages may be asserted
against the State for violation of the Equal Protection
and Search and Seizure Clauses of the State Constitution.

The rights embodied in sections 11 and 12 were first constitutionalized when our present Constitution was adopted in 1938 but the principles expressed in those sections were hardly new. The Equal Protection Clause of the Fourteenth Amendment had been thoroughly debated and adopted by Congress and ratified by our Legislature after the Civil War, and the concepts underlying it are older still. Indeed, cases may be found in which this Court identified a prohibition against discrimination in the Due Process Clauses of earlier State Constitutions, clauses with antecedents traced to colonial times (*see, e.g., People v. King*, 110 N.Y. 418, 18 N.E. 245; Charter of Liberties and Privileges, 1683, § 15, reprinted in 1 Lincoln, Constitutional History of New York, at 101).

The prohibition against unlawful searches and seizures originated in the Magna Carta and has been a part of our statutory law since 1828. The civil cause of action was fully developed in England and provided a damage remedy for the victims of unlawful searches at common law (*see, Huckle v. Money*, 2 Wils. 205, 95 Eng. Rep. 768 [1763]; *Wilkes v. Wood*, Lofft 1, 98 Engl. Rep. 489 [1763]; *Entick v. Carrington*, 19 State Tr. 1029, [1558-1774] All ER Rep. 41 [1765]).

Thus, there is historical support for the claimants' contention that the rights guaranteed by these two provisions have common-law antecedents warranting a tort remedy for invasion of the rights they recognize. Indeed, the availability of a civil suit for damages sustained as the result of a constitutional violation was contemplated by the delegates to the Constitutional Convention of 1938. They did not consider whether one was desirable — they assumed a civil remedy already existed. At least that is so with respect to Section 12. The debates over the proposed exclusion of evidence unlawfully obtained in criminal proceedings make that abundantly clear.[9]

Prior to the Convention of 1938, Judge Cardozo had written an opinion for the Court of Appeals holding that evidence obtained in violation of the search and seizure clause of the Civil Rights Act could be used against the defendant in a criminal trial. The defendant's remedy

---

[9] Exclusion is now mandated by the application of the Fourth Amendment applying to the States through the Fourteenth Amendment (*Mapp v. Ohio*, 367 U.S. 643, 81 S.Ct. 1684, 6 L.Ed.2d 1081).

for the wrong, he said, was a civil suit for damages (*see, People v. Defore*, 242 N.Y. 13, 19, 150 N.E. 585, *cert. denied* 270 U.S. 657, 46 S.Ct. 353, 70 L.Ed. 784). Based upon Cardozo's statement, the Convention delegates assumed that damages were unavailable to the victim of unconstitutional action and they used that argument to help persuade the Convention that exclusion was unnecessary to deter official misconduct (*see,* 1 Rev. Record of N.Y. Constitutional Convention, 1938, at 416, 425, 459). These debates reveal that the concept of damages for constitutional violations was neither foreign to the delegates nor rejected by them. That the Convention adopted the equal protection provision without similarly discussing the damage remedy does not establish that the delegates disfavored it nor does it foreclose our consideration of that relief.

Moreover, implying a damage remedy here is consistent with the purposes underlying the duties imposed by these provisions and is necessary and appropriate to ensure the full realization of the rights they state (*see, Bivens, supra*, 403 U.S., at 406, 91 S.Ct., at 2010 [Harlan, J., concurring]; *see also, Cort v. Ash*, 422 U.S. 66, 78, 95 S.Ct. 2080, 2088, 45 L.Ed.2d 26: Restatement [Second] of Torts § 874A, comment d). The analysis is not unlike `that which the Supreme Court and this Court have used to find a private right of action based upon certain regulatory statutes and is consistent with the rule formulated by the Restatement (*see, e.g., Merrill Lynch, Pierce, Fenner & Smith v. Curran*, 456 U.S. 353, 374-377, 102 S.Ct. 1825, 1837-1838, 72 L.Ed.2d 182; *Cannon v. University of Chicago*, 441 U.S. 677, 717, 99 S.Ct. 1946, 1968, 60 L.Ed.2d 560; *Chotapeg. Inc. v. Bullowa*, 291 N.Y. 70, 73-74, 50 N.E.2d 548; *Abounader v. Strohmeyer & Arpe Co.*, 243 N.Y. 458, 154 N.E. 309; *cf. CPC Intl. v. McKesson Corp.*, 70 N.Y.2d 268, 519 N.Y.S.2d 804, 514 N.E.2d 116; *see generally*, Restatement [Second] Torts § 874A; *see generally*, Burnham, *op. cit.;* and *see*, Wells and Eaton, *op. cit.*).

The provisions clearly define duties and impose them on government officers and employees. Section 11 is divided into two parts.[10] The first sentence directs that "[n]o person shall be denied the equal protection of the laws of this state or any subdivision thereof." The provision was intended to afford coverage as broad as that provided by the Fourteenth Amendment to the United States Constitution (*see, Dorsey v. Stuyvesant Town Corp.*, 299 N.Y. 512, 530, 87 N.E.2d 541, 2 Rev. Record of N.Y. State Constitutional Convention, 1938, at 1065). The section imposes a clear duty on the State and its subdivisions to ensure that all persons in the same circumstances receive the same treatment (*see, Davis v. Passman, supra* [implying a Federal cause of action based on the Fifth Amendment for denial of equal protection]).

The remainder of Section 11 prohibits discrimination. It is implicit in the language of the provision, and clear from a reading of the constitutional debates, that this part of the section was not intended to create a duty without enabling legislation but only to state a general principle recognizing other provisions in the Constitution, the existing Civil Rights Law or statues to be later enacted (*see, Dorsey v. Stuyvesant Town Corp., supra*, at 531, 87 N.E.2d 541; 2 Rev. Record of N.Y. Constitutional Convention, 1938, at 1069, 1144; *id*, Vol. 4, at 2626-2627). The Legislature subsequently implemented those guarantees by provisions of various statutes which regulate the conduct of both State officers and private individuals (*see, e.g.*, Executive Law § 290 *et seq.* [Human Rights Law]; Civil Rights Law § 40 *et seq.*; Labor Law § 220-e).

The language of Section 12 imposes a duty regulating the conduct of police officials.[11] It is consistent with the search and seizure provisions found in the Federal Constitution and the Constitutions of other States. Though a similar provision is found in the earlier

---

[10] "§ 11. [Equal protection of laws; discrimination in civil rights prohibited].

"No person shall be denied the equal protection of the laws of this state or any subdivision thereof. No person shall, because of race, color, creed or religion, be subjected to any discrimination in his civil rights by any other person or by any firm, corporation, or institution, or by the state or any agency or subdivision of the state." (N.Y. Const., art. 1, § 11.)

[11] "12. [Security against unreasonable searches, seizures and interceptions]

"The right of the people to be secure in their persons, houses, papers and effects, against unreasonable searches and seizures, shall not be violated, and no warrants shall issue, but upon probable cause, supported by oath or affirmation, and particularly describing the place to be searched, and the persons or things to be seized." (N.Y. Const., art. I, § 12.)

enacted Section 8 of the Civil Rights Law, the constitutional section is self-executing.

[11] These sections establish a duty sufficient to support causes of action to secure the liberty interests guaranteed to individuals by the State Constitution independent of any common-law tort rule. Claimants alleged that the defendant's officers and employees deprived them of the right to be free from unlawful police conduct violating the Search and Seizure Clause and that they were treated discriminatorily in violation of the State Equal Protection Clause. The harm they assert was visited on them was well within the contemplation of the framers when these provisions were enacted for fewer matters have caused greater concern throughout history than intrusions on personal liberty arising from the abuse of police power. Manifestly, these sections were designed to prevent such abuses and protect those in claimants' position. A damage remedy in favor of those harmed by police abuses is appropriate and in furtherance of the purpose underlying the sections.

Nor should claimants' right to recover damages be dependent upon the availability of a common-law tort cause of action. Common-law tort rules are heavily influenced by overriding concerns of adjusting losses and allocating risks, matters that have little relevance when constitutional rights are at stake. Moreover, the duties imposed upon government officers by these provisions address something far more serious than the private wrongs regulated by the common law. To confine claimants to tort causes of action would produce the paradox that individuals, guilty or innocent, wrongly arrested or detained may seek a monetary recovery because the complaint fits within the framework of a common-law tort, whereas these claimants, who suffered similar indignities, must go remediless because the duty violated was spelled out in the State Constitution.

Damages are a necessary deterrent for such misconduct. The remedies now recognized, injunctive or declaratory relief, all fall short. Claimants are not charged with any crime as a result of their detention and thus exclusion has no deterrent value. Claimants had no opportunity to obtain injunctive relief before the incidents described and no ground to support an order enjoining future wrongs. For those in claimants' position "it is damages or nothing" (*see, Bivens,* 403 U.S., at 410, 91 S.Ct., at 2012, *supra* [Harlan, J., concurring]). The damage remedy has been recognized historically as the appropriate remedy for the invasion of personal interests in liberty, indeed, damage remedies already exist for similar violations of the Federal Constitution. Those created by Congress and the Supreme Court, however, fail to reach State action though it is on the local level that most law enforcement functions are performed and the greatest danger of official misconduct exists. By recognizing a narrow remedy for violations of sections 11 and 12 of article I of the State Constitution, we provide appropriate protection against official misconduct at the State level.

VI.

A number of observations are in order about the dissent.

Judge Bellacosa cites several cases dealing with legal defenses which the State may interpose to avoid liability (*see, e.g., Arteaga v. State of New York,* 72 N.Y.2d 212, 532 N.Y.S.2d 57, 527 N.E.2d 1194 [legislative or judicial immunity]; *Tarter v. State of New York,* 68 N.Y.2d 511, 510 N.Y.S.2d 528, 503 N.E.2d 84 [quasijudicial or discretionary actions]; *Weiss v. Fote,* 7 N.Y.2d 579, 200 N.Y.S.2d 409, 167 N.E.2d 63 [same]; *Miller v. State of New York,* 62 N.Y.2d 506, 510, 478 N.Y.S.2d 829, 467 N.E.2d 493 [special duty]; *Merced v. City of New York,* 75 N.Y.2d 798, 552 N.Y.S.2d 96, 551 N.E.2d 589 [same]; *Steitz v. City of Beacon,* 295 N.Y. 51, 64 N.E.2d 704 [same]; *Sharapata v. Town of Islip,* 56 N.Y.2d 332, 452 N.Y.S.2d 347, 437 N.E.2d 1104 [immunity from punitive damages or "remedial immunity"]; *see generally,* Prosser and Keeton, Torts, at 1032 *et seq.* [5th ed.]). These defenses, sometimes referred to loosely as "immunities," should not be confused with sovereign immunity. The immunity waived by Section 8 of the Act is the historic immunity derived from the State's status as a sovereign and protects the State from suit. The defenses the dissent refers to are based on the special status of the defendant as a governmental entity. The State is amenable to suit but may nevertheless assert these grounds to avoid paying damages for some tortious conduct because, as a matter of policy, the courts have foreclosed liability (*see, Arteaga v. State, supra; Weiss v. Fote, supra*). The cited cases have little to do with the jurisdictional issue before us and, notably, in each of them the Court entertained jurisdiction and decided the matter on the basis of the defense asserted.

Addressing a related concept, the dissent has equated immunity with the special duty rule. The special duty rule holds that a plaintiff cannot recover against a municipality for failure to supply police protection or similar services absent a special relationship between the plaintiff and the police or municipality (citing *Merced v. City of New York*, 75 N.Y.2d 798, 552 N.Y.S.2d 96, 551 N.E.2d 589, *supra; Steitz v. City of Beacon*, 295 N.Y. 51, 64 N.E.2d 704, *supra*). The duty to supply police protection or similar services is a duty owed to the public at large and, as such violation of those duties does not create civil liability absent some special relationship or undertaking by the government in favor of the plaintiff (*compare, Merced v. City of New York, supra, with De Long v. County of Eric*, 60 N.Y.2d 296, 469 N.Y.S.2d 611, 457 N.E.2d 717). The duties imposed by the Constitution, which are at issue here, secure rights and privileges to individuals, as individuals. A breach of these duties is actionable under existing law. The question before us is whether a civil action for damages may be based upon the duties imposed by sections 11 and 12 of article I of the State Constitution.

The dissent also criticizes the majority for creating new respondeat superior liability against the State. It asserts this action is contrary to the holding of the Supreme Court in *Monell* (dissenting opn., at 205, at 243 of 652 N.Y.S.2d, at 1149 of 674 N.E.2d). It is the statute that imposes vicarious liability on the State, however, not this Court (*see,* Court of Claims Act § 9[2]). Our decisions have uniformly recognized this fact (*see, Jackson v. State of New York*, 261 N.Y. 134, 184 N.E. 735, *supra; Liubowsky v. State of New York*, 260 A.D. 416, 23 N.Y.S.2d 633, *affd.* 285 N.Y. 701, 34 N.E.2d 385; *Robison v. State of New York*, 263 A.D. 240, 32 N.Y.S.2d 388, later appeal 266 A.D. 1054, 45 N.Y.S.2d 725, *affd.* 292 N.Y. 631, 55 N.E.2d 506).

The authorities cited in the dissent for the proposition that the State may not be liable vicariously simply do not support any such rule. Indeed, in *Becker v. City of New York*, 2 N.Y.2d 226, 159 N.Y.S.2d 174, 140 N.E.2d 262 [holding that the waiver of immunity contained in the Court of Claims Act applies not only to the State but also to its subdivisions under the rule in *Bernardine v. City of New York*, 294 N.Y. 361, 62 N.E.2d 604], we held the State had waived its immunity from respondeat superior liability and specifically recognized that the State and its subdivisions were liable for the acts of their employees (*id.*, at 236, 159 N.Y.S.2d 174, 140 N.E.2d 262). Accordingly, we reversed a judgment in favor of the City which had held otherwise and sent the case back for a new trial. In *Welch v. State of New York*, 203 A.D.2d 80, 610 N.Y.S.2d 21 the Appellate Division held, as we do here, that the State cannot be liable on the basis of respondeat superior in a Section 1983 action under the rule in *Morell v. New York City Dept. of Social Servs.*, 436 U.S. 658, 98 S.Ct. 2018, 56 L.Ed.2d 611, *supra*. A Section 1983 action is controlled by the Federal statute which limits liability to actions taken "under color" of State law, i.e., as a matter of governmental policy or custom (*see, Monell, supra*, 436 U.S., at 691-692, 98 S.Ct., at 2036). A plaintiff seeking to recover on the basis of respondeat superior simply does not come within the terms of Section 1983. The *Welch* decision, based as it is on the Federal enabling statute, is inapposite to the action here based on the State Constitution and governed by the State statutes waiving immunity and imposing respondeat superior liability for actions of officers and employees.

[12] Nor is there any reason why the State should not be vicariously liable for constitutional torts by its officers or employees acting in the course of their employment. The State is answerable for the conduct of its officers who commit common-law torts, such as assault and false imprisonment, even if they are joined with constitutional torts and there is no reason why, if constitutional torts are actionable, the State should not be similarly liable in the absence of a common – law tort. Indeed, whether the delegates to the Constitutional Convention contemplated liability of the State or of individuals for violating a constitutional duty – and the dissent acknowledges that individuals may be liable (*see, e.g.,* dissenting opn., at 208, at 245 of 652

N.Y.S.2d, at 1151 of 674 N.E.2d) — is irrelevant. If individuals may be liable, then the State is liable because the Legislature, by enacting sections 8 and 9(2) of the Court of Claims Act, has determined that the State is answerable for the wrongs of its officers and employees.

The dissent maintains that there is little deterrent value in holding the State responsible for the torts of its officers and employees; the individual wrongdoer should pay, it says, not the State. Preliminarily, it should be noted that claimants assert liability against the State based upon inadequate training and supervision by the State as well as liability based on the individual officers' conduct. Moreover, in many cases the State will be secondarily liable for the employees' act because it has assumed the obligation to defend and indemnify them (*see,* Public Officers Law § 17).

But aside from those considerations, the State is appropriately held answerable for the act of its officers and employees because it can avoid such misconduct by adequate training and supervision and avoid its repetition by discharging or disciplining negligent or incompetent employees. Moreover, there is no reason why the deterrent value of holding the State answerable for an actionable assault by one of its employees is warranted but the deterrent value of holding it liable for an employee's constitutional tort is not. Thus, contrary to the reasoning of *Federal Deposit Ins. Corp. v. Meyer,* 510 U.S. 471, 114 S.Ct. 996, 127 L.Ed.2d 308, *supra* and *Bivens,* 403 U.S. 388, 91 S.Ct. 1999, 29 L.Ed.2d 619, *supra,* that liability should be confined to the individual wrongdoer, there is merit to imposing liability on the party who is ultimately responsible — and who the wrongdoer will often blame for ordering or directing the conduct complained of, the State.

Bivens and *Federal Deposit Ins. Corp.* must be seen in context. The Federal Government was not sued in *Bivens* because it was immune from suit (*see, Bivens, supra,* 403 U.S., at 410, 91 S.Ct., at 2012 [Harlan, J., concurring]; *see also,* Bandes, *op. cit.,* at 342). In *Federal Deposit Ins. Corp. v. Meyer,* 510 U.S. 471, 114 S.Ct. 996, 127 L.Ed.2d 308, *supra,* the injured party tried to hold an agency of the Federal Government liable because the individual defendant enjoyed immunity from liability in a *Bivens* action. There is no similar problem here because the State has waived immunity for the acts of its officers and employees (*see*, Court of Claims Act § 9[2]) and this provision distinguishes the case before us from the Federal cases limiting liability to individuals.

It should be noted that Congress could not, in view of the Eleventh Amendment of the United States Constitution, overrule a State's claimed sovereign immunity from suit in Federal or State courts (*see*, *Pennhurst State School & Hosp. v. Halderman*, 465 U.S. 89, 98-99, 104 S.Ct. 900, 906-907, 79 L.Ed.2d 67; *Quern v. Jordan*, 440 U.s. 332, 342, 99 S.Ct. 1139, 1145-1146, 59 L.Ed.2d 358). The dissent, relying on the restraint sometimes evident in Supreme Court decisions involving constitutional torts, fails to recognize that these concerns of federalism underlie much of the Supreme Court's reluctance to expand the relief available under Section 1983 and thereby unduly interfere with States' rights. On this point, Professor Shapo, partially quoted in the dissent (at 204-205, at 242-243 of 652 N.Y.S.2d, at 1148-1149 of 674 N.E.2d), stated that the *"federal judiciary should tread warily in utilizing a civil damage remedy against local law enforcement officers"* (*see*, Shapo, *Constitutional Tort: Monroe v. Pape, and the Frontiers Beyond*, 60 Nw. U. L. Rev. 277, 325 [emphasis added]). In *Paul v. Davis*, 424 U.S. 693, 96 S.Ct. 1155, 47 L.Ed.2d 405, the Supreme Court, concerned about the reach of the Due Process Clause in Section 1983 actions, declined to use the Fourteenth Amendment to create "a font of [Federal] tort law" and impose it on the States (*id.*, at 701, 96 S.Ct., at 1160-1161; *see also*, Whitman, Constitutional Torts, *op. cit.*, at 10). The Court did, however, acknowledge that actions could be based upon the specific constitutional guarantee against unreasonable searches and seizures (*id.*, at 700-701, 96 S.Ct., at 1160-1161). We recognize here a similar cause of action based on our State Constitution.

The dissent expresses the concern that recognition of a damage remedy here will result in the courts being deluged with lawsuits and seriously jeopardize the public treasury. Our decision does not hold that every tort by a government employee is actionable, or that those which may be will be actionable under all circumstances.

Claimants must still establish that their constitutional rights have been violated and that a damage remedy is available to them. Perhaps the most effective answer to the dissent's contention, however, was expressed by Justice Harlan in his concurring opinion in *Bivens*:

> "Judicial resources, I am well aware, are increasingly scarce these days. Nonetheless, when we automatically close the courthouse door solely on this basis, we implicitly express a value judgment on the comparative importance of classes of legally protected interests. And current limitations upon the effective functioning of the courts arising from budgetary inadequacies should not be permitted to stand in the way of the recognition of otherwise sound constitutional principles" (403 U.S., at 411, 91 S.Ct., at 2012).

The point is that no government can sustain itself, much less flourish, unless it affirms and reinforces the fundamental values that define it by placing the moral and coercive powers of the State behind those values. When the law immunizes official violations of substantive rules because the cost or bother of doing otherwise is too great, thereby leaving victims without any realistic remedy, the integrity of the rules and their underlying public values are called into serious question. A damage remedy for constitutional torts depriving individuals of their liberty interests is the most effective means of deterring police misconduct, it is appropriate to the wrong and it is consistent with the measure by which personal injuries have historically been regulated.

Finally, it should be noted that sovereign immunity — although originally a common — law doctrine defined solely by the courts — is now a creature of statute. Thus, it is within the power of the Legislature to redefine the jurisdiction of the Court of Claims if it sees fit to do so.

Accordingly, Claims 7 and 8, insofar as they state causes of action based upon sections 11 and 12 of article I of the State Constitution, are sustained.

VII.

The Remaining Causes of Action

[13]   Claimants have discontinued Claim 6 alleging a conspiracy and Claim 10 alleging a violation of Public Officers Law article 6-A. Additionally, we need not address the sufficiency of Claim 9, asserting a cause of action based on Section 40-c of the Civil Rights Law, because claimants acknowledge in their brief that damages are not available to them from the State on that cause of action.[12] The claim under Section 8 of the Civil Rights Law is duplicative of the constitutional claim based on Section 12 of article I of the Constitution and should be dismissed for that reason. Finally, the Court of Claims has jurisdiction over Claim 11, asserting a cause of action for negligent training and supervision.

In sum, the Court of Claims has jurisdiction of all the causes of action asserted in this claim. Claims 7 and 8 insofar as they allege claims based upon violations of article I, §§ 11 and 12 of the New York State Constitution and Claim 11 alleging a claim for negligent training and supervision are facially sufficient to state causes of action against defendants and should be reinstated. The remaining causes of action were properly dismissed.

Accordingly, the order of the Appellate Division should be modified, without costs, and the case remitted to the Court of Claims for further proceedings in accordance with this opinion and, as modified, affirmed.

BELLACOSA, Judge (dissenting).

Because the dismissal of this entire case against the State by the Court of Claims and the Appellate Division is the justifiable result, I respectfully dissent. The synopsis for my vote to affirm is:

---

[12] Civil Rights Law § 40-c states various discriminatory acts and the claim defendants violated Penal Law § 240.25 is asserted purely on the reference to it as a discriminatory act in Civil Rights Law § 40-c.

- The New York Constitution grants the Legislature the responsibility to define the subject-matter jurisdiction of the Court of Claims;

- Duly considered and enacted statutes expressly prescribe and implement that power and forbid implied State liability;

- The judicial inferential interpretive method is neither supportable nor suitable for the resolution of this preanswer dispute, in which the Court of Appeals promulgates new subject-matter jurisdiction for a court of limited powers and recognizes new remedies and causes of action against the State;

- The constitutional tort theory and nomenclature should not be equated and subsumed within conventional tort doctrines, as a route to resolving fundamental subject-matter jurisdiction and sovereign immunity issues, affecting New York's jurisprudence in and for the Court of Claims.

The following reasoning and documentation of authorities is necessary because of the complex nature of this dispute and the extensive majority explication. The reinstatement of constitutional tort claims in the Court of Claims invests that court with inferred subject-matter jurisdiction, which the Legislature has not seen fit to confer expressly. The means used to find an implied legislative authorization to achieve the end result substitutes for a quintessentially legislative prerogative.

This Court, "[b]y recognizing" what it tries to minimize as a "narrow remedy for violations of sections 11 and 12 of article I of the State Constitution" so as to "provide appropriate protection" under a distinct "'species of tort liability'" (majority opn., at 178, 192, at 226, 235 of 652 N.Y.S.2d, at 1132, 1141 of 674 N.E.2d), requires the State itself to answer for alleged "official" wrongdoings. The exposure includes the stigma of societal fault and the payment of unknown sums of public funds, not only for this case but also for innumerable others certain to be improvised within its precedential repertoire.

The significant and sharp controversy grows out of a race-based, community-wide police sweep, as part of an investigation stemming from a serious crime committed in an upstate college town. As Judge Hanifin stated, however, in his comprehensive and cogent trial court opinion: "This Court [of Claims] cannot expand its jurisdiction based on the emotional content of the issues presented to it" (*Brown v. State of New York*, Ct. Cl., Mar. 17, 1994, claim No. 86979. *affd.* 221 A.D.2d 681, 633 N.Y.S.2d 409). The focus in this Court, too, should remain solely on the statutory construction question posed by this case: whether the State itself may be sued in the New York State Court of Claims for affronts classified as "constitutional torts."

I conclude that the analysis and result that ultimately tap the deep reservoir of State responsibility are fundamentally flawed as a matter of law and history. No sustainable root and nexus have been sufficiently identified to overcome the twin towers of State protection expressly reflected in the statutorily specified subject-matter jurisdiction of the Court of Claims and limited surrender of sovereign immunity. Instead, a wide web of words within the statutory interpretation method has been used to discern a route to the desired result, while traditional respect for and proper distribution of power between legislative and judicial branches are deflected. On the other hand, though I express a dissenting viewpoint, I also recognize and respect the cogency and reasonableness that is reflected in the decision my colleagues reach in this complicated case.

I. JURISDICTION AND SOVEREIGN IMMUNITY

A.

The first general statutory provision for claims against the State related to the operation of the Erie Canal (L. 1817, c. 262). Through many statutory permutations, the modern Court of Claims was constituted in 1915 (L. 1915, chs. 1, 100). Its essential jurisdiction prescribed that *"[i]n no case shall any liability be implied against the state*, and no award shall be made on any claim against the state except upon such legal evidence as would establish liability against an individual or corporation in a court of law or equity" (Code Civ. Pro.§ 264 [emphasis added]).

In *Smith v. State of New York*, 227 N.Y. 405, 125 N.E. 841, this Court held that the State had not waived its sovereign immunity except as expressly surrendered and, thus, preserved the unrelinquished sovereign immunity of the State (*id.*, at 409-410, 125 N.E. 841). The operative construction canon is that "[s]tatutes in derogation of the sovereignty of a state must be *strictly construed* and a waiver of immunity from liability must be *clearly expressed*. * * * In the absence of a legislative enactment *specifically* waiving this immunity, the state cannot be subjected to a liability therefor" (*Smith v. State, supra*, at 410, 125 N.E. 841 [emphasis added]). The majority now ordains a new canon, using a judicial inference method not to fill a natural or legislative interstice, but to discover a vaguely unexplored universe of extensive tort exposure against the State, triable in a court of limited jurisdiction.

B.

The history of the exclusive legislative authority with respect to the investiture of jurisdiction in the Court of Claims is plainly expressed in New York Constitution, article VI,§ 9 (originally added in 1925 as art. VI,§ 23). It states that "[t]he court shall have jurisdiction to hear and determine claims against the state or by the state against the claimant or between conflicting claimants *as the legislature may provide*" (N.Y. Const., art. VI,§ 9 [emphasis added]; *see*, Court of Claims Act § 9).

Court of Claims Act § 9(2) precisely lists the subject-matter jurisdiction in words and structure that indicate a careful consideration by the Legislature of the categories and circumscriptions of claims to which the State's waiver of immunity would also apply. This Court, interpreting the State's statutory post-*Smith* waiver of immunity (L. 1929, ch. 467), noted that "[i]t includes only claims which appear to the judicial mind and conscience to be such as the *Legislature may declare* * * * the State should satisfy" (*Jackson v. State of New York*, 261 N.Y. 134, 138, 184 N.E. 735, *rearg. denied*

261 N.Y. 637, 185 N.E. 771 [emphasis added]). The legislative history accompanying recodification of the Court of claims Act shows no intention, understanding or contemplation to sweep the State's assumption of liability into uncharged and open waters as are at issue in this case (*see*, Bill Jacket, L. 1939, ch. 860, Mem. of James Barrett, Presiding Judge of the Ct. Cl., at 2; Mem. of Senator Feinberg, at 2-3).

Section 9 of the Court of Claims Act provides, "[t]he court shall have jurisdiction * * * 2. To hear and determine a claim of any person, corporation or munic- ipality against the state * * * for the torts of its officers or employees while acting as such officers or employees, providing the claimant complies with the *limitations* of this article" (*id.* [emphasis added]). Correspondingly, the State's waiver of sovereign immunity is contained in Section 8 of the Act, which provides: "The state hereby waives its immunity from liability and action and hereby assumes liability and consents to have the same determined in accordance with the same rules of law as applied to actions in the supreme court *against individuals or corporations*" (Court of Claims Act § 8 [emphasis added]). It is the interplay and application of the various constitutional and legislative declara- tions, with their evident and express limitations that ought to govern this controversy, not speculative attri- butions of implied and assumed legislative intent.

The plaintiff's predicate argument is that the Leg- islature, through its use of the word "torts" implied an all-encompassing corral of wrongs. For starters and contrary to this theory, traditional tort law is not an undefinable, limitless arena of wrongs. Rather, the word of art reflects "[t]he civil action for tort * * * is commenced and maintained by the injured person, and its primary purpose is to compensate for the damage suffered, *at the expense of the wrongdoer*: (Prosser and Keeton, Torts § 2, at 7 [5th ed.]). Professor Prosser also notes the realistic and sensible limitation that "[i]t does not lie within the power of any judicial system to remedy all human wrongs" (Prosser, *op. cit.*, § 4, at 23). Indeed, the word "tort," for subject-matter jurisdictional pur- poses, should be viewed and determined discretely within that universe and context.

A core feature of the defendant State's more nuanced argument, moreover, is not that the word "tort" is frozen like a fossil in time as of the original enactment of Court of Claims' jurisdiction. That is a strawman argument posed to overcome the cogent State position on virtually all points and authorities. Indeed, the focus in this statutory construction exercise should remain fixed on the proposition that the term "tort," for these jurisdictional purposes, pertains only to those claims reasonably understood by the enactors, as part of the common-law tradition, developed within the tort root rubric and jurisprudence (*see, e.g., Bovsun v. Sanperi*, 61 N.Y.2d 219, 473 N.Y.S.2d 357, 461 N.E.2d 843; *Battalla v. State of New York*, 24 N.Y.2d 980, 302 N.Y.S.2d 813, 250 N.E.2d 224, affg, 26 A.D.2d 203, 272 N.Y.S.2d 28). The State's argument should prevail by a reasonable interpretation of the governing statute, the history of its enactment, and this Court's restrained interpretation of it and its own powers in this regard. A transformative redefinition and expansion into a fundamentally different juridical genre gives the holding of the instant case breadth-taking dimensions. Moreover, that approach ignores the well-established discipline that subject-matter jurisdiction, groundbreaking new remedies and their policy and practical ramifications, are matters appropriately within the legislative purview and, thus, not within some generalized supervisory or inferential adjudicative role of the courts (*see generally*, Gershman, *Supervisory Power of the New York Courts*, 14 Pace L. Rev. 41 [1994]).

This Court, following the general canon that any waiver of immunity by the State is to be narrowly construed, has further noted that the waiver of immunity (§ 8) and grant of jurisdiction (§ 9) are not absolute and open-ended (*Weiss v. Fote*, 7 N.Y.2d 579, 585-587, 200 N.Y.S.2d 409, 167 N.E.2d 63). Relevant and analogous precedents illustratively point out that the "State waived that immunity which it had enjoyed solely by reason of its sovereign character," but that "the State retained its immunity for those governmental actions requiring expert judgment or the exercise of discretion" (*Arteaga v. State of New York*, 72 N.Y.2d 212, 215-216, 532 N.Y.S.2d 57, 527 N.E.2d 1194; *see, Tarter v. State of New York*, 68 N.Y.2d 511, 518-519, 510 N.Y.S.2d 528, 503 N.E.2d 84).

Significantly, in *Sharapata v. Town of Islip*, 56 N.Y.2d 332, 452 N.Y.S.2d 347, 437 N.E.2d 1104, this Court emphasized that it *"is hard to believe that any attempt to include punitive damages* [in Court of Claims Act § 8] *would not have induced lively legislative debate*, contemporary State history preceding the formulation of Section 8 gives no indication that the matter ever evoked any legislative interest" (*id.*, at 337, 452 N.Y.S.2d 347, 437 N.E.2d 1104 [emphasis added]). The Court stated that:

> "[T]he twin justifications for punitive damages — punishment and deterrence — are hardly advanced when applied to a governmental unit. As [then] Justice TITONE realistically put it in his opinion below, it would be anomalous to have "the persons who bear the burden of punishment, i.e., the taxpayers and citizens," constitute "the self-same group who are expected to benefit from the public example which the granting of such damages supposedly makes of the wrongdoer'" (*Sharapata v. Town of Islip, supra,* 56 N.Y.2d, at 338-339, 452 N.Y.S.2d 347, 437 N.E.2d 1104, *affg.* 82 A.D.2d 350, 441 N.Y.S.2d 275).

The holding of the instant case disregards the usefulness garnered from the parallel purposes and pertinent guidance reflected in the analysis of the punitive damages issue in respect to the constitutional tort "species" (majority opn., at 178, 192, at 226, 235 of 652 N.Y.S.2d, at 1132, 1141, of 674 N.E.2d).

In *Steitz v. City of Beacon*, 295 N.Y. 51, 64 N.E.2d 704, this Court, in interpreting the application of Court of Claims Act § 8 to municipalities, noted that "[a]n intention to impose upon the city the crushing burden of such [liability] *should not be imputed to the Legislature in the absence of language clearly designed to have that effect*" (*id.*, at 55, 65 N.E.2d 704 [emphasis added]). Yet, that is precisely what the stretched and attenuated analysis does in the instant case (*see, Weiss v. Fote*, 7 N.Y.2d 579, 586-587, 200 N.Y.S.2d 409, 167 N.E.2d 63, *supra*). It is, after all, the unique governmental police power that is involved here, not some ordinary form of individual or corporate tortious act. For that reason, among others, the claims asserted in this case are nowhere near "sufficiently similar" (majority opn., at 183, at 230 of 652 N.Y.S.2d, at 1136 of 674 N.E.2d) to traditionally recognized individual or corporate tort conduct, the limiting and qualifying phrase of the statute itself.

The refined interpretation previously accorded to the State's waiver of immunity is also reflected in the handling of torts involving members of the State militia. In *Goldstein v. State of New York*, 281 N.Y. 396, 24 N.E.2d 97, the Court noted that "if the word 'officers' is given its broad meaning it would include every officer engaged in performing a duty placed upon him by law, including the Governor, judges, members of the Legislature and all others occupying an official position in the State. *Such an interpretation of the statute would lead to an absurd conclusion"* (*id.,* at 405, 24 N.E.2d 97 [emphasis added]). The claim was dismissed because the State militia were not within the meaning of the State's waiver of sovereign immunity (*id.,* at 406, 24 N.E.2d 97; *see also, Newiadony v. State of New York*, 276 App.Div. 59, 93 N.Y.S.2d 24). In 1953, the Legislature amended the Court of Claims Act (L. 1953, ch. 343) adding Section 8-a by which the State waived its immunity for such torts. Notably, the Legislature — not the courts — expressly expanded this jurisdictional reach into that category, as it also did after *Smith* (*see, Jackson v. State of New York*, 261 N.Y. 134, 138, 184 N.E. 735, *supra*). The Legislature does not leave these substantial definitional duties and demarcations to chance, implication, the fertile inferential method or to other entities, because definiteness and distribution of powers are important societal and jurisprudential values. It knows well how to be very plain about such matters in fulfilling its up-to-now distinctive responsibility.

The Legislature has never contemplated the common-law word "tort" to include the kind of extensive constitutional domain advanced here because, not only did it not exist until recently; but also had the notion occurred or been presented to that legislative body, the policy debates would surely and necessarily have been "lively" indeed (*Sharapata v. Town of Islip*, 56 N.Y.2d 332, 337, 452 N.Y.S.2d 347, 437 N.E.2d 1104, *supra*).

When the Legislature enacted Section 12(1) of the Court of Claims Act — "In no case shall any liability be *implied* against the state" (emphasis added) — it could not have contemplated this kind of substantial judicial exertion, promulgating new substantive remedies.

II. CONSTITUTIONAL TORTS

A.

The coined term "constitutional tort" is used generally to refer to civil damage actions initiated under 42 U.S.C.§ 1983. In *Monroe v. Pape*, 365 U.S. 167, 81 S.Ct. 473, 5 L.Ed.2d 492, claimants whose Federal rights were infringed by acts of police officers wee allowed to sue individual officers. They could not sue the municipal employer because Congress did not intend municipalities to be "persons" under 42 U.S.C.§ 1983 (*id.*, at 187, 191, 81 S.Ct., at 484, 486). (I agree, by the way, that the 42 U.S.C.§ 1981 claims remain properly dismissed, but for the more fundamental reason of lack of subject-matter jurisdiction.) The Court in *Monroe* never expressly or impliedly recognized new species of claims embraced in the classification or common-law history of tort, especially for purposes like local subject-matter jurisdiction.

The origin of the sobriquet "constitutional tort," which so overridingly drives the analysis of the instant case, is found in the title of a law review article that sets forth to "analyze the jurisprudential development of a *federal statutory remedy*" (Shapo, *Constitutional Tort: Monroe v. Pape, and the Frontiers Beyond*, 60 Nw. U. L. Rev. 277, 323-324 [1965] [emphasis added]; *see*, Burnham, *Separating Constitutional and Common-Law Torts: A Critique and a Proposed Constitutional Theory of Duty*, 73 Minn. L. Rev. 515, n. 2 [1989]). Substance and the essence of the claims ought to control, however, not catchy nomenclature for law review titles, later conveniently utilized as semantical shorthand. Chief Judge Fuld, in *Morrison v. National Broadcasting Co.*, 19 N.Y.2d 453, 280 N.Y.S.2d 641, 227 N.E.2d 572, *revg.* 24 A.D.2d 284, 266 N.Y.S.2d 406, in a related context, quoting a well-established rubric, stated: "'We look for the realty, and the essence of the action and not its

mere name'" (*id.*, at 459, 280 N.Y.S.2d 641, 227 N.E.2d
572, quoting *Brick v. Cohn-Hall-Marx Co.*, 276 N.Y. 259,
265, 11 N.E.2d 902).

The proposition that "constitutional torts" are equivalent to or derive their essential nature from any
common-law tort antecedent, worthy of present-day Court
of Claims cognizance without legislative authorization,
finds no support anywhere. Quite the contrary, in
describing a "constitutional tort," Professor Shapo
states "[i]t is *not quite a private tort*, yet contains
tort elements; it is *not 'constitutional law*,' but employs
a constitutional test. Because of this interesting
amalgam, serious questions arise about the measurement
of the substantive right" (Shapo, *op. cit.*, at 324).
Indeed, the wordsmith himself warns that the "judiciary
should tread warily in utilizing a civil damage remedy
against local law enforcement officers [I now add a
fortiori, against the State itself], where much that is
vital to the case grows uniquely from the local situation"
(Shapo, *op. cit.*, at 325). Serious confusion is sown
from intermingling distinct doctrinal sources, as is
evident in the spiral of this case, which has no Federal
statutory source and no equivalent or corresponding State
statutory enabling predicate.

In *Monell v. New York City Dept. of Social Servs.*,
436 U.S. 658, 98 S.Ct. 2018, 56 L.Ed.2d 611, when the
Supreme Court overruled *Monroe* and declared that municipalities could be liable under 42 U.S.C.§ 1983, it
added that "a municipality cannot be held liable *solely*
because it employs a tortfeasor — or, in other words,
*a municipality cannot be held liable under § 1983 on a
respondeat superior theory*" (*Monell v. New York City
Dept. of Social Servs., supra*, at 691, 98 S.Ct., at 2036
[emphasis added]). This key analytical link is rejected
by the majority.

Finding accountability for the first time within this
uniquely governmental police power context, against the
express admonitions of *Monell* and by the classification
of such acts as within this State's surrender of sovereign
immunity, fails to credit and respect this important
historical and precedential limitation. Ironically, now
the effect of the State's limited waiver of liability
in Court of Claims Act § 8 makes "the State and its

subdivisions liable, under the doctrine of *respondeat superior*, for the negligent acts of its paid employees" in pure constitutional police power circumstances, never contemplated, debated or legislatively enacted (*Becker v. City of New York*, 2 N.Y.2d 226, 235, 159 N.Y.S.2d 174, 140 N.E.2d 262; *see also, Welch v. State of New York*, 203 A.D.2d 80, 81, 610 N.Y.S.2d 21). Since that is not what the State statutorily surrendered, the upending of this key limitation by something interpretatively elevated and recognized as constitutional tort actions against the State and swept into the Court of Claims' limited charter is severely unsettling (*see,* Prosser, *op. cit.,* § 69, at 500).

This Court has emphasized, "'[t]he municipal corporation is different. It is not organized for any purpose of gain or profit, but it is a legal creation engaged in carrying on government and administering its details for the general good and as a matter of public necessity'" (*Sharapata v. Town of Islip*, 56 N.Y.2d 332, 337, 452 N.Y.S.2d 347, 437 N.E.2d 1104, *supra*, quoting *Costich v. City of Rochester*, 68 App.Div. 623, 631, 73 N.Y.S. 835). Under the majority's holding, the State may well be deprived of traditional defenses, although those would be available to a government defendant in a Federal constitutional rights-statutory source case (*see, Collins v. Harker Hgts.*, 503 U.S. 115, 121-122, 112 S.Ct. 1061, 1066-1067, 117 L.Ed.2d 261; *Monell v. New York City Dept. of Social Servs.*, 436 U.S. 658, 691, 98 S.Ct. 2018, 2036, 56 L.Ed.2d 611, *supra*). Indeed, most of the rules limiting tort responsibility — to which the State's waiver of sovereign immunity and immunity-like protections would otherwise apply — are rendered inoperative and impotent when claims trace their origins to the superior ordinances of constitutional dimension, without any specific statutory enablement or qualifying features.

The United States Supreme Court itself — with apt instruction concerning the instant question — has noted differences between "constitutional torts" and common-law torts (*see, Paul v. Davis*, 424 U.S. 693, 701, 96 S.Ct. 1155, 1160-1161, 47 L.Ed.2d 405 [refused to make Due Process Clause "*a font of tort law*" and noted the "constitutional shoals" confronting any attempt to derive a body of tort law from Federal civil rights statutes (emphasis added)]; *Carey v. Piphus*, 435 U.S. 247, 258-259,

98 S.Ct. 1042, 1049-1050, 55 L.Ed.2d 252; *Daniels v. Williams*, 474 U.S. 327, 332, 106 S.Ct. 662, 665-666, 88 L.Ed.2d 662; *Farmer v. Brennan*, 511 U.S. 825, 837-838, 114 S.Ct. 1970, 1978-1979, 128 L.Ed.2d 811.) In analyzing constitutional tort violations the Supreme Court has "emphasize[d] the separate character of the inquiry into the question of municipal responsibility and the question of whether a constitutional violation occurred" (*Collins v. Harker Hgts.*, 503 U.S. 115, 122, 112 S.Ct. 1061, 1067, 117 L.Ed.2d 261, *supra*). That some of these notions spring from federalism restraints is not my point; the flat refusal of the majority to heed home-grown constraints, well-established and tested within the historical development and this State's own constitutional, jurisprudential and statutory disciplines, is the point of my departure and puzzlement (majority opn., at 195, at 237 of 652 N.Y.S.2d, at 1143 of 674 N.E.2d).

Further, in "constitutional tort" exertions, claimants are not required to satisfy other thresholds limiting municipal liability, like the "special duty rule" applicable to "ordinary" common-law tort actions against governments. In this respect, this Court has held and cautioned that:

> "Absent this requirement, a municipality would be exposed to liability every time one of its citizens was victimized by crime and the municipality failed to take appropriate action although notified of the incident — so vast an expansion of the duty of protection should not emanate from the judicial branch. 'Before such extension of responsibilities should be dictated by the indirect imposition of tort liabilities, there should be a legislative determination that that should be the scope of public responsibility'" (*Kircher v. City of Jamestown*, 74 N.Y.2d 251, 259, 544 N.Y.S.2d 995, 543 N.E.2d 443, quoting *Riss v. City of New York*, 22 N.Y.2d 579, 582, 293 N.Y.S.2d 897, 240 N.E.2d 860; *see, Miller v. State of New York*, 62 N.Y.2d 506, 478 N.S.2d 829, 467 N.E.2d 493: *Merced v.*

> City of New York, 75 N.Y.2d 798, 552
> N.Y.S.2d 96, 551 N.E.2d 589; see also,
> Schuster v. City of New York, 5 N.Y.2d
> 75, 180 N.Y.S.2d 265, 154 N.E.2d 534).

For these additional reasons, the Federal statutory "constitutional tort" concept should not be merged and subsumed within the meaning of "tort," as specifically delineated by Court of Claims Act § 8 — unless other appropriate protections for the State are somehow balanced into the equation. This is not done, additionally fortifying the wisdom of keeping the nuanced weighing within the legislative gambit.

No intimation appears anywhere that the Legislature ever thought of, no less enabled or implemented, this new overflowing "font" of State tort liability (*see, Paul v. Davis*, 424 U.S. 693, 701, 96 S.Ct. 1155, 1160-1161, 47 L.Ed.2d 405, *supra*). To infer that it did fundamentally disregards differences of kind and essence, not just degree. The comprehensive treatment and redefinition of the word "tort" in New York for jurisdictional purposes transforms the distinctive and unique qualities, nature, source, purpose and history between conventional torts and constitutional torts.

B.

Plaintiffs are also successful in achieving a result that declares "constitutional torts" as within the class of claims to which the State has surrendered its immunity in the Court of Claims. This conclusion is seriously questionable, as this Court has never before conflated a constitutional duty into such a pervasive and comprehensive tort theory for damages, energized by sheer inference. Notably, claimants are not asking for or receiving some equivalent remedy existing and recognized elsewhere, as the prayer for relief is generously characterized (majority opn., at 179, at 227 of 652 N.Y.S.2d, at 1133 of 674 N.E.2d.) This relief, as acknowledged throughout the majority opinion, is an additionally recognized remedy, despite forthright acknowledgments of express and analytical limitations. Moreover, it is a remedy in New York only and against the State of New York alone.

I do not doubt for a moment that the Court of Claims possesses jurisdiction over "constitutional tort" type claims when they share a recognized common-law lineage or specifically authorized root such as assault, false arrest, trespass, malicious prosecution and many more (*Bovsun v. Sanperi*, 61 N.Y.2d 219, 473 N.Y.S.2d 357, 461 N.E.2d 843, *supra*; *Battalla v. State of New York*, 24 N.Y.2d 980, 302 N.Y.S.2d 813, 250 N.E.2d 224, *supra*). Thus, "[b]y virtue of its waiver of [sovereign] immunity, the State of New York has provided a means by which the citizen can obtain satisfaction in such cases and, at the same time, has afforded protection to itself from unjustified claims based on groundless accusations" (McNamara, Jr., *The Court of Claims: Its Development and Present Role in the Unified Court System*. 40 St. John's L. Rev. 1, 40 [1965]).

In this respect, however, the reliance on *People v. Defore*, 242 N.Y. 13, 150 N.E. 585, *cert. denied* 270 U.S. 657, 46 S.Ct. 353, 70 L.Ed. 784 for the proposition that this Court in the early part of this century recognized a private damages action for search and seizure against the State is as curious in its application to this case as it is disturbing for its wider implications (*see*, Pitler, *Independent State Search and Seizure Constitutionalism: The New York State Court of Appeals' Quest for Principled Decisionmaking*, 62 Brook. L. Rev. 1, *seriatim*, and at 13-14, 86-97, 152-185 [1996]). In *Defore*, this Court rejected suppression of illegally acquired evidence as an inappropriate remedy (*People v. Defore*, *supra*, at 25, 28, 150 N.E. 585). Judge Cardozo commented that direct remedies were available to a person subjected to an illegal search, stating that "*[t]he officer* might have been resisted, or *sued for damages*, or even prosecuted for oppression" (*id.*, at 19, 150 N.E. 585 [emphasis added]). The majority in the instant case treats Judge Cardozo's dictum about a private right of action as if it applied against the State — indeed such an assumption is also attributed to the 1938 Constitutional delegates. This mischaracterization is registered frequently in the majority opinion and is reflected in numerous citations that also do not support the enlarged proposition for which they are advanced.

The delegates to the 1938 Constitutional Convention, in considering proposals to adopt the Federal exclusionary rule, appreciated that the civil damages remedy, merely theoretically posed in *Defore*, was available only against the offending officer (*see*, Statement by Harold Riegelman, 1 Rev. Record of N.Y. Constitutional Convention, 1938, at 416 ["The State forbids an unreasonable search. The State officer disobeys the injunction. *He* can and should be punished and made to answer in damages" (emphasis added)]; *see also*, Pitler, *op. cit.*, at 94). These authorities demonstrate that absolutely no basis whatsoever is presented for even a tenuous speculation that Judge Cardozo or the 1938 Constitutional delegates ever imagined a direct action against the State itself for money damages resulting solely from constitutional provisions deemed self-executing.

The same Constitutional Convention, relied upon now to infer a direct action damage remedy against the State, expressly rejected — after sharp and lively debate — not only an exclusionary remedy, but also a forfeiture of office remedy (1 Rev. Record of N.Y. Constitutional Convention, 1938, at 578). These express rejections were not done because of the hypothetical dictum of Judge Cardozo concerning a personal civil action, but for fundamental, doctrinal and jurisprudential reasons that were fully debated (*see*, Statement of Harry E. Lewis, Chairman, Bill of Rights Comm., 1 Rev. Record of N.Y. Constitutional Convention, 1938, at 577-578 [opposed office forfeiture amendment to Search and Seizure Clause because it "is necessarily legislative"]). This history and specificity prove, it seems to me, that the civil remedy announced today was never on any delegate's or legislator's mind or list, nor should it be assumed to have been (*see*, Pitler, *op. cit.*, at 157). The gap between express rejection of considered remedies and the discovery of this new additional one is not bridgeable by statutory construction. The subject-matter jurisdiction and cause of action, now simply inferred, results after forthright judicial debate, but in the entirely wrong deliberative forum.

The shock to legislators and delegates would be all the more profound in the modern light of the unquestionable acceptance of the exclusionary rule. It does not take much imagination to conjure up the operational impact of all filings of motions to suppress being accompanied by notices of claims. This prospect would surely have been "foreign" to legislators and delegates, and the practical and procedural ramifications would have concerned them, to put it mildly, that every grant of suppression — as one example only — carried the potential of a virtually certain civil filing and of monetary recovery in the Court of Claims against the State.

Moreover, if such a vein of tort exposure were to be carefully explored through the full labyrinth of the New York Constitution, some would surely and legitimately also raise questions of judicial impact and resources, weighed against the proportionate societal benefit to be obtained (*see,* Pitler, *op. cit.,* at 54 ["To Cardozo, the far-reaching consequence of the exclusionary rule — that it would free some of the most serious criminal offenders — *was simply too high a price to pay, at least until the Legislature spoke with a clearer voice*" [emphasis added]]). This case engenders a different price where, again, the Legislature has not expressed, implied or spoken in a clear voice.

The reflections and debate on that feature alone, in the customary legislative fashion, would provoke a reality check on the policy implications of seeding new tort clouds so generously (*see,* Eisenberg and Schwab, *The Reality of Constitutional Tort Litigation,* 72 Cornell L. Rev. 641, 694 [1987]; Rosen, *The Bivens Constitutional Tort: An Unfulfilled Promise,* 67 N.C. L. Rev. 337, 343 [1989]). If this portentous decision were considered and made by the Legislature — as I believe it should be — the judicial impact and resources questions would engender not only serious, respectful debate, but also robust differences of views and particularized, qualified expressions of law, effected through the statutory enactment method, not the inferential, adjudicative interpretive model.

C.

The plaintiffs alternatively argue for an implied private right of action against the State based on violations of article I, §§ 11 and 12 of the New York State Constitution. Heavy reliance is placed upon *Bivens v. Six Unknown Fed. Narcotics Agents,* 403 U.S. 388, 91 S.Ct. 1999, 29 L.Ed.2d 619. *Bivens* held that an individual government agent or officer could be held liable in

Federal court for monetary damages as a result of a violation of the Fourth Amendment to the United States Constitution (*id.*, at 397, 91 S.Ct., at 2005). Whether the government was liable under an implied right of action based on the violation of a constitutional right, however, was not confronted.

That question was presented to the Court in *Federal Deposit Ins. Corp. v. Meyer*, 510 U.S. 471, 114 S.Ct. 996, 127 L.Ed.2d 308. The Supreme Court stated that it "implied a cause of action against federal officials in *Bivens* in part *because a direct action against the Government was not available*" (*id.*, at 485, 114 S.Ct., at 1005 [emphasis added]). The Court noted that "the purpose of *Bivens* is to deter *the officer.* * * * If we were to imply a damages action directly against federal agencies, thereby permitting claimants to bypass qualified immunity, there would be no reason for aggrieved parties to bring damages actions against individual officers" (*id.*, at 485, 114 S.Ct., at 1005). Finally, the Supreme Court instructively noted that:

> "Here, unlike in *Bivens*, there are 'special factors counselling hesitation' in the creation of a damages remedy. *Bivens*, 403 U.S., at 396 [91 S.Ct., at 2004-2005]. *If we were to recognize a direct action for damages against federal agencies, we would be creating a potentially enormous financial burden for the Federal Government.* * * * *[D]ecisions involving '"federal fiscal policy"' are not ours to make* [citation omitted]. *We leave it to Congress to weigh the implications of such a significant expansion of Government liability*" (*id.*, at 486, 114 S.Ct., at 1005-1006 [emphasis added]).

Using *Bivens* as a source of authority for what is wrought by the instant holding, yet ignoring its critical limitations as reflected in *Federal Deposit Ins. Corp. v. Meyer, supra*, sidesteps the New York question. The Supreme Court's deference to Congress to fix some boundaries of government liability and fiscal accountability closely parallels this Court's consistent respect for the Legislature's prerogatives in these matters under explicit direction from this State's Constitution. New York now reverses its own prudent history, bucks the Federal trend and even discards the useful guiding lights of many sibling States (*see, e.g., Figueroa v. State of Hawaii,* 61 Haw. 369, 382, 604 P.2d 1198, 1205-1206; *Hunter v. City of Eugene,* 309 Or. 298, 302-303, 787 P.2d 881, 883-884).

Assuming that there was any justification — and none has been persuasively shown — for this Court to combine a transmuted *Bivens* analysis and the Restatement (Second) of Torts as a joint sourcing for the groundbreaking decision, such an approach should nevertheless be rejected as unnecessary as well as unauthorized. The New York Legislature and the courts of this State already provide effective and adequate remedies for infringements of constitutional rights. Although the range and reach of some of these remedies may be debatable, the policy choices and degree features are not appropriate justification for this Court's magnification of the subject-matter jurisdiction of the Court of Claims — a constitutionally delegated legislative power. Indeed, the suggestion that absent this added judicial remedy against the State some amorphous kind of immunity attaches to the underlying conduct is a chimera; claims in this very controversy remain pending against potentially responsible individuals in appropriate general jurisdiction forums, like State Supreme Court and the Federal courts (*see, Brown v. City of Oneonta,* 916 F.Supp. 176 [N.D.N.Y.], 911 F.Supp. 580 [N.D.N.Y.], 160 F.R.D. 18 [N.D.N.Y.]).

Given the breadth of the Legislature's specific enactments providing mechanisms to enforce constitutionally guaranteed rights, including the Civil Rights Law (*see, e.g.,* Civil Rights Law § 40-c; Executive Law § 290 *et seq.*), a judicially created *Bivens* remedy against the State in the Court of Claims is a significant policy overreach. Indeed, the specificity of remedial legislative enactments and enablements, under standard statutory construction principles, punctuates the argument for some reasonable restraint against the *carte blanche* invitation inscribed by today's bold precedent.

As a further example, the Human Rights Law regime is rendered substantially unnecessary and the unanimous rationale underlying *Koerner v. State of New York*, 62 N.Y.2d 442, 449, 478 N.Y.S.2d 584, 467 N.E.2d 232 is simultaneously made superfluous. If this particular Human Rights Law provision is no barrier to the wider adjudicative interpretive web, then neither are other New York State constitutional provisions under the extended self-executing principle (*compare*, Pitler, *op. cit.*, at 49-54 [analysis of Civil Rights Law history]).

The plaintiffs assert that the Equal Protection and Search and Seizure Clauses of the State Constitution, though essentially "self-executing," require no specific legislative enablement to energize the Court of Claims remedy and empowerment they now obtain for themselves as a class and for others precedentially. While "it is now presumed that constitutional provisions are self-executing" (*People v. Carroll*, 3 N.Y.2d 686, 691, 171 N.Y.S.2d 812, 148 N.E.2d 875), no precedent from this Court suggests that a private damages remedy may be inferred and promulgated in the loose and unlimited fashion spread under this large and general umbrella.

Inferring subject-matter jurisdiction, in any event, is a far reach even from recognizing a new cause of action, itself no small matter. One would expect strict and rigorous standards to justify inferring — i.e., attributing to the authorized policymaker, the Legislature — first, a whole new subject-matter jurisdiction for a limited specialized court, and, second, a whole new category of tort liability (not just a duty element). One searches in vain for rigorous analysis and reliable sources justifying such dual, high hurdle attribution (*compare*, *Matter of Tap Elec. Contr. Serv. v. Hartnett*, 76 N.Y.2d 164, 169, 556 N.Y.S.2d 988, 556 N.E.2d 427 ["courts are reluctant to infer preemption * * * in the absence of an express indication * * * of an intent to do so"], citing *Hillsborough County v. Automated Med. Labs.*, 471 U.S. 707, 717-718, 105 S.Ct. 2371, 2377-2378, 85 L.Ed.2d 714).

Just as importantly, the judicial rationalization of a need to declare an additional private damages remedy against the State itself, derived from the State Constitution and several other more attenuated and elliptical sources, does not equate to and certainly does not warrant the conferral of subject-matter jurisdiction over such a claim in a court of limited, special powers. The State Constitution itself in article VI,§ 9 gives the Legislature the sole authority to determine the monetary claims that may be brought and adjudicated against the State. The cause of action that the majority concedes it needs to "imply" in order to give it life — because it surely is not legislatively expressed — should not prevail over an explicit constitutional imperative and specifically delegated authorization to the Legislature. In any event, such a cause of action should not generally be inferred absent clear, pertinent and unequivocal evidence of the drafters' intent (*see, CPC Intl. v. McKesson Corp.*, 70 N.Y.2d 268, 276-277, 519 N.Y.S.2d 804, 514 N.E.2d 116). There is none such in this case.

Whatever the history of the unreasonable search and seizure protection inspires for the long reach of this case, nothing — not one iota of data — can be identified to support a corresponding constitutional tort grounded in the Equal Protection Clause. This prong of the holding concededly represents the farthest extension of all, with extraordinary precedential implications for a host of constitutional claims reduced to torts that may now enter the wells and dockets of Court of Claims courthouses.

In sum, the lower courts correctly held that the Legislature did not invest the Court of Claims with subject-matter jurisdiction and that the State has not consented to stand accountable in that court for alleged violations by its officers and employees of these constitutional police power exertions. This result and rationale should be upheld whether the claims are the egregious type as alleged to have occurred here or the myriad variety of others interspersed throughout the New York State Constitution. Thus, I vote to affirm the order of the Appellate Division in its entirety.

KAYE, C.J., and TITONE, SMITH and CIPARICK, JJ. concur with SIMONS, J.

BELLACOSA, J., dissents in part and votes to affirm in a separate opinion.

LEVINE, J., taking no part.

Order modified, etc.

chapter thirteen

What to expect after an incident

"Everyone has a plan until they get hit."

Mike Tyson

*"Expect only 5% of an intelligence report to be accurate.
The trick of a good commander is to isolate the 5%."*

Douglas MacArthur

After an incident of workplace violence "all hell is going to break loose!" Think about it! Your workplace has just been traumatized by a violent incident. Employees have been injured or killed. Every law enforcement agency in the area is photographing and taping your workplace. Television reporters swarm around your facility interviewing anyone who will talk to them and beaming the information around the world. Corporate is calling wanting to know what is going on, and anyone who is anyone is calling or stopping at your facility. Crowds of on-lookers block traffic around your facility, and employees mill around the parking area. In essence, all control has been lost!

Post-incident planning is essential in order to protect not only your human assets and property assets but also your company's efficacy assets. Post planning is "making the best of a bad situation." After a workplace violence incident, the company is going to incur losses....the only question is how much!

What kind of losses should a post-incident plan or response plan be prepared to address? What kind of assets are at risk? In evaluating the assets of most companies, the assets tend to lie in the following categories:

- Human assets
- Equipment or technology assets
- Financial assets
- Building or structural assets
- Company reputation and prestige
- Intellectual assets

Conversely, in a workplace violence incident the following kinds of losses can be anticipated:

- Loss of production
- Loss of sales
- Repair or replacement of equipment
- Workers' compensation costs
- Litigation costs
- Agency fines
- Defense costs
- Damage to building structure
- Drop in stock price (i.e., financial worth of company)
- Loss of loan status (i.e., cost more for operational capital)
- Loss of prestige in business community
- Loss of marketability of products
- Loss of future workforce (i.e., people don't want to work for your company)
- Loss of "brain power" or intellectual capabilities
- Loss of current workforce (i.e., employees leaving)
- Stigma on the company name (i.e., what does "Bhopal" or "Exxon Valdez" bring to mind?)
- Loss of potential future earnings and sales

In developing the written Workplace Violence and Response Plan, careful attention should be provided to assessing the assets to be protected and the anticipated losses in order to properly develop a plan of action to minimize the losses, gather the information necessary for future defense, and to safeguard remaining assets. A broad-base analysis can be utilized anticipating the type of potential risks and, as noted previously, this program may be developed in conjunction with existing emergency and disaster preparedness plans.

Types of Possible Situations

- Domestic dispute
- Employee with a weapon
- Ex-employee with a gun
- Equipment sabotage
- Fire or arson
- Intentional chemical release
- Employee with weapon and hostages
- Employee with multiple weapons
- Employee with bomb
- Bomb threat
- Cyber terrorism
- Sabotage of computer system

- Sabotage of water system
- Sabotage of food system
- Fighting between employees without weapons
- Stalking of employee by current employee
- Stalking of employee by other individual
- Extortion threatening violence

Types of Potential Plans

- Internal evacuation (i.e., move employees within the facility)
- External evacuation (i.e., move employees outside of the facility)
- Highway or traffic control
- Location of command center
- Communications

Possible Components of a Workplace Violence Preparedness and Response Plan

- Written policies
- Written plans
- Communication systems
- Command center locations
- Emergency telephone directory
- Alarm system
- Evacuation routes and plans
- Accounting for employees
- First aid and triage area
- Medical transport via ground and air
- Protection of records/computers
- Reporting to management
- Reporting to union
- Reporting to shareholders
- Contractors on site contracts
- Coordination with public sector
- Media control
- Emergency lighting
- Sprinkler systems
- Disaster kits
- Medical supplies
- Shut down procedures
- Responsibilities and accountability
- Mock drills
- Traffic control
- Investigation and documentation equipment
- Controlling litigation
- Contact procedures for insurance entities
- Clean up and repair equipment

The response section of the Workplace Violence Prevention and Response Program should prepare for every possible contingency in order that the proper personnel are selected and trained, equipment is available and in working order, and all facets of the response plan are addressed. Remember, members of your management team and employees will be under a tremendous amount of stress following a workplace violence incident and may not be "thinking straight." Preplanning and preparation will permit these individuals to function in this very difficult situation to preserve and minimize the damages.

Workplace violence response kits

As is often utilized in emergency and disaster preparedness programs, assembly of the equipment and documents that may be needed following a workplace violence incident or other emergency situation is essential. Workplace violence incidents can happen at any time, and assembly of these vital components following a workplace violence incident may not be feasible. To this end, assembly of response kits prior to the incident is a low-cost "insurance policy" to ensure proper gathering of important documentation which can be utilized in future litigation or related activities. A response kit can include the following:

- Copy of written programs
- Copy of appropriate forms
- Testing equipment
- Camera, film, batteries
- Videocamera, tape, batteries
- Pens, notepads, and other accessories
- Tape player, tapes, batteries
- Law enforcement and local emergency service telephone numbers
- Notification list, including critical stress debriefing personnel
- Laptop computer
- Nylon carrying case

In addition to preparing for workplace violence incidents which happen on site, special consideration within the Workplace Violence Prevention and Response Program may also be provided for employees who work outside the operational area, such as executives and sales personnel. These employees may be in another company's facilities or even have stopped for a cup a coffee at a convenience store when a workplace violence incident happens. In the eyes of the law, these employees are "working;" thus, it is the same as if these individuals were in their office. Appropriate education and training may be applicable for these employees.

An area often overlooked in preparing a Workplace Violence Prevention and Response program is the shareholders. This is a greater area of importance today than ever before, given the number of retirees, pension funds, mutual funds, and average households who now own stock in companies. These individuals have risked their retirement savings on the profitability of your company. Additionally, these individuals and their brokers are acquiring information via television, wire services, and computer services regarding your company virtually immediately after a workplace violence incident. If these individuals feel that their investment is in jeopardy because of the workplace violence incident, they may sell their stock to protect their investment. If a large number of shareholders sell their stock, the price of the stock will go down. With the stock price falling, the value of your company also falls. This lowering in value can have a dramatic trickle down effect on the company for a substantial period of time. Prudent professionals may want to consider this important area as part of the response component of the overall program including the following:

- Be prepared for sudden departure of investors when the news breaks
- Provide prepared information to stock brokers to calm their nerves
- Provide timely information to shareholders
- Schedule interviews by executives with CNNFN, CNBC, and related networks.
- Provide internet information on your web site
- Keep shareholders informed after an incident
- Provide timely media releases
- Provide prepared sound bites for media
- Put the proper "spin" on releases

A vitally important area that should be addressed in the Workplace Violence Prevention and Response Program is the control of the media. Given today's technology with on-site news vehicles uplinking the news immediately to the television station and transference to global stations such as CNN, footage of a workplace violence incident in No Where, USA, will be seen in the living rooms of the world within hours if not minutes. This type of news coverage and videotape footage has an impact and creates an impression of your company in the minds of your shareholders, your competitors, Congress, the legal community, and the general public. This impact is dynamic in so far as the general public has even created a new term from the workplace violence incidents at the U.S. Postal Service..."going postal" on someone, meaning assaulting or aggressively attacking. Preparing to address the media after a workplace violence incident could include:

- Preparing sound bite information about your company
- Preparing packets of company literature
- Identifying a media center on location and requiring the media to remain in this area
- Selecting a spokesperson
- Screening all comments through legal or other corporate levels
- Never using "No Comment"

- Preparing proper identification of company officials
- Preparing proper security at center
- Preparing background for camera background
- Select proper "image" clothing for spokesperson
- Ensure media remains at the media center
- Locate employees away from the media
- Make the media's job easy — remember their deadlines
- Don't release family information
- Expect the unexpected

In addition to the enforcement personnel and the media, various governmental agencies will be required to investigate the workplace violence scene. These agencies, such as the Occupational Safety and Health Administration, may require notification in fatality or multiple injury situations, and virtually all governmental agencies can issue monetary penalties. To prepare for the arrival of governmental agencies, the Workplace Violence Prevention and Response Program should include the following:

- Know the requirements and limitations
- Designated company liaison
- Identify the type of inspection
- Prepare your evidence along with the inspector (i.e., photos, testing, etc.)
- Prepare your forms and documents; have documents reviewed by legal department and corporate
- Prepare for criminal investigation in a fatality or multiple-injury situation
- Notification procedures (8 hours for OSHA)
- Prepare for district attorney (criminal rules)
- Protect trademark and patent information
- Prepare for requests for internal documents

If the investigation is criminal in nature, consideration should be given to the following:

- Preserve Constitutional rights
- Have legal counsel present
- Know criminal law rules
- Remember *Miranda* rights
- Provide no documents voluntarily
- State or federal laws being used?
- Company's position?
- Individual position?

Given our litigious society, the Workplace Violence Prevention and Response Plan may need to include preparation for subsequent litigation by involved parties. In preparing for litigation consideration may be given to the following areas:

- Careful documentation of incident scene
- Acquisition of medical records
- Review of personnel records
- Contact with local bar association

In summation, every aspect of the response and recovery process should be anticipated — from the proper cleanup and disposal of the bodily fluids to the information provided to shareholders. Following a workplace violence incident, the way in which a company manages the situation can have a definite long-term bearing on the lives of all directly or indirectly involved. In essence, the company wants to be prepared to make the best of a bad situation. If properly managed, time will heal the wounds. If improperly managed, the situation can be disastrous. Preplanning and preparation are the keys, and there is no substitute.

chapter fourteen

Other selected readings and studies

"Education is no longer thought of as a preparation for adult life, but as a continuing process of growth and development from birth until death."

Stephen Mitchell

"The future of civilization is, to a great extent, being written in the classrooms of the world today."

Milton L Smith

Workplace violence issues and protective measures are still emerging in the American workplace. Professionals should keep abreast with the most recent research in this important area in order to ensure every possible measure is taken to minimize or eliminate the risk of workplace violence. To assist in researching workplace violence issues, below is a group of selected studies and readings in alphabetical order.

"Attacks Up Against Judges, Lawyers," *National Law Journal* (November, 1997).

AFSCME (American Federation of State, County, and Municipal Employees) and AFL-CIO [1995]. Hidden violence against women at work. *Women in Public Service* 5(fall):1–6

Alexander BH, Franklin GM, Wolf ME [1994]. The sexual assault of women at work in Washington State, 1980 to 1989. *Am J Public Health* 84:640–642.

Amandus HE [1994]. Status of NIOSH research on prevention of robbery-related intentional injuries to convenience store workers. *Questions and Answers in Lethal and Non-Lethal Violence: Proceedings of the Third Annual Workshop of the Homicide Research Working Group.* Washington, D.C.: National Institute of Justice.

Amandus HE, Hunter RD, James E, Hendricks S [1995]. Reevaluation of the effectiveness of environmental designs to reduce robbery risk in Florida convenience stores. *J Occup Environ Med* 37(6):711–717.

Applegate, Jane, "Small Firms Not Immune To Workplace Violence," *Lexington Herald Leader*, (Sunday, Sept. 10, 1995 at p.8.).

"As Workplace Violence Episodes Increase, Employers Face Myriad Of Legal Issues," *Daily Labor Report*, (February 14, 1994).

Bachman R [1994]. Violence and theft in the workplace. In *U.S. Department of Justice Crime Data Brief*. Washington, D.C.: U.S. Government Printing Office, NCJ–148199.

Barab J [1995]. Workplace violence: how labor sees it. *New Solutions*, spring issue.

Barrier, M. "The Enemy Within," *Nation's Business*, Feb 1995 p18-22.

Bell CA [1991]. Female homicides in United States workplaces, 1980–1985. *Am J Public Health* 81:729–732.

Bell CA, Stout NA, Bender TR, Conroy CS, Crouse WE, Myers JR [1990]. Fatal occupational injuries in the United States, 1980 through 1985. *JAMA* 263(22):3047–3050.

Bensimon, Helen Frank "Violence in the Workplace," *Training and Development Magazine*, p 27 (Jan 1994).

Bensley L, Nelson N, Kaufman J, Silverstein B, Kalat J [1993]. Study of assaults on staff in Washington State Psychiatric Hospitals. Olympia, WA: State of Washington Department of Labor and Industries.

Block R, Felson M, Block C [1984]. Crime victimization rates for incumbents of 246 occupations. *Sociol Soc Res* 69:442–451.

Borwegen W [1995]. Violence as a preventable occupational hazard: a labor perspective. *New Solutions*, spring issue.

Bowie V [1989]. *Coping with Violence: A Guide for the Human Services*. Sydney, Australia: Karibuni Press.

Brierley B [1996]. Personal communication on February 7, 1996, between B. Brierley of the IACP/Dupont Kevlar Survivors' Club and Lynn Jenkins, Division of Safety Research, National Institute for Occupational Safety and Health, Centers for Disease Control and Prevention, Public Health Service, U.S. Department of Health and Human Services.

Bureau of Justice Statistics (BJS) [1994]. Criminal victimization in the United States, 1992: a national crime victimization survey report. Washington, D.C.: U.S. Department of Justice, Bureau of Justice Statistics, NCJ–145125, p. 79.

Bureau of Justice Statistics (BJS) [1993]. Employment and earnings. Washington, D.C.: U.S. Department of Labor, Bureau of Labor Statistics, January issue.

Bureau of Justice Statistics (BJS) [1994a]. Annual survey of occupational injuries and illnesses. Washington, D.C.: U.S. Department of Labor, Bureau of Labor Statistics. Unpublished database.

Bureau of Justice Statistics (BJS) [1994b]. National census of fatal occupational injuries, 1993. Washington, D.C.: U.S. Department of Labor, Bureau of Labor Statistics, BLS News, USDL–94–384.

Bureau of Justice Statistics (BJS) [1994c]. Violence in the workplace comes under closer scrutiny. Issues in labor statistics. Washington, D.C.: U.S. Department of Labor, Bureau of Labor Statistics, Summary 94–10.

Bureau of Justice Statistics (BJS) [1994d]. Work injuries and illnesses by selected characteristics, 1992. Washington, D.C.: U.S. Department of Labor, Bureau of Labor Statistics.

Bureau of Justice Statistics (BJS) [1995]. National census of fatal occupational injuries, 1994. Washington, D.C.: U.S. Department of Labor, BLS News, USDL–95–288.

Cabral R [1995]. Workplace violence: viable solutions under collective bargaining. *New Solutions*, spring issue.

Carmel C, Hunter M [1989]. Staff injuries from inpatient violence. *Hosp Community Psychiatry* 40:41–46.

Castillo DN [1994]. Nonfatal violence in the workplace: directions for future research. In *Questions and Answers in Lethal and Non-Lethal Violence: Proceedings of the Third Annual Workshop of the Homicide Research Working Group*. Washington, D.C.: National Institute of Justice.

Castillo DN, Jenkins EL [1994]. Industries and occupations at high risk for work-related homicide. *J Occup Med* 36:125–132.

Castelli, Jim, "NIOSH Condemns WorkPlace-Murder..." *Safety and Health*, (March, 1993 at p. 77).

Cauldron, Shari, "Working Scared," *Legal Assistant Today*, (November/December, 1995 at p. 24).

"Causes of Workplace Deaths in 1996," *Occupational Hazards*, (October, 1997 at p. 22).

CDC (Centers for Disease Control and Prevention) [1994]. Occupational injury deaths of postal workers — United States, 1980–1989. *MMWR* 43(32):587, 593–595.

Census of Fatal Occupational Injuries, Bureau of Labor Statistics, U.S. Department of Labor, August 1994.

Chapman SG [1986]. *Cops, Killers and Staying Alive: The Murder of Police Officers in America*. Springfield, IL: Charles C Thomas.

Chelimsky E, Jordan FC, Russell LS, Strack JR [1979]. Security and the small business retailer. Washington, D.C.: National Institute of Law Enforcement and Criminal Justice, Law Enforcement Assistance Administration, Department of Justice.

Civil Service Employees Association [1993]. A matter of life and death: worksite security and reducing risks in the danger zone. Albany, NY: Civil Service Employees Association, Inc.

Collins JJ, Cox BG [1987]. Job activities and personal crime victimization: implications for theory. *Soc Sci Res* 16:345–360.

Collins, Pam and Southerland, Mittie, "Workplace Violence: A Recent Case Analysis," *Presented at Academy of Criminal Justice Sciences*, Chicago, Illinois, (March, 1994).

Comment, "The Creation of CommonLaw Rule: Fellow Servant Rule, 1837-1860, p. 132 *University of Pennsylvania Law Review* 579 (1984).

Crime Prevention Division [1990]. Safety tips for the taxi driver and the for-hire vehicle driver. New York, NY: City of New York Police Department.

Crow W, Bull JL [1975]. *Robbery Deterrence: An Applied Behavioral Science Demonstration*. La Jolla, CA: Western Behavioral Sciences Institute.

Davis H [1987]. Workplace homicides of Texas males. *Am J Public Health* 77:1290–1293.

Davis J, Honchar PA, Suarez L [1987]. Fatal occupational injuries of women, Texas 1975–1984. *Am J Public Health* 77:1524–1527.

Erickson R [1991]. Convenience store homicide and rape. In *Convenience Store Security: Report and Recommendations*. Alexandria, VA: National Association of Convenience Stores.

FBI [1992]. Killed in the line of duty: a study of selected felonious killings of law enforcement officers. Washington, D.C.: U.S. Department of Justice, Federal Bureau of Investigation.

FBI [1994]. Uniform crime reports for the United States, 1993. Washington, D.C.: U.S. Department of Justice, Federal Bureau of Investigation.

Friedman RJ, Framer MB, Shearer DR [1988]. Early response to post-traumatic stress. EAP Digest, September/October issue.

Ginsburg, Roy A., "Employment Law — Courts Around the Country Are Addressing Whether Employers May Be Liable for Negligent Hiring, Supervision and Retention of Employees Who Harm Others," *National Law Journal*, (July 8, 1996 at B-5-6).

Goetz RR, Bloom JD, Chenell SL, Moorhead JC [1991]. Weapons possession by patients in a university emergency department. *Ann Emerg Med* 20:8–10.

Goodman RA, Jenkins EL, Mercy JA [1994]. Workplace-related homicide among health care workers in the United States, 1980 through 1990. *JAMA* 272(21):1686–1688.

Hales T, Seligman P, Newman SC, Timbrook CL [1988]. Occupational injuries due to violence. *J Occup Med* 30:483–487.

Jama 100 years ago: assaults upon medical men [1992]. *JAMA* 267(22):2987.

Jenkins EL [1994]. Occupational injury deaths among females: the U.S. experience for the decade 1980 to 1989. *Ann Epidemiol* 4(2): 146–151.

Jenkins EL [1996]. Workplace homicide: industries and occupations at high risk. *Occup Med State of Art Reviews* 11(2):219–225.

Jenkins EL, Layne LA, Kisner SM [1992]. Homicide in the workplace: the U.S. experience, 1980–1988. *AAOHN J* 40:215–218.

Jenkins EL, Kisner SM, Fosbroke DE, Layne LA, Stout NA, Castillo DN et al. [1993]. Fatal injuries to workers in the United States, 1980–1989: a decade of surveillance; national profile. Washington, D.C.: U.S. Government Printing Office, DHHS (NIOSH) Publication No. 93–108.

Johnson, D., Kiehlbauch, J., Kurutz, J., "Workplace Violence Scenario for Supervisors," *HR Magazine*, p. 63 Test 67.

"Judiciary Panel to Examine WorkPlace Violence," *Daily Labor Report*, (December 2, 1993).

Keep N, Gilbert P [1992]. California Emergency Nurses Association's informal survey of violence in California emergency departments. *J Emerg Nurs* 18:433–442.

Kennish, John W., "Violence in the Workplace," *Professional Safety*, (June, 1995 at p.34).

Kraus JF [1987]. Homicide while at work: persons, industries, and occupations at high risk. *Am J Public Health* 77:1285–1289.

Kurlan, Warren., "Workplace Violence," *Risk Management*, p. 76 (June 1993).

Lavoie FW, Carter GL, Danzi DF, Berg RL [1988]. Emergency department violence in United States teaching hospitals. *Ann Emerg Med* 17:1227–1233.

Levin PF, Hewitt JB, Misner ST [1992]. Female workplace homicides: an integrative research review. *AAOHN J* 40:229–236.

Lipscomb JA, Love CC [1992]. Violence toward health care workers: an emerging occupational hazard. *AAOHN J* 40:219–228.

Liss GM, Craig CA [1990]. Homicide in the workplace in Ontario: occupations at risk and limitations of existing data sources. *Can J Public Health* 81:10–15.

Lusk SL [1992]. Violence experienced by Nurse's aides in nursing homes: an exploratory study. *AAOHN J* 40:237–241.

Lynch JP [1987]. Routine activity and victimization at work. *J Quantitative Criminol* 3: 283–300.

Mahoney BS [1991]. *Victimization of Pennsylvania Emergency Department Nurses in the Line of Duty* [dissertation]. State College, PA: Pennsylvania State University.

Martin, Justin, "It's Murder in the Workplace," *Fortune Magazine*, (August 9, 1993 at p.12).

"Murder On The Job," *Personnel Journal*, (February, 1992, at p.72).

National Institute for Safety and Health (NIOSH), Violence in the workplace: risk factors and prevention strategies, *Current Intelligence Bulletin* 57, (July 5, 1996).

Nelson N [1995]. *Violence in Washington Workplaces,* 1992. Olympia, WA: Washington State Department of Labor and Industries, Technical report 39–1–1995.

NIOSH [1992]. Homicide in U.S. workplaces: a strategy for prevention and research. Morgantown, WV: U.S. Department of Health and Human Services, Public Health Service, Centers for Disease Control, National Institute for Occupational Safety and Health, DHHS (NIOSH) Publication No. 92–103.

NIOSH [1993]. NIOSH Alert: request for assistance in preventing homicide in the workplace. Cincinnati, OH: U.S. Department of Health and Human Services, Public Health Service, Centers for Disease Control and Prevention, National Institute for Occupational Safety and Health, DHHS (NIOSH) Publication No. 93–109.

NIOSH [1995]. National Traumatic Occupational Fatalities (NTOF) Surveillance System. Morgantown, WV: U.S. Department of Health and Human Services, Public Health Service, Centers for Disease Control and Prevention, National Institute for Occupational Safety and Health. Unpublished database.

Norris D [1990]. Violence Against Social Workers: The Implications for Practice. London, England: Jessica Kingsley Publishers.

Northwestern National Life [1993]. Fear and violence in the workplace: a survey documenting the experience of American workers. Minneapolis, MN: Northwestern National Life.

O'Carroll PW, Mercy JA [1989]. Regional variation in homicide rates: why is the West so violent? *Violence and Victims* 4(1):17–25.

Office of the Attorney General [1991]. Study of safety and security requirements for "at-risk businesses." Tallahassee, FL: Office of the Attorney General, State of Florida, Department of Legal Affairs.

Oregon Department of Consumer and Business Services [1994]. Violence in the workplace, Oregon, 1988 to 1992: a special study of workers' compensation claims caused by violent acts. Salem, OR: The Department Information Management Division.

"OSHA Now Assesses Employer Liability," *HR Focus,* (June, 1995 at p.7).

"OSHA to Clarify Recordkeeping On Violence," *Daily Labor Report,* (January 18, 1995).

Petersen, Donald J. and Massengill, Douglas, "The Negligent Hiring Doctrine — A Growing Dilemma for Employers," 15 *Employee Relations Law Journal* at 419(1) (1989-90).

"Postal Service Offering Employee Assistance, Hiring Programs to Address Workplace Violence," *Daily Labor Report,* (August 6, 1993).

"Preventing Deaths and Injuries of Adolescent Workers," Center For Disease Control and Prevention, U.S. Department of Health and Human Services, (May, 1995).

Public Law 91–596: Occupational Safety and Health Act of 1970.

Reeves FL, Hall JA [1990]. Taxicab operators' safety program. Atlanta, GA: Department of Public Safety, Bureau of Taxicabs and Vehicles for Hire.

Rowett CR [1986]. Violence in social work: a research study of violence in the context of local authority social work. Cambridge, England: University of Cambridge Institute of Criminology.

"Safety and Health: NIOSH Explores How Best to Study Stress, Violence in Post Offices," *Government Employees Relations Report, Federal News,* (December 06, 1993).

Safety Shields Task Force, Manitoba Taxicab Board [1990]. Report and recommendations: taxi safety shields. Manitoba, Canada: The Task Force.

Schreiber FB [1991]. 1991 national survey of convenience store crime and security. In *Convenience Store Security: Report and Recommendations.* Alexandria, VA: National Association of Convenience Stores.

Seligman PJ, Newman SC, Timbrook CL, Halperin WE [1987]. Sexual assault of women at work. Am J Ind Med 12:445–450.

Shepherd E, ed. [1994]. *Violence in Health Care: A Practical Guide to Coping with Violence and Caring for Victims.* New York, NY: Oxford University Press.

State of California [1993]. Guidelines for security and safety of health care and community service workers. Sacramento, CA: Division of Occupational Safety and Health, Department of Industrial Relations.

Stout NA [1992]. Occupational injuries and fatalities among health care workers in the United States. *Scand J Work Environ Health* 18(Suppl 2):88–89.

Stout NA, Jenkins EL, Pizatella TJ [1996]. Occupational injury mortality rates in the United States: changes from 1980 to 1989. *Am J Public Health* 86:73–77.

Sullivan C, Yuan C [1995]. Workplace assaults on minority health and mental health care workers in Los Angeles. *Am J Public Health* 85(7):1011–1014.

"Surgeon General Elders Urges Employers to Take Steps to Curb WorkPlace Violence," *Daily Labor Report,* (December 17, 1993).

Taking Steps to Prevent Workplace Violence, *Occupational Hazards,* (October, 1997 at p.66-68).

Thomas JL [1992]. Occupational violent crime: research on an emerging issue. *J Safety Res* 23:55–62.

Toscano G, Weber W [1995]. Violence in the workplace. Compensation and working conditions. Washington, D.C.: U.S. Department of Labor, Bureau of Labor Statistics.

Toscano G, Windau J [1994]. The changing character of fatal work injuries. *Monthly Labor Review* 117(10):17–28.

Uchida CD, Brooks LW, Koper C [1987]. Danger to police during domestic encounters: assaults on Baltimore county police, 1984–1986. *CJPR* 2:357–371.

"Violence in the Workplace Comes Under Closer Scrutiny," *Issues In Labor Statistics,* U.S. Department of Labor, (August, 1994).

Windau J, Toscano G [1994]. Workplace homicides in 1992. Compensation and working conditions, February 1994. Washington, D.C.: U.S. Department of Labor, Bureau of Labor Statistics.

Windau, Janice, Toscano, Guy, Murder Inc. — Homicide in the American workplace, *Business in Society,* (Spring, 1994 at p.58).

"Worked to Death," *The Houston Chronicle,* (Fall, 1994 p.1-16).

"Worker Remembers Postal Shootings," *Richmond Register,* (August 19, 1996 at p.B3).

"Workplace Deaths Reported by 12% of HR Managers: Private Sector Rate at 3.5%," *Daily Labor Report,* (April 14, 1994).

"Workplace Murders Provoke Lawsuits and Better Security," *The New York Times,* (February 14, 1994 at p.B4).

"Workplace Violence Generates Two Kinds of Torts," *National Law Journal,* (April 17, 1995 at p.C32).

Workplace Violence: OSHA says guidelines will target the retail and healthcare sectors, 1995 *DLR* p. 16 (BNA January 23, 1995).

Zimmering FE, Zuehl J [1986]. Victim injury and death in urban robbery: a Chicago study. *J Leg Studies* 15(1):1–40.

Case Studies[1]

Introduction

The call comes in.

Someone's been shot — there is a fight going on — someone's been threatened — someone's being stalked by an ex-boyfriend — someone's threatening suicide — someone wants to put a stop to the "bullying" behavior that's been going on in his office.

These are just a few examples of the types of incidents reported.

How each agency responds to these reports will differ, not only among agencies but also within each agency, because the circumstances surrounding each situation are different. Even in agencies that are highly structured and have well-thought-out procedures in place, the response will necessarily depend on:

- The nature of the incident,
- The circumstances surrounding the incident,
- Who is available to respond, and
- Who has the skills needed to deal with the particular situation.

What has been learned from agencies' many years of experience is that the most effective way to handle these situations is to take a team approach, rather than having one office handle a situation alone. In some cases of workplace homicides, it became apparent that the situation got out of control because personnel specialists did not inform security about a problem employee, or coworkers were not warned about the threatening behavior of an ex-employee, or one agency specialist felt he had to "go it alone" in handling the situation.

Agencies should have plans in place ahead of time so that emergency and non-emergency situations can be dealt with as soon as possible. However, it is also necessary to build the maximum amount of flexibility possible into any plan.

---

[1] These case studies are from the recently published "Dealing with Workplace Violence: A Guide for Agency Planners," Published by the Office of Personnel Management.

Basic concepts

Since agencies and situations differ, specific steps or procedures to follow on a Government-wide basis would be inappropriate and impractical. However, there are some basic concepts that all agencies should keep in mind when formulating their strategy to address workplace violence.

- Respond promptly to immediate dangers to personnel and the workplace.
- Investigate threats and other reported incidents.
- Take threats and threatening behavior seriously; employees may not step forward with their concerns if they think that management will dismiss their worries.
- Deal with the issue of what may appear to be frivolous allegations (and concerns based on misunderstandings) by responding to each report seriously and objectively.
- Take disciplinary actions when warranted.
- Support victims and other affected workers after an incident.
- Attempt to bring the work environment back to normal after an incident.

How to use the case studies

The case studies presented in this chapter are derived from real life situations that have arisen in Federal agencies. They are intended to provide assistance to agency planners as they develop workplace violence programs and assess their readiness to handle these types of situations. It should be noted that, in some of the case studies, the circumstances have been modified to make them better learning tools.

As you read the case studies, keep in mind that there is no one correct way to handle each situation. The case studies should not be taken as specific models of how to handle certain types of situations. Rather, they should be a starting point for a discussion and exploration of how a team approach can be instituted and adapted to the specific needs and requirements of your agency.

Questions for discussion

The case studies are intended to raise questions such as:

1. Do we agree with the approach the agency took in the case study?
2. If not, why wouldn't that approach work for us?
3. Do we have adequate resources to handle such a situation?

Questions for program evaluation

Establish a system to evaluate the effectiveness of your response in actual situations that arise so that you can change your procedures if necessary.

Ask the following questions after reviewing each of the case studies and after planning how your agency would respond to the same or a similar situation:

1. Does our workplace violence program have a process for evaluating the effectiveness of the team's approach following an incident?
2. Would our written policy statement and written procedures limit our ability to easily adopt a more effective course of action in the future, if an evaluation of our response showed that a change in procedures was necessary?
3. Do we have plans to test our response procedures and capability through practice exercises and preparedness drills and change procedures if necessary?

Case study 1 — a shooting

The incident

The report comes in: Two employees have been killed in the workplace and two have been wounded. A witness has called 911 and the police and ambulances have arrived. The perpetrator (an agency employee) has been taken into custody, the victims are being sent to the hospital, and the police are interviewing witnesses and gathering evidence.

Response

In this situation, the agency's crisis response plan called for the immediate involvement of:

1. A top management representative
2. A security officer
3. An employee relations specialist
4. An EAP counselor
5. An official from the public affairs office

Top management representative. The manager, an Assistant Director of a field office with 800 employees, coordinated the response effort because she was the senior person on duty at the time. In addition to acting as coordinator, she remained available to police throughout the afternoon to make sure there were no impediments to the investigation.

She immediately called the families of the wounded and assigned two other senior managers to notify the families of the deceased. She also arranged for a friend of each of the deceased coworkers to accompany each of the managers. She took care of numerous administrative details, such as authorizing expenditures for additional resources, signing forms, and making decisions about such matters as granting leave to coworkers. (In this case, the police evacuated the building, and employees were told they could go home for the rest of the day, but that they were expected to return to duty the following day.)

To ensure a coordinated response effort, she made sure that agency personnel involved in the crisis had cell phones for internal communication while conducting their duties in various offices around the building.

Security staff. The security staff assisted the police with numerous activities such as evacuating the building.

Employee Relations Specialist. The employee relations specialist contacted the agency's Office of General Counsel and Office of Inspector General and alerted them to the situation so that they could immediately begin to monitor the criminal proceedings. He made a detailed written record of the incident, but he did not take statements from witnesses because it could have impeded the criminal prosecution of the case. He also helped the supervisor draft a letter of proposed indefinite suspension pending the outcome of the criminal proceedings. He worked closely with the prosecutor's office to obtain relevant information as soon as it was available so the agency could proceed with administrative action when appropriate.

Employee Assistance Program (EAP) counselor. The agency had only one EAP counselor on duty at the time. However, in prior planning for an emergency, they had contracted with a local company to provide additional counselors on an "as needed" basis. The one EAP counselor on duty called the contractor and four additional counselors were at the agency within an hour. The counselors remained available near the scene of the incident to reassure and comfort the employees.

The agency EAP counselor arranged for a series of Critical Incident Stress Debriefings (CISD) to take place two days later (see Chapter 10 for a discussion of CISD) and also arranged for two contract EAP counselors to be at the workplace for the next week to walk around the offices inquiring how the employees were doing and to consult with supervisors about how to assist employees to recover. She made contact with the wounded employees and their family members, as well as family members of the deceased, to offer EAP services.

Public Affairs Officer. The Public Affairs Officer handled all aspects of press coverage. She maintained liaison with the media, provided an area for reporters to work, and maintained a schedule of frequent briefings.

Questions for the agency planning group

1. How would your agency have obtained the services of additional EAP counselors?
2. How would employees be given information about this incident?
3. Who would clean up the crime scene?
4. Would you relocate employees who worked in the area of the crime scene?

5. What approach would your agency take regarding granting administrative leave on the day of the incident and requests for leave in the days/weeks following the incident?
6. How would you advise management to deal with work normally assigned to the victims/perpetrator?
7. What support would your agency provide to supervisors to get the affected work group(s) back to functioning?

Case study 2 — viciously beating and wounding a coworker

The incident

The following incident was reported to the agency's Incident Response Team. A female employee had broken off a romantic relationship with a male coworker, but he wouldn't leave her alone. She finally had a restraining order served to him at work. After the police officer served the restraining order, the perpetrator totally lost control and entered the woman's cubicle. He hit her; she fell from her chair. While she was on the floor, he broke a soda bottle and cut her face with the broken glass. While this was going on, coworkers heard the commotion and called the police. The perpetrator fled the scene before police arrived and the victim was transported to the hospital.

Response

The Incident Response Team immediately implemented the following plan.

Security. The Security officer worked with hospital security to ensure that the victim got around-the-clock security while she was in the hospital. He ensured that the hospital staff knew not to give out any information about the victim to callers. He gave the victim advice, reading material, and a video on personal safety. He made sure the perpetrator's card key was inactivated, and he had pictures of the perpetrator made for the building guards. He coordinated efforts with local police.

Employee Assistance Program (EAP). The EAP counselor visited the victim in the hospital and ensured that she was being seen regularly by a social worker on the hospital staff. She worked with the victim's colleagues to help them be supportive of the victim when she came back to work. The EAP counselor visited the worksite to let coworkers know she was available to them.

Employee Relations. The employee relations specialist contacted the agency's Office of General Counsel and Office of Inspector General and alerted them to the situation so that they could begin to monitor the criminal proceedings. He helped the supervisor develop a notice of proposed indefinite suspension using the crime provision.

Union. The union was fully supportive of the agency's efforts to help the victim. Since both the victim and the perpetrator were bargaining unit employees, the union was aware of its role to represent all employees in the bargaining unit. In this particular case, because of the viciousness of the attack, union officials were reluctant to take the case through the entire negotiated grievance procedure. In addition, realizing that this could happen to other employees, the union officials obtained brochures on stalking from their national headquarters and invited an expert speaker on the subject to a chapter meeting.

Supervisor. The employee's supervisor obtained all the necessary forms and assisted the employee in filing an Office of Workers Compensation Programs (OWCP) claim to pay for hospital and medical costs. The supervisor and the employee's coworkers visited her in the hospital, kept in touch with her during her convalescence, and kept her up-to-date on news from the office.

Agency Attorney. An agency attorney maintained contact with the local prosecutor's office.

Resolution

The police caught and arrested the perpetrator after about 10 days. The agency proposed and effected a removal action against the perpetrator based on a charge of "wounding a coworker." He did not appeal the action.

The employee remained hospitalized for two days and then went to the home of a friend until the perpetrator was apprehended. She remained at home for another two weeks before returning to work. Her OWCP claim was accepted. She continues to stay in touch with the Employee Assistance Program counselor who had visited her at the hospital and assisted her during her time away from the office. The counselor referred her to a support group for battered women, and she finds it very helpful.

The perpetrator was found guilty and received jail time. After jail time was served, and at the suggestion of an agency attorney, the court forbade the perpetrator to contact the victim or the agency as one of the conditions of probation. The security officer alerted security guards and discussed security precautions with the victim, ensuring that there would be an effective response if the perpetrator violated this restriction.

Questions for the agency planning group

1. Who at your agency would monitor the proceedings of the criminal case, e.g., to be aware of the situation if the perpetrator got out of jail on bail or probation?
2. Does your security office establish liaison with and keep in contact with agency or local law enforcement authorities in order to coordinate efforts in these type of cases?
3. Do you have a procedure in place for cleaning up the crime scene after investigators are finished examining it?
4. Would employees at your agency know who to call in an emergency—for example, 911, the Federal Protective Service, in-house security, or in-house law enforcement?

Case study 3 — a suicide attempt

The incident

A member of the agency's Incident Response Team received a frantic call from an employee saying that her coworker just left her office muttering about the final straw—you all won't have me to push around any more. She said she's been worried for weeks about the possibility of her coworker committing suicide and knows now she should have called earlier. The staff member who took the call told the employee to see if she could find her coworker and remain with her. Help was on its way.

Response

For incidents involving suicide threats, the agency's plan was to call local police if there seemed to be an immediate danger and, if not enough was known about the situation, to contact security and the Employee Assistance Program (EAP) counselor to do an immediate assessment of the situation.

The team member who took the initial call first contacted security who immediately located the two employees. The EAP counselor could not be reached immediately, so the team member called an employee in the Human Resources (HR) department who had earlier volunteered to help out in emergency situations (she had been trained in her community in dealing with suicide attempts).

The HR specialist arrived at the distressed employee's office within two minutes of the call. The employee was crying at this point and saying she was beyond help. *It'll be over soon.* The HR specialist recognized what was happening and asked the security officer to call police and an ambulance and tell them there was a suicide attempt. After calling the police, the security officer went outside to meet the emergency workers and direct them to the scene. The HR specialist then learned from the woman that she had swallowed 10 pills an hour earlier. The police and ambulance were on the scene within three minutes of the call and the woman was hospitalized.

The HR specialist contacted the employee's family and then prepared a report of the incident. The EAP counselor consoled and supported the coworker who had initially called the Incident Response Team.

Emergency treatment was successful, and the employee was admitted to the hospital's psychiatric unit. The EAP counselor and HR specialist stayed in touch with the employee and supported her in planning her return to work. She returned to work four weeks later, functioning with the help of anti-depressant medication and twice-weekly psychotherapy sessions.

With the employee's consent, the Employee Assistance Program (EAP) counselor arranged a meeting involving the employee, her supervisor, and the Human Relations specialist to coordinate her treatment and work activities. The supervisor agreed to adjust the employee's work schedule to fit her therapy appointments as a reasonable accommodation, and the supervisor provided guidance on procedures and medical documentation requirements for leave approval. The counselor, supervisor, and employee agreed on a plan for getting the employee immediate emergency help if she should feel another crisis coming on.

Resolution

Two years later, the employee is doing well, working a normal schedule, and continues to be a productive employee. She no longer takes anti-depressant medication, but she stays in touch with both her psychiatrist and the EAP counselor.

Questions for the agency planning group

1. Do you agree with the agency's approach in this case?
2. Does your agency have back-up plans for situations where key team members are not available?
3. Has your agency identified employees with skills in handling emergencies?
4. Does your workplace violence policy and training encourage employees to report incidents at an early stage?
5. Does your workplace violence policy and training encourage employees to seek guidance with regard to problems that trouble them even though they don't understand the nature of the problem?
6. If the employee had left the building before emergency personnel arrived, does your plan provide for contacting the appropriate authorities?

Case study 4 — stalking

The incident

A supervisor called the Employee Relations office to request a meeting of the workplace violence team for assistance in handling a situation he's just learned about. He was counseling with one of his employees about her frequent unscheduled absences, when she told him a chilling story of what she's been going through for the past year. She broke up with her boyfriend a year ago and he's been stalking her ever since. He calls her several times a week (she hangs up immediately). He shows up wherever she goes on the weekends and just stares at her from a distance. He often parks his car down the block from her home and just sits there. He's made it known he has a gun.

Response

This agency's plan calls for the initial involvement of security, the Employee Assistance Program (EAP), and Employee Relations in cases involving stalking.

The security officer, the EAP counselor, and employee relations specialist met first with the supervisor and then with the employee and supervisor together. At the meeting with the employee, after learning as much of the background as possible, they gave her some initial suggestions.

- Contact her local police and file a report. Ask them to assess her security at home and make recommendations for improvements.
- Log all future contacts with the stalker and clearly record the date, time, and nature of the contact.
- Let voice mail screen incoming phone calls.
- Contact her own phone company to report the situation.
- Give permission to let her coworkers know what was going on (she would not agree to do this).
- Vary her routines, e.g., go to different shops, take different routes, run errands at different times, report to work on a variable schedule.

The team then worked out the following plan:

1. The **Employee Relations** specialist acted as coordinator of the response effort. He made a written report of the situation and kept it updated. He kept the team members, the supervisor, and the employee apprised of what the others were doing to resolve the situation. He also looked into the feasibility of relocating the employee.
2. The **Security** officer immediately reported the situation to the local police. With the employee's consent, she also called the police where the employee lived to learn what steps they could take to help the employee. She offered to coordinate and exchange information with them. The security officer arranged for increased surveillance of the building and circulated photos of the stalker to all building guards with instructions to detain him if he showed up at the building. She brought a tape recorder to the employee's desk and showed her the best way to tape any future voice mail messages from the stalker. She also contacted the agency's phone company to arrange for its involvement in the case.
3. The **Employee Assistance Program** counselor provided support and counseling for both the employee and the supervisor throughout the time this was going on. He suggested local organizations that could help the employee. He also tried to convince her to tell coworkers.
4. The **Union** arranged to sponsor a session on stalking in order to raise the consciousness of agency employees about the problem in general.

After a week, when the employee finally agreed to tell coworkers what was going on, the EAP counselor and security jointly held a meeting with the whole work group to discuss any fears or concerns they had and give advice on how they could help with the situation.

Resolution

In this case, the employee's coworkers were supportive and wanted to help out. They volunteered to watch out for the stalker and to follow other security measures recommended by the security specialist. The stalker ended up in jail because he tried to break into the employee's home while armed with a gun. Security believes that the local police were able to be more responsive in this situation because they had been working together with agency security on the case.

Questions for the agency planning group

1. Do you agree with the agency's approach in this case?
2. What would you do in a similar situation if your agency doesn't have security guards?
3. What would you do if coworkers were too afraid of the stalker to work in the same office with her?
4. What would you do if/when the stalker gets out of jail on bail or out on probation?
5. Would your Office of Inspector General have gotten involved in this case, e.g., coordinated agency efforts with local law enforcement agencies?

Case study 5 — a domestic violence situation

The incident

A team member, the employee relations specialist, receives a phone call from an employee. She reports that she has just finished a long conversation with a friend and coworker, a part-time employee, who revealed to her that she is a victim of domestic violence. The coworker, a single mother, relates that she had always considered her friend a very fortunate person—attractive, personable, and, because of her husband's success in business, able to devote a good deal of her time to her beautiful home and two small children.

Finding her in tears after a phone call from home, the employee had encouraged her coworker to go for a walk at lunch time and talk about whatever was wrong. To her surprise, she learned that the woman's husband has been abusing her since their first child was born. He is careful to injure her only in ways that do not leave visible signs, and she feels sure no one would ever believe her word against his. The family's assets, even "her" car, are all in his name, and her part-time salary would not be enough for herself and the children to live on. Further, he has threatened to kill her if she ever leaves him or reveals the truth.

After talking with the employee, the coworker agreed to let the situation be reported to the workplace violence team.

Response

The **Employee Relations** specialist agreed to meet with both employees immediately. The abused woman asked to have her friend along and, at the employee relations specialist's suggestion, gave her permission to explain the situation to the two employees' supervisor. After interviewing her in a caring, supportive way to get basic information, she asked other team members, the security director and the Employee Assistance Program (EAP) counselor, to join her in analyzing the situation. Then she met with the abused employee, her friend (at her request), and her supervisor to report on the team's recommendations.

The **EAP** counselor arranged for the abused woman to see another counselor, who had an open appointment that same day, for counseling and referral to the community agencies that could help her.

The counselor referred her to a comprehensive shelter for victims of abuse. She explained the comprehensive services the shelter could offer her: a safe place to stay with her children, advice on how to get out of her home situation safely, legal advice, and much other helpful information. At first, the employee was afraid to change the status quo. After several meetings with the Employee Assistance Program counselor and encouraging talks with her friend, she agreed to talk with the shelter staff. Her friend drove her to the meeting. They worked with her to develop a safe plan for leaving home with her children.

The employee asked the workplace violence team to coordinate with the shelter staff. After discussing her plan with them, the **Security** director identified that right after she left home would be a high risk period and arranged for a guard to be at the workplace during that time. He supplied photographs of the husband to the guard force.

With the woman's consent, the supervisor and security director discussed the situation with coworkers, shared the picture with them, and explained what they should do in various contingencies. At the meeting one coworker began complaining about danger to herself. The friend argued persuasively that, *This could happen to any of us. Would you rather we stick together, or leave one another to suffer alone?* This rallied the group, and the complainer decided to go along with the others.

The **Supervisor** agreed to use flexitime and flexiplace options to make the employee more difficult to find. Not only would she be working a different schedule; she would report to a suburban telecommuting center instead of the agency's central office.

The supervisor explained to the employee that she would like very much to have her on board full time, as she was an excellent worker, but that there was no position available. However, she encouraged her to seek a full time job, and made phone calls to colleagues in other departments to develop job leads for her. One of her professional associates offered to allow the employee to use their organization's career transition center, which had excellent job search resources, and was located in a different part of town from her normal worksite.

Resolution

The employee executed her plan for leaving home and moved to the shelter with her children. She worked with an attorney to obtain financial support and to begin divorce proceedings. She often had times of doubt and fear but found the shelter staff very supportive. Her coworkers encouraged her to call daily with reports on her progress.

The husband appeared at the office only once, a few days after his wife moved into the shelter. He shouted threats at the security guard, who calmly called for backup from the local police. Fearing for his reputation, he fled the scene before police could arrive. The guard force continued to monitor any efforts by the husband to gain entry to the building.

Six months later, the employee has obtained a full-time position at a nearby office within the same agency. She discovered that they also had a workplace violence team and made them aware of her situation, just in case she should need their help. She and her children have moved into an apartment. The children are seeing a child psychologist, recommended by the Employee Assistance Program counselor, to help them make sense of an upsetting situation, and she attends a support group for battered women. Her friend from her former office has helped her with encouragement, support, and suggestions on how to handle the stresses of single parenthood.

Questions for the agency planning group

1. Are your team members knowledgeable about domestic violence?
2. What do you think about the role of the friend? How would you encourage agency employees to support coworkers in these types of situations?
3. Does your agency have access to career transition services to help in these types of situations?
4. Has your planning group identified someone knowledgeable about restraining/protective orders to discuss with the employee the pros and cons of obtaining one?

Case study 6 — a threat

The incident

At a smoking break with one of his colleagues from down the hall, an employee was reported to have made this statement, I like the way some employees handle problems with their supervisors—they eliminate them. One of these days I'm going to bring in my gun and take care of my problem. The employee who heard the statement reported it to his supervisor, who in turn reported it to his supervisor, who called a member of the workplace violence team.

When the supervisor of the employee who allegedly made the threat was contacted, he reported that, several months earlier, the same employee had responded to his casual question about weekend plans by saying, **I'm going to spend the weekend in my basement with my guns practicing my revenge.** At that time, the supervisor had warned the employee that such talk was unacceptable at work and referred the employee to the Employee Assistance Program (EAP).

Response

In the case of a threat where there does not appear to be an imminent danger, the agency's plan called for the **Employee Relations** specialist to conduct an immediate preliminary investigation and for the team to meet immediately afterward to look at the available evidence and strategize a preliminary response.

That afternoon, the employee relations specialist interviewed the employee who heard the threat, that employee's supervisor, the supervisor of the employee who made the threat, and subsequently the employee who allegedly made the threat. The employee who made the threat denied saying any such thing. There were no witnesses. The supervisor expressed concern for his and his staff's safety. He said that this morning's incident confirmed his earlier concerns. Based on comments from supervisors and the employee who made the threat, the employee relations specialist recommended that a more thorough investigation be done.

At the meeting where the employee relations specialist's findings were discussed, the following people were present: the first- and second-level supervisor of the employee who allegedly made the threat, an Associate Director of the agency, the agency security officer, the employee relations specialist, the EAP counselor, and an attorney from the General Counsel's Office.

One of the team members recommended that the employee be given a counseling memo and referred to the Employee Assistance Program. The consensus of the others, however, based on the employee relations specialist's verbal report, was to recommend to the supervisor that the employee be placed on excused absence pending an investigation and that he be escorted from the premises.

The **Security Officer** and the employee's second-level **Supervisor** went together to give the alleged threatener a letter that simply stated *This is to inform you that effective immediately you will be in a paid, non-duty status, pending an agency determination regarding your actions on June 10. You are required to provide a phone number where you can be reached during working hours.* They also took away his identification badge and office keys, and escorted him to the building exit.

The team consulted with the agency's Office of Inspector General which arranged for a criminal investigation to be conducted. The **Criminal Investigator** interviewed all of the employee's coworkers and two other employees who the coworkers indicated had knowledge of this employee's prior statements against his supervisors. He then interviewed the alleged threatener.

The criminal investigator checked to see if the employee had a police record. He did not. The investigator also checked his workplace to see if he had any weapons at the office or if he had any written material of a threatening nature. The search of his workplace found nothing of consequence.

The investigative report showed that the employee told his coworkers on several occasions that he had no respect for his supervisor and that he thought that threatening him was an effective way to solve his problems with him. Signed statements indicated that he bragged about knowing how to get his way with his boss.

The prosecutor's office, after receiving the investigative report, made a determination that it would not prosecute the case and informed management that they could proceed with administrative action. The team recommended a proposed removal action since the evidence showed that the employee was using threats to intimidate his supervisor.

Resolution

The second-level supervisor proposed a removal action based on a charge of "threatening a supervisor." A top manager who had not been directly involved in the case insisted that the agency enter into a settlement agreement that would give the employee a clean Standard Form (SF) 50. However, based on the particular facts in this case, the team convinced him that he was not solving any problems by settling the case in this way and was, in fact, just transferring the problem to another unsuspecting employer. The top manager finally agreed and the employee was removed from Federal service.

> **Note:** Even though the agency did not settle the case, and did, in fact, effect a removal action, the employee was soon hired by another agency anyway. The new agency never checked his references and is now experiencing the same intimidating behavior from the employee.

Questions for the agency planning group

1. What would your agency have done about checking references before hiring this employee?
2. What do you think would have been the risks of settling the case with a clean SF 50?
3. How would your agency have handled the case if the key witness (i.e., the employee who heard the threat) was not credible?

Case study 7 — veiled threats

The incident

A team member took a phone call from a supervisor who said, **One of my employees said this morning that he knows where my kids go to school. I know that doesn't sound like much to you, but if you saw the look in his eyes and heard the anger in his voice, you'd know why I need your help in figuring out what to do.**

Response

The team member who took the call heard more details about the incident and then set up a meeting with the supervisor who made the report, a security specialist, an employee relations specialist, and an Employee Assistance Program (EAP) counselor.

At the meeting, the **Supervisor** who made the report told the team that the employee who said that he knows where his kids go to school has been engaging in intimidating behavior against him for the past year since he became his supervisor. The supervisor had spoken with him on several occasions to let him know that his behavior was unacceptable. He also had given him a written warning along with a written referral to the EAP.

Because the office was in a General Services Administration controlled building, the **Security** specialist then called the regional office of the Federal Protective Service (FPS). The FPS contacted the threat assessment unit of the state police, who agreed to assign a threat assessment consultant to assist the agency. In a phone consultation with the team, the **Threat Assessment Consultant** suggested that the team arrange for an immediate investigation by an investigator who was experienced in workplace violence cases. The investigator should explore the following areas:

1. What further background information can be learned about the relationship between the supervisor and alleged threatener?
2. What is the relationship between the supervisor and his other employees and coworkers?
3. Have there been problems of a similar nature with the alleged threatener's previous supervisors? If so, how were they resolved or handled? If there were problems with previous supervisors, were they similar to or different from the current situation?
4. What are the alleged threatener's relationships with coworkers? Might there be other potential victims? Are there also interpersonal problems between the alleged threatener and other employees?
5. Are there unusually stressful problems in the life of the alleged threatener, e.g., divorce, financial reversal, or any other recent significant traumatic event?

6. Does anyone else feel threatened based on their interaction with the alleged threatener?
7. Does the alleged threatener have access to weapons? Has he recently acquired weapons?

The threat assessment consultant scheduled another telephone consultation with the team for three days later. He also suggested that the investigator not interview the alleged perpetrator until after the next phone consultation.

The investigation was conducted immediately by a **Professional Investigator** and the team reviewed the investigative report prior to the next phone conversation with the threat assessment consultant. The report contained statements by the employee's supervisor about veiled threats the employee had made, such as *If you give me that assignment, you'll be sorry, I know where you live*, and *I see you every day on your way to work* (the employee lives at the opposite end of town from the supervisor). Also in the investigative report was a transcript (and a tape recording) of two voice mail messages that the supervisor found intimidating—one in which the employee said he needed annual leave that day to go for target practice and another one in which he said he couldn't come to work that day because he had to go hunting.

Again, the supervisor's statement showed that he considered the employee's tone of voice to be intimidating and said that, on the day previous to each of these phone calls, the employee had acted as though he was angry about new assignments the supervisor had given him. The supervisor said he has taken several precautions as a result of the threats. For example, he told his children to take precautions, installed dead-bolt locks at his home, and asked the local police to do a security survey of his home. In addition to the investigative report, the security office obtained a police record showing a misdemeanor conviction for spousal abuse several years earlier.

Participating in the phone consultation with the threat assessment consultant was the workplace violence team, the second-line supervisor, and the director of the office. The purpose of the consultation was to:

• Analyze the information contained in the investigative report,
• Determine what additional information was needed,
• Determine whether to interview the alleged perpetrator,
• Help the team members organize their thinking about how to proceed with the case, and
• Discuss a range of options that could be taken.

The threat assessment consultant recommended that the investigator interview three coworkers, the employee's ex-wife who had been abused, and subsequently the alleged threatener. The purpose of the interview with the alleged threatener would be to corroborate what was said by the others and get his explanation of why he made the statements. The interview would also communicate to him that this kind of conduct has been noticed, troubles people, and is not condoned. He advised that security measures, including having a security officer in the next room, be taken when the alleged threat-ener was interviewed. The threat assessment consultant also gave the team guidance in the preservation of evidence, such as written material and tape recordings, and in the documentation of all contacts.

During the interview, the alleged threatener made what the investigator believed were several additional veiled threats against the supervisor. He even behaved in a way that led the investigator to be concerned about his own safety.

Based on the findings of the investigation, the threat assessment con-sultant concluded that the employee presented behaviors that showed that a real possibility existed that the employee, if pushed, might carry out some of his threats toward the supervisor and his family. He expressed concern that, if the employee continued to work in the same office, the situation could escalate. Management decided to place the employee on excused absence for the safety of the threatened supervisor.

The threat assessment consultant worked with team members to develop a plan for ongoing security. For example, he suggested the team identify one member to coordinate case management, recommended monitoring any further communication between the employee and other agency employees (e.g., any phone calls, any email messages, any showing up at residences, etc. were to be reported to the case manager). He recommended that security officials be in the area, though not visible, whenever meetings were held with the employee. The threat assessment consultant remained available for telephone consultation as the team carried out the plan.

Resolution

Despite the agency's concerns that any agency action might trigger an action against the supervisor's family, the agency removed the employee based on a charge of threatening behavior. The agency's analysis considered the cred-ibility of the supervisor and employee, and the information and evidence gathered. The employee did not appeal the removal action.

The agency security officer gave the supervisor advice on personal safety and discussed with him the pros and cons of obtaining a restraining order for his family. The security officer also helped the supervisor get in touch with the local office of victim assistance for additional ideas on ways to protect his family. The threat assessment consultant also spoke with the supervisor and suggested that he may want to go to the school, school bus driver, and neighbors and make them aware of the problem and the threat-ener's appearance (show them his picture). The reason for involving/alert-ing the school and neighbors would be to encourage them to report any suspicious activities to the police. He also talked to the supervisor about police involvement and discussed filing criminal charges. If the police said the situation was not serious enough to file criminal charges, he suggested finding out from the police what was serious enough to warrant an arrest.

For example, he could explore with police what would constitute a pattern of behavior that might be considered serious enough to pursue action under the state's stalking or harassment statute.

Questions for the agency planning group

1. If this incident were reported at your agency, would you have used a criminal investigator or administrative investigator to conduct the initial investigation?
2 If your agency has a criminal investigative service, have you discussed the feasibility of involving agency criminal investigators at an early stage in the process of dealing with threatening behavior, i.e., in situations where threatening behavior does not yet rise to the level of a crime?
3. Has your agency identified a threat assessment professional to whom you could turn for assistance if the need arose?
4. Do you have a good source for keeping up with Merit Systems Protection Board case law on charges and threats?
5. If this happened at your agency, and the threatening behavior continued, what would you do to protect the supervisor and his family?

Case study 8 — a threat

The incident

*A visibly upset male employee cornered a female employee in her office, and said quietly and slowly that she **will pay with her life** for going over his head to ask about his work. The male employee then stared at his coworker with his hands clenched rigidly at this side before leaving the office and slamming the door behind him. The female employee, fearful and shaken, reported this to her supervisor, who immediately reported the incident to the director of Employee Relations.*

Response

The agency's response plan calls for involvement of Employee Relations, Security and the Employee Assistance Program (EAP) in cases involving threats. Immediately following the report to the response team, the **Security Officer** contacted the female employee to assist her in filing a police report on the threat and to discuss safety measures that she should be taking. The victim was also referred to the EAP, where she received brief counseling and educational materials on handling severe stress.

An investigation was immediately conducted by an investigator from **the Office of Inspector General.** In her statement, the female employee repeated what she had reported to the supervisor earlier about the threat. In his statement, the male employee stated that, on the day in question, he had been upset about what he felt were some underhanded activities by the female employee and his only recollection about the conversation was that he made a general statement like *You'll pay* to her. He stated that this was not a threat, just an expression. The investigation showed that the employee had several previous incidents of intimidating behavior which had resulted in disciplinary actions.

Resolution

After reviewing the results of the investigation, the supervisor proposed a removal action, finding that the female employee's version of the incident was more credible. In his response to the proposed notice, the employee brought in medical documentation that said he had a psychiatric disability of Post Traumatic Stress Disorder, which caused his misconduct, and he requested a reasonable accommodation. The deciding official consulted with an agency attorney and employee relations specialist who explained that nothing in the Rehabilitation Act prohibits an agency from maintaining a workplace free of violence or threats of violence. Further, they explained that a request for reasonable accommodation does not excuse employee misconduct nor does it shield an employee from discipline. The deciding official determined that removal was the appropriate discipline in this case. The employee did not appeal the action.

Questions for the agency planning group

1. Do you agree with the agency's approach in this case?
2. If this situation occurred at your agency, would you have involved law enforcement early in the process?
3. Who would conduct the investigation at your agency?
4. What else would your agency have done to protect the employee?
5. Would you have requested more medical documentation from the employee?
6. Would you have considered the Merit Systems Protection Board case law on mitigation when determining the penalty?

Case study 9 — a threat made during an EAP counseling session

The incident

When the employee first contacted the in-house Employee Assistance Program (EAP) counselor, he said that he had been referred by his supervisor because of frequent tardiness and his inability to complete his assignments on time. He complained of listlessness, lack of interest in his job, and inability to sleep. He expressed boredom with his job, unhappiness over the fact that his career was at a standstill, and eagerness to retire and move to another state. The counselor referred the employee to a psychiatrist for evaluation. The employee agreed to sign releases so the counselor could contact both his supervisor and the psychiatrist.

Feedback from the employee's supervisor confirmed the information he had provided to the EAP counselor. For the past year, his arrival time at work had been erratic and his productivity was down. The psychiatrist diagnosed depression, pre-scribed an anti-depressant, and referred the employee for psychotherapy. Several weeks later, the supervisor called the EAP counselor to report that, although the employee was no longer tardy, he often came in looking disheveled; coworkers complained that his speech and manner were sometimes bizarre; and he bragged of drinking large amounts of alcohol each evening.

The counselor immediately called the employee and asked him to come in for a follow-up visit. He agreed and appeared late that afternoon in a euphoric state. He said that, although he was sleeping only a few hours each night, he had never felt better in his life and had decided against psychotherapy. He admitted to having an occasional glass of wine but denied heavy drinking. The counselor encouraged him to return to the psychiatrist for re-evaluation but he refused.

The employee was in a talkative mood and began to reminisce about his Federal career—first his early successes, then recent disappointments, such as being passed over repeatedly for promotions and failure to receive any type of recognition. As he continued, he revealed in a matter-of-fact tone that he had been spending his evenings planning revenge on his managers because they had treated him unfairly for many years and they deserved to be punished. He believed he had planned the "perfect murder" and that he would never be caught.

Thinking at first that he was just venting his frustration, the counselor ques-tioned him further and quickly realized that he was very serious. She urged him to call his psychiatrist immediately and he again refused but said he would "think about calling" in a day or two. As soon as the employee left her office, the counselor called the psychiatrist and asked whether he viewed the employee's statement as a threat. The psychiatrist said he believed it was a serious threat and recommended that she take immediate action. The EAP counselor called the police and agency officials and informed them about the situation.

Response

The following morning when the employee reported to the office, he was met by the local police. A police officer brought him to the community's emergency services clinic for an evaluation and subsequently transported him to the hospital. He remained in the hospital for several weeks. Following discharge, he remained at home for several more weeks, during which time agency management held many discussions with his treating and consulting physicians. It was finally decided that the employee would be allowed to return to work, and not removed from his position, on the condition that, as long as he remained an employee of the agency, he would continue in psychotherapy, remain on medication as prescribed, refrain from alcohol and other drug abuse, and be seen on a regular basis by a psychiatric consultant to the agency. The employee agreed to the plan, often known as a last chance agreement.

Resolution

Although coworkers had been concerned about the employee's strange behavior and had seen him removed from the premises by the police, several had visited him in the hospital and were supportive of his return to the office. He worked his remaining years with no further problems, then retired and moved to another state.

Questions for the agency planning group

1. Do you agree with the agency's approach in handling this case?
2. Would you have let the employee back to work after his hospitalization? What information would you need to make this determination?
3. What safety precautions would your agency take if you did/did not take him back?
4. What should the EAP counselor have done if he denied making the threat?
5. Would your agency have proposed disciplinary action prior to the last chance agreement?

Case study 10 — threats made by an ex-employee

The incident

*The first incident report that came in to the agency's newly formed workplace violence team was from a field office. Two months after an employee retired on disability retirement, he began threatening his ex-supervisor. He knocked on his ex-supervisor's apartment door late one evening. He left threatening statements on the supervisor's home answering machine, such as **I just wanted to let you know I bought a gun.** On one occasion, a psychiatrist called the supervisor and the agency's security office and told them that the ex-employee threatened to murder his ex-supervisor. The psychiatrist said the threat should be taken seriously especially because he was drinking heavily. A coworker received an anonymous letter stating, **It is not over with [name of supervisor].** Each time a threat was reported, the agency's security office would take extra measures to protect the supervisor while at the workplace and the supervisor would report the incident to the local police. Each time, the supervisor was informed that the police were unable to take action on the threats because they did not rise to a criminal level. The supervisor spoke with the county magistrate about a restraining order, but again was told the threats did not rise to the level required to obtain a restraining order.*

Response

The workplace violence team held a conference call with the threatened supervisor, the director of the office, and the security chief of the field office. They suggested the following actions.

Recommendations for the **Security Officer:**

- Confirm the whereabouts of the ex-employee and periodically reconfirm his whereabouts.
- Meet with local police to determine whether the ex-employee's behavior constitutes a crime in the jurisdiction and whether other applicable charges (such as stalking or harassment) might be considered. Ask if the police department has a threat assessment unit or access to one at the state level. Ask police about contacting the U.S. Postal Service for assistance in tracing the anonymous letter (18 USC 876).
- Meet with the psychiatrist who called the agency and ask him to send a letter to the chief of police reporting the threats. Also, inform the psychiatrist about the ex-employee's behavior and discuss whether or not involuntary hospitalization might be an option. Attempt to establish an ongoing dialog with the psychiatrist and try to get a commitment from him to share information about the case to the extent allowed by confidentiality.
- Provide periodic updates to the threatened supervisor on the status of the case, actions taken, and actions being contemplated.
- Provide support and advice to the threatened supervisor, including telephone numbers and points of contact for local telephone company, local law enforcement, and local victim assistance organizations.

Recommendations for the **Director of the Field Office:**

- Meet with security and police to consider options (and their ramifications) for encouraging the ex-employee to cease and desist his threatening activities.
- Provide support to the supervisor by encouraging the supervisor to utilize the Employee Assistance Program.

Recommendations for the threatened **Supervisor:**

- Keep detailed notes about each contact with the threatener. Give copies of all the notes to the police. (They explained to the supervisor that in all probability, each time he went to the police, it was treated like a new report, and thus, as individual incidents, they did not rise to the level of a crime.)
- Contact the phone company to alert them to the situation.
- Tape record all messages left on the answering machine.
- Contact the local office of victim assistance for additional ideas.

Resolution

Contact with the local police confirmed that each report had been taken as a new case. When presented with the cumulative evidence, in fact, the ex-employee's behavior did rise to the level of stalking under state law. The police visited the ex-employee and warned him that further threats could result in an arrest. At the threatened supervisor's request, the county magistrate issued a restraining order prohibiting personal contact and any communication. Two months after the restraining order was issued, the ex-employee was arrested for breaking the restraining order. The agency security office and the supervisor kept in contact with the police about the case to reduce any further risk of violence.

Questions for the agency planning group

1. Do you think the agency's approach in this case was adequate to protect the supervisor?
2. Have you already established liaison with appropriate law enforcement authorities to ensure that situations such as this get the proper attention from the beginning?
3. What would your agency do if the psychiatrist refused to get involved? Are you familiar with any laws in your state requiring mental health professionals to protect potential victims when threats have been made?
4. How would you continue to monitor the threatener's activities after he is released from jail?
5. What would your agency do if the case continued without the ex-employee being arrested?

Case study 11 — threats from non-employees

The incident

The agency's new workplace violence team receives a call from a small field office. The office staff consists of three employees, two of whom spend much of their workday outside of the office. All three employees have had close calls in the past in dealing with violent individuals. On two occasions, clients who came into the office lost their tempers because they received answers they did not like. Several times the employees who conducted their business outside the office were the targets of threats and aggressive behavior. How can you help us out here in the field? they asked the workplace violence team.

Response

Presented with this problem, the workplace violence team consulted with the following organizations:

- The local law enforcement agency in the jurisdiction where the field office was located;
- Several Federal law enforcement agencies, including the Federal Protective Service;
- Other Federal Government agencies that had small field offices and/or employees who spent most of their workday outside the office;
- The National Victims' Center;
- Prevention units of State Police in several states where the agency had field offices.

Resolution

The agency implemented the following plan not only for the office that made the initial request, but for many of their other field offices as well.

- Install a panic button in the office which is connected to a security service.
- Install a video camera (with an audio component) in the public service area to record any incident that occurs in the office.
- Reconfigure office furniture, especially in public service areas, to maximize security (e.g., rearrange the office furniture and dividers to give the appearance that the employee is not alone).
- Train all employees in personal safety techniques.
- Provide back-up for employees in the field when a threatening situation is suspected.
- Provide employees with copies of the laws regarding harassment, threats, and stalking in their states.
- Provide employees with lists of state and local organizations that can assist them in preventing violence and in dealing with potentially violent situations.
- Arrange for regional and field offices to develop and maintain liaison with state and local law enforcement agencies.
- Establish a system/procedure for employees in the field to check in periodically throughout the day, e.g., an employee would call and say *I'm entering the Jones residence, and I will call you back in 30 minutes.*
- Provide cellular phones, personal alarms, and other safety devices, as appropriate, to employees in the field.

Questions for the agency planning group

1. Do you agree with the agency's approach in this case?
2. What more could have been done?

Case study 12 — intimidation

The incident

A supervisor reported to a Human Resources (HR) specialist that he recently heard from one of his employees (alleged victim) that another one of his employees (alleged perpetrator) has been intimidating him with his "in your face" behavior. The alleged perpetrator has stood over the alleged victim's desk in what he perceived as a menacing way, physically crowded him out in an elevator, and made menacing gestures. The supervisor stated that the alleged perpetrator was an average performer, somewhat of a loner, but there were no behavior problems that he was aware of until the employee came to him expressing his fear. He said that the employee who reported the situation said he did not want the supervisor to say anything to anyone, so the supervisor tried to observe the situation for a couple of days. When he didn't observe any of the behavior described, he spoke with the alleged victim again and told him he would consult with the Crisis Management Team.

Response

In cases involving intimidation, this agency's crisis response plan called for involvement of Human Resources (HR) and the Employee Assistance Program (EAP) (with the clear understanding that they would contact other resources as needed). The first thing the HR specialist did was to set up a meeting for the next day with the supervisor, an EAP counselor, and another HR specialist who was skilled in conflict resolution.

At that meeting, several options were discussed. One was to initiate an immediate investigation into the allegations, which would involve interviewing the alleged victim, any witnesses identified by the alleged victim, and the alleged perpetrator. Another suggestion offered by the EAP counselor was that, in view of the alleged victim's reluctance to speak up about it, they could arrange a training session for the entire office on conflict resolution (at which time the EAP counselor could observe the dynamics of the entire work group). The EAP counselor noted that conflict resolution classes were regularly scheduled at the agency. The supervisor also admitted that he was aware of a lot of tensions in the office and would like the EAP's assistance in resolving whatever was causing them.

After discussing the options, the supervisor and the team decided to try the conflict resolution training session before initiating an investigation.

At the training session, during some of the exercises, it became clear that the alleged victim contributed significantly to the tension between them. The alleged victim, in fact, seemed to contribute significantly to conflicts not only with the alleged perpetrator, but with his coworkers as well. The alleged perpetrator seemed to react assertively, but not inappropriately, to the alleged victim's attempts to annoy him.

Resolution

Office tensions were reduced to minimum as a result of the training session and follow-up work by the Employee Assistance Program. The employee who initially reported the intimidation to his supervisor not only realized what he was doing to contribute to office tensions, but he also actively sought help to change his approach and began to interact more effectively with his coworkers. He appreciated getting the situation resolved in a low-key way that did not cause him embarrassment and began to work cooperatively with the alleged perpetrator. The alleged perpetrator never learned about the original complaint, but he did learn from the training session more effective ways to interact with his coworkers. This incident took place over a year ago, and the agency reports that both are productive team players.

Questions for the agency planning group

1. Do you agree with the agency's approach in this situation?
2. Can you think of other situations that could be addressed effectively through an intervention with the work group?
3. In what kinds of situations would this approach be counter-productive?
4. Can you envision a scenario where using the group conflict resolution session to get at any individualized problem might have a negative, rather than a positive, effect?
5. Has your agency conducted employee training in conflict resolution, stress management, dealing with hostile persons, etc.?

Case study 13 — intimidation

The incident

An employee called a member of the agency crisis team for advice, saying that a coworker was picking on her, and expressing fear that something serious might happen. For several weeks, she said, a coworker has been making statements like, You actually took credit for my work and you're spreading rumors that I'm no good. If you ever get credit for my work again, that will be the last time you take credit for anybody's work. I'll make sure of that. She also said that her computer files have been altered on several occasions and she suspects it's the same coworker. When she reported the situation to her supervisor, he tried to convince her that there was no real danger and that she's blowing things out of proportion. However, she continued to worry. She said she spoke with her union representative who suggested she contact the agency's workplace violence team.

Response

The agency's plan called for the initial involvement of employee relations and the Employee Assistance Program (EAP) in situations involving intimidation. The **Employee Relations** specialist and the **EAP** counselor met with the **Supervisor**. He told them he was aware of the situation, but that the woman who reported it tended to exaggerate. He knew the alleged perpetrator well, had supervised him for years, and said, *He just talks that way; he's not really dangerous.* He gave examples of how the alleged perpetrator is all talk and not likely to act out. One example occurred several months earlier when he had talked to the alleged perpetrator about his poor performance. The employee had become agitated and accused the supervisor of being unfair, siding with the other employees, and believing the rumors the coworkers were spreading about him. He stood up and in an angry voice said, *You better start treating me fairly or you're going to be the one with the problem.* He then stormed out of the room, saying, *Don't ever forget my words.* The supervisor reasoned that, since he's always been this way, he's not a real threat to anyone.

During the initial meeting, the team encouraged the supervisor to take disciplinary action but he didn't believe it was appropriate. They asked him to at least sign a written statement about these incidents. He was reluctant to make any kind of written statement and could not be persuaded by their arguments to do so.

The employee relations specialist conducted an investigation. The results confirmed continuing intimidating behavior on the part of the alleged perpetrator. In interviews with the coworkers of the victim, they confirmed the menacing behavior of the perpetrator and several felt threatened themselves. None was willing to sign affidavits. The investigator also found a witness to the incident where the supervisor had been threatened. As the alleged perpetrator had left the supervisor's office and passed by the secretary's desk, he had said, *He's an (expletive) and he better watch himself.* Although he did not directly threaten the secretary, she also was intimidated by the perpetrator and said he often acts in a menacing fashion. However, the secretary was also unwilling to sign an affidavit.

After confirming the validity of the allegations, but with the supervisor refusing to take action, and the only affidavit being from the employee who originally reported the situation, the team considered three courses of action:

1. Arrange for the reassignment of the victim to a work situation that eliminated the current threatening situation;
2. Report the situation to the second line supervisor and recommend that she propose disciplinary action against the perpetrator; and
3. Locate an investigator with experience in workplace violence cases to conduct interviews with the reluctant witnesses. The investigator would be given a letter of authorization from the director of the office

stating the requirement that employees must cooperate in the investigation or face disciplinary action.

The team located an **Investigator,** who was experienced in workplace violence cases, from a nearby Federal agency and worked out an interagency agreement to obtain his services. During the investigation, he showed the letter of authorization to only one employee and to the supervisor, since he was able to persuade the others to sign written affidavits without resorting to showing them the letter.

The agency **Security** specialist met with the perpetrator to inform him that he was to have no further contact with the victim. He also met with the victim to give her advice on how to handle a situation like this if it were to happen again. In addition, he recommended a procedure to the team that would monitor computer use in the division. This action resulted in evidence showing that the employee was, in fact, altering computer files.

Resolution

The first-line supervisor was given a written reprimand by the second-line supervisor for failing to take proper action in a timely manner and for failing to ensure a safe work environment. He was counseled about the poor performance of his supervisory duties. The alleged perpetrator was charged with both disruptive behavior and gaining malicious access to a non-authorized computer. Based on this information, he was removed from Federal Service.

Questions for the agency planning group

1. Would supervisory training likely have resulted in quicker action against the perpetrator?
2. Do you have other approaches for convincing a recalcitrant supervisor to take action?
3. Do you have other approaches for convincing reluctant witnesses to give written statements?
4. Are you aware of the problems associated with requiring the subject of an investigation to give statements? (See page 27 for a discussion of this issue.)
5. If you had not been able to convince the reluctant witnesses to give written statements, and you only had the one affidavit to support the one incident, do you think this would have provided your agency with enough evidence to take disciplinary action? If so, what type of penalty would likely be given in this case?

Case study 14 — *frightening behavior*

The incident

A supervisor contacts the Employee Relations Office because one of his employees is making the other employees in the office uncomfortable. He said the employee does not seem to have engaged in any actionable misconduct but, because of the agency's new workplace violence policy, and the workplace violence training he had just received, he thought he should at least mention what was going on. The employee was recently divorced and had been going through a difficult time for over two years and had made it clear that he was having financial problems which were causing him to be stressed out. He was irritable and aggressive in his speech much of the time. He would routinely talk about the number of guns he owned, not in the same sentence, but in the same general conversation in which he would mention that someone else was causing all of his problems.

Response

At the first meeting with the supervisor, the **Employee Relations** specialist and **Employee Assistance Program** (EAP) counselor suggested that, since this was a long-running situation rather than an immediate crisis, the supervisor would have time to do some fact-finding. They gave him several suggestions on how to do this while safeguarding the privacy of the employee (for example, request a confidential conversation with previous supervisors, go back to coworkers who registered complaints for more information, and, if he was not already familiar with his personnel records, pull his file to see if there are any previous adverse actions in it). Two days later they had another meeting to discuss the case and strategize a plan of action.

The **Supervisor's** initial fact-finding showed that the employee's coworkers attributed his aggressive behavior to the difficult divorce situation he had been going through, but they were nevertheless afraid of him. The supervisor did not learn any more specifics about why they were afraid, except that he was short-tempered, ill-mannered, and spoke a lot about his guns (although, according to the coworkers, in a matter-of-fact rather than in an intimidating manner).

After getting ideas from the employee relations specialist and the EAP counselor, the supervisor sat down with the employee and discussed his behavior. He told the employee it was making everyone uncomfortable and that it must stop. He referred the employee to the EAP, setting a time and date.

Resolution

As a result of counseling by the supervisor and by the Employee Assistance Program counselor, the employee changed his behavior. He was unaware that his behavior was scaring people. He learned new ways from the EAP to deal with people. He accepted the EAP referral to a therapist in the community to address underlying personal problems. Continued monitoring by the supervisor showed the employee's conduct improving to an acceptable level and remaining that way.

Questions for the agency planning group

1. Do you agree with the agency's approach in this case?
2. Can you think of other situations that would lend themselves to this kind of low-key approach?
3. Does your agency have effective EAP training so that supervisors are comfortable in turning to the EAP for advice?

Case study 15 — Frightening Behavior

The incident

Several employees in an office went to their supervisor to report an unusual situation which had occurred the previous day. An agency employee from a different building had been in and out of their office over a seven-hour period, remarking to several people that "the Government" had kept her prisoner, inserted microphones in her head to hear what she was thinking, and tampered with her computer to feed her evil thoughts. She also said that her doctors diagnosed her as paranoid schizophrenic, but that they are wrong about her. She made inflammatory remarks about coworkers, and made statements such as Anybody in my old job who got in my way came down with mysterious illnesses.

Response

The **employee relations** specialist, who took the report, immediately informed the employee's supervisor about the incident. She learned from the employee's supervisor that until a few months ago, the employee performed adequately, but had always seemed withdrawn and eccentric. However, her behavior had changed (it was later learned that she had stopped taking her medication) and she often roamed around the office, spending an hour or more with any employee she could corner. Several employees had reported to the supervisor that they were afraid she might hurt them. She also learned that a former supervisor had previously given the employee a reprimand and two counseling memoranda for inappropriate language and absence from the worksite. The former supervisors had repeatedly offered her leave for treatment as a reasonable accommodation but she had refused.

Upon the recommendation of the employee relations specialist, the employee was placed on excused absence pending further agency inquiry and response, with a requirement to call in daily. The employee turned in her ID and building keys and the building guards were informed of the situation.

The employee relations specialist, who was a trained investigator, conducted interviews with both the employees who filed the reports and with the employee's coworkers. She found that most of the employees were afraid of the woman and had reluctantly filed the report. They said that the woman made repeated statements that the doctors called her paranoid schizophrenic, but that she was fine except for the devices implanted in her body and her computer. Employees in the woman's own office refused to give written statements because they were so afraid.

The employee relations specialist then set up a meeting with the woman's first- and second-line supervisor, the director of personnel, the legal office, the director of security, the agency's medical officer, and the Employee Assistance Program (EAP) counselor. The following options were raised:

- Propose an indefinite suspension pending an investigation (option rejected because the agency already had all the information it needed about the incident).
- Reassign or demote the employee to another office (option rejected because the reported conduct was too serious).
- Propose suspension based on her day-long frightening and disruptive comments and conduct (option rejected because the reported conduct was too serious).
- Order a medical examination to determine whether the employee was fit for duty (option rejected because the employee was not in a position with medical standards or physical requirements).
- Offer a medical examination (option rejected because supervisor already tried it several times).
- Offer her leave for treatment (option rejected because supervisor already tried it several times).

The team recommended that the supervisor issue a proposal to remove based on the events in the other office, i.e., her day-long frightening and disruptive comments and conduct. The notice would also reference the earlier counseling memos and the reprimand which placed her on notice concerning her absence from her office and inappropriate behavior.

The supervisor presented her with the proposed removal with the employee relations specialist present and a security officer standing by in an adjoining office. Three weeks later, the employee and her brother-in-law came in for her oral reply to the proposed notice. She denied making any of the statements attributed to her. Her brother-in-law asked the deciding official to order her to go for a psychiatric examination, but he was told that regulations prohibited the agency from doing so. The employee did not provide any additional medical documentation.

Resolution

The agency proceeded with a removal action based on her disruptive behavior. Once her brother-in-law realized that her salary and health benefits would soon cease, he was able to convince her to go to the hospital for the help she needed and to file for disability retirement. The agency assisted her with the filing of the forms with the Office of Personnel Management. The disability retirement was approved which provided her with income and a continuation of medical coverage.

Questions for the agency planning group

1. Do you agree with the agency's approach in handling this case?
2. Does your employee training direct employees to call security or 911 in emergency situations?
3. Is your team knowledgeable about accessing appropriate community resources for emergency situations?
4. What if the employee had not been willing and able to apply for disability retirement herself? Would your agency have filed for disability retirement on the employee's behalf?
5. Does your agency's supervisory training encourage early intervention in cases of this type?

Case study 16 — disruptive behavior

The incident

After workplace violence training was conducted at the agency, during which early intervention was emphasized, an employee called the Employee Assistance Program (EAP) member of the workplace violence team for advice on dealing with his senior coworker. He said the coworker, who had been hired at the GS-14 level six months earlier, was in the habit of shouting and making demeaning remarks to the employees in the office. The senior coworker was skilled in twisting words around and manipulating situations to his advantage. For example, when employees would ask him for advice on a topic in his area of expertise, he would tell them to use their own common sense. Then when they finished the assignment, he would make demeaning remarks about them and speak loudly about how they had done their work the wrong way. At other times, he would demand rudely in a loud voice that they drop whatever they were working on and help him with his project. The employee said he had attempted to speak with his supervisor about the situation, but was told not to make a mountain out of a mole hill.

Response

The **EAP Counselor** met with the employee. The employee described feelings of being overwhelmed and helpless with regard to the situation. The demeaning remarks were becoming intolerable. The employee believed that attempts to resolve the issue with the coworker were futile. The fact that the supervisor minimized the situation further discouraged the employee. By the end of the meeting with the EAP, however, the employee was able to recognize that not saying anything was not helping and was actually allowing a bad situation to get worse.

At a subsequent meeting, the EAP counselor and the employee explored skills to address the situation in a respectful, reasonable, and responsible manner with both his supervisor and the abusive coworker. She suggested using language such as:

Please lower your voice.
I don't like shouting.
I don't like it when you put me down in front of my peers.
It's demeaning when I am told that I am...
Don't point your finger at me.
I want to have a good working relationship with you.
I recognize how important your project is. I will help you this afternoon when
*   I have finished the job I am working on.*

The employee learned to focus on his personal professionalism and responsibility to establish and maintain reasonable boundaries and limits by using these types of firm and friendly "I statements," acknowledging that he heard and understood what the supervisor and coworker were saying, and repeating what he needed to communicate to them.

After practicing with the Employee Assistance Program counselor, the employee was able to discuss the situation again with his supervisor. He described the situation in non-blaming terms, and he expressed his intentions to work at improving the situation. The supervisor acknowledged that the shouting was annoying, but again asked the employee not to make a mountain out of a mole hill. The employee took a deep breath and said, *It may be a mole hill, but never the less it is affecting my ability to get my work done efficiently.* He requested that the supervisor be ready to intervene if the behavior continued. Finally, the supervisor stated that he did not realize how disruptive the situation had become and agreed to monitor the situation.

The next time the coworker raised his voice and demanded that the employee drop what he was doing and work on his project, the employee used his newly acquired assertiveness skill and stated in a calm and quiet voice. *I don't like to be shouted at. Please lower your voice.* After a moment of silence he continued, *I recognize how important this is to you. I will take care of it this afternoon after I have finished the job I am working on.* When the coworker started shouting again, the employee restated in a calm voice, *I don't like being shouted at. Please lower your voice.* The coworker stormed away.

Meanwhile, the **Supervisor** began monitoring the situation. He noted that the abusive coworker's interactions had improved with the newly assertive employee, but continued to be rude and demeaning toward the other employees. The supervisor consulted with the EAP and employee relations specialist for assistance, recognizing that the discord was increasing. The EAP counselor told him, *Generally, people don't change unless they have a reason to change.* The counselor added that the reasons people change can range from simple "I statements," such as those suggested above, to disciplinary actions. The **Employee Relations** specialist discussed disciplinary options with the supervisor.

The supervisor then met with the abusive coworker who blamed the altercations on the others in the office. The supervisor remembered that his goal was to put an end to the abusive behavior and that placing blame was often counterproductive to achieving this goal. He therefore responded, *I understand the others were stressed. I'm glad you understand that shouting, speaking in a demeaning manner, and rudely ordering people around is unprofessional and disrespectful. It is unacceptable behavior and will not be tolerated.* During the meeting, he also referred the employee to the Employee Assistance Program (EAP).

The coworker continued his rude and demeaning behavior to the other employees in spite of the supervisor's efforts. The others, after observing the newly acquired confidence and calm of the employee who first raised the issue, requested similar training from the EAP. The supervisor met again with the EAP counselor and employee relations specialist to strategize next steps.

Resolution

When all of the employees in the office started using assertive statements, the abusive coworker became more cooperative. However, it took a written reprimand, a short suspension, and several counseling sessions with the EAP counselor before he ceased his shouting and rude behavior altogether.

Questions for the agency planning group

1. Does your workplace violence training include communication skills to put a stop to disruptive behavior early on (including skills for convincing reluctant supervisors to act)?
2. How would your agency have proceeded with the case if the coworker had threatened the employee who confronted him?
3. What recourse would the employee have had if the supervisor had refused to intervene?

Appendix A:
Table of cases and cites

29 C.F.R. 1910.5(a)(1)

29 U.S.C. Sect. 654(a)(1)

Acands, Inc. v. Godwin, 340 Md. 334, 667 A.2d 116, Prod.Liab.Rep. (CCH) P 14,416 (Md., Oct 18, 1995) (NO. 23 SEPT.TERM. 1994)

Alexander v. Com., Dept. of Public Welfare, 137 Pa.Cmwlth, 342, 586 A2d 475 (Pa.Cmwlth., Jan 24, 1991) (NO. 1033 C.D. 1990)

Allen v. Tyson Foods, Inc., 121 F.3d 642, 74 Fair Empl.Prac.Cas. (BNA) 1694, 11 Fla. L. Weekly Fed. C. 505 (11th Cir.(Ala.) Sep 10, 1997) (NO. 96-6593)

Alvarez v. Dart Industries, Inc., 55 Cal.App.3rd 91, 127 Cal.Rptr. 222 (1976)

American Justice Ins. Reciprocal v. Ohio Twp. Ass'n Risk Management Authority, 121 F.3d 707 (Table, Text in WESTLAW), Unpublished Disposition, 1997 WL 415317 (6th Cir. (Ohio), Jul 22, 1997) (NO. 96-3615)

Amvest Mortgage Corp. v. Antt, Cal.App.4th, 1997 WL 672689 (Cal.App. 1 Dist., Oct 30, 1997) (NO. A074966))

Anderson v. U.S., F.3d, 1997 WL 690066 (9th Cir.(Wash.), Nov 04, 1997) (NO.96-35768)

Argonne Apartment House Company v. Garrison, 42 F.2d 605 (D.C. Cir. 1930)

Archibeque v. Moya, 116 N.M. 616, 866 P.2d 344 (N.W., Dec 15,1993) (NO. 21, 359)

Armstrong v. Chrysler Corp., 972 F.Supp. 1085, 156 L.R.R.M. (BNA) 2768 (E.D.Mich., Jul 10, 1997) (NO. 96-CV-50251-AA)

Baker v. Turner Const. Co., 200 A.D.2d 525, 607 N.Y.S.2d 10 (N.Y.A.D. 1 Dept., Jan 25, 1994) (NO. 50915)

Ballard's Administratrix v. Louisville & Nashville Railroad Company, 128 Ky. 826, 110 S.W. 296 (1908)

Barry v. Young Men's Christian Ass'n of Metropolitan Hartford, Inc., 1995 WL 27433 (Conn.Super., Jan 17, 1995) (NO. CV94-053 2 37)

Battle v. Philadelphia Housing Authority, 406 Pa.Super. 578, 594 A.2d 769 (Pa.Super., Aug 12, 1991) (NO. 2826 PHILA. 1990)

Bedford v. City of Mandeville. 1997 WL 543103 (E.D.La., Sep 04, 1997) (NO. CIV. A. 96-0737)

Bias v. IPC Intern. Corp., 107 F.3d 865 (Table, Text in WESTLAW), Unpublished Disposition, 1997 WL 100925 (4th Cir.(Md.), Mar 06, 1997)

Biernacki v. Carter, 1996 WL 727396 (N.D.111., Dec 16, 1996) (NO. 95 C 1694)

Bishop v. Miche, 943 P.2d 706, (Wash.App. Div. 1, Sep 22, 1997) (NO. 37844-9-I)

Braswell v. Braswell, 410 S.E.2d 897

Breiland v. Advance Circuits, Inc., F.Supp, 1997 WL 580598, 7 A.D. Cases 619 (D.Minn., Sep 16, 1997) (NO. 4-96-660 (DSD/JMM))

Broad v. Patico Corp., N.Y.S.2d, 1997 WL 619753 (N.Y.A.D. 2 Dept., Oct 06, 1997) (NO. 96-09862)

Brown v. City of Tuscon, 122 F.3d 1070 (Table, Text in WESTLAW), Unpublished Disposition, 1997 WL 556365 (9th Cir.(Ariz.), Aug 29, 1997) (NO. 94-16984)

Bryant v. Livigni, and National Super Markets, Inc., 619 N.E.2d 550

Bowers v. Cook County, 1997 WL 461095 (N.D.Ill., Aug 08, 1997) (NO. 96 C 6114)

Burke v. Fitzgerald, 1997 WL 600087 (Conn.Super., Sep 22, 1997) (NO. CV 950322083S)

Burkhart v. Washington Metropolitan Area Transit Authority, 112 F.3d 1207, 324 U.S. App.D.C. 241, 133 Lab.Cas. P 58,249, 47 Fed. R. Evid. Serv. 142, 6 A.D. Cases 1333, 10 NDLR P 40 (D.C.Cir., May 16, 1997) (NO. 96-7163)

Calhoun v. Jumer, 1997 WL 651145 (Ill.App. 4 Dist., Oct 21, 1997) (NO. 4-97-0243)

Calvillo v. Pacific Gas and Elec. Co., 1997 WL 488120, 10 NDLR P 260 (N.D.Cal., Aug 11, 1997) (NO. C 96-2704-FMS)

Callaway v. New Mexico Dept. of Corrections, 117 N.M. 637, 875 P.2d 393 (N.M.App., Mar 29, 1994) (NO. 14,525)

Cancellier v. Federated Department Stores, 672 F.2d 1312 (CA-9, 1982)

Cannizzaro v. Neiman Marcus, Inc., 1997 WL 538740 (N.D.Tex., Aug 20, 1997) (NO. CIV.A. 3:96-CV-0934-P)

Cannizzaro v. Neiman Marcus, Inc., F.Supp., 1997 WL 622787 (N.D.Tex., Aug 20, 1997) (NO. CIV.A. 3:96-CV-0934)

Carraro v. Director, Employment Sec. Div., 54 Ark.App. 210, 924 S.W.2d 819 (Ark.App., Jun 26, 1996) (NO. E 95-154)

Carter v. Skokie Valley Detective Agency, LTD., 628 N.E.2d 602

City of Huntington Beach v. City of Westminster, 57 Cal.App.4th 220, 66 Cal.Rptr.2d 826, 97 Cal. Daily Op. Serv. 6743, 97 Daily Journal D.A.R. 10,933 (Cal.App. 4 Dist., Aug 21, 1997) (NO. G015484)

Clark v. City of Ithaca, 235 A.D.2d 746, 652 N.Y.S.2d 819 (N.Y.A.D. 3 Dept., Jan 16, 1997) (NO. 76627)

Cintron v. Sysco Food Services of Chicago, Inc., 1997 WL 457500 (N.D. 111., Aug 07, 1997) (NO. 96 C 1632)

Columbia Aluminum Corp. v. United Steelworkers of America, Local 8147, 922 F.Supp. 412, 151 L.R.R.M. (BNA) 2778 (E.D. Wash., Oct 19, 1995) (NO. CY-95-3098-AAM)

Contreras v. Suncast Corp., 1997 WL 598120 (N.D.Ill., Sep 19, 1997) (NO. 96 C 3439)

Cummings v. Koehnen, 568 N.W.2d 418 (Minn., Aug 28, 1997) (NO. C6-96-1118)

D'Alessandro v. Westfall, 972 F.Supp. 695 (W.D.N.C., Mar 24, 1997) (NO. CIV. 4:95CV70)

Daugherty v. Delta Lloyd's Ins. Co. of Houston, 1996 WL 640611, No Publication, (Tex.App.-Hous. (14 Dist.), Nov 07, 1996) (NO. 14-95-00049 CV)

Davis v. Civil Service Com'n, 55 Cal.App.4th 677, 64 Cal.Rptr.2d 121, 97 Cal. Daily Op. Serv. 4291, 97 Daily Journal D.A.R. 7145 (Cal.App. 2 Dist., May 22, 1997) (NO. E 95-154)

Davis v. Hennepin County, 559 N.W.2d 117 (Minn.App., Feb 11, 1997) (NO. C7-96-1841)

De Los Santos v. Cambridge Tankers, Inc., 114 F.3d 1193 (Table, Text in WESTLAW), Unpublished Disposition, 1997 WL 289422 (9th Cir. (Cal.), May 29, 1997) (NO. 96-15258)

District of Columbia v. Tinker, 691 A.2d 57 (D.C.App., Mar 13, 1997) (NO. 93-CV-1020)

District of Columbia v. Walker, 689 A.2d 40 (D.C.App., Feb 06, 1997) (NO. 93-CV-113, 94-CV-1309, 94-CV-14)

Doe v. Clyde-Green Springs Exempted Village Schools, 1997 WL 586748 (Ohio App. 6 Dist., Sep 19, 1997) (NO. S-96-019)

Doe v. Hartz, 970 F.Supp. 1375 (N.D.Iowa, Jun 23, 1997) (NO. C 96-4091-MWB)

Doe v. Neal, 1994 WL 702876 (E.D.Pa., Dec 12, 1994) (NO. CIV. a 94-4294)

Doe v. Young, 656 So.2d 569, 20 Fla. L. Weekly D1420 (Fla.App. 5 Dist., Jun 16 1995) (NO. 95-616)

Donovan v. Mt. Ida College, 1997 WL 259522 (D.Mass., Jan 03, 1997) (NO. CIV.A. 96CV10289RGS)

Doud v. Las Vegas Hilton Corp., 109 Nev. 1096, 864 P.2d 796 (Nev., Nov 29, 1993) (NO. 23513)

Drawhorn v. Texaco Chemical Co., 887 S.W.2d 510 (Tex.App.-Beaumont, Nov 23, 1994) (NO. 09-93-117 CV))

Dullard v. Bereley Associates Co., 606 F.2d 890 (2nd Cir.(N.Y.), Aug 27, 1979) (NO. 731, 732, 733, 734, 78-7457, 78-7512, 78-7560, 78-7561)

Dunn v. Pacific Employers Ins. Co., 332 N.C. 129, 418 S.E.2d 645 (N.C., Jul 17, 1992) (NO. 139PA91)

Edwards v. Burgess, 96-2064 (La.App 4 Cir. 10/3/97), So.2d, 1997 Wl 606994 (La.App. 4 Cir., Oct 01, 1997) (NO. 96-CA-2064)

Eller v. Allen, 623 So.2d 545, 18 Fla. L. Weekly D1719 (Fla.App. 5 Dist., Aug 06, 1993) (NO. 93-119)

Ernst v. Parkshore Club Apartments Ltd. Partnership, 863 F.Supp. 651 (N.D. Ill., Aug 24, 1994) (NO. 93 C 1207)

Estate of Michael v. Suburban Power Piping Corp., 1997 WL 127197 (Ohio App. 8 Dist., Cuyahoga County, Mar 20, 1997) (NO. 70873)

Firestone Textile Company v. Meadows, 666 S.W.2d 730 (Ky. 1983)

Fortenberry v. United Air Lines, 1997 WL 457499 (N.D.Ill., Aug 07, 1997) (NO. 96 C 3198)

Fritz v. Guida-Fenton Opthalmology Associates, 1997 WL 614592 (Conn.Super., Sep 18, 1997) (NO. CV 970398021S)

Fulford v. ITT Rayonier, Inc., 676 F.Supp. 252 (S.D. Ga., Nov 12, 1987)

Fund v. Hotel Lenox of Boston, Inc., 418 Mass. 191, 635 N.E.2d 1189 (Mass, Jul 07, 1994) NO. S-6492)

G.B. Goldman Paper Co. v. United Paperworkers Intern. Union, Local 286, 957 F.Supp. 507, 154 L.R.R.M. (BNA) 2489, 133 Lab. Cas. P 11, 830 (E.D.Pa., Feb 03, 1997) (NO. CIV. A. 95-7319)

Gee v. Southwest Airlines, 110 F.3d 1400, 65 USLW 2665, 97 Cal. Daily Op. Serv. 3962, 97 Daily Journal D.A.R. 4580, 97 Daily Journal D.A.R. 6779 (9th Cir.(Cal.), Apr 04, 1997) (NO. 95-17175, 95-56278, 95-36117, 95-36188)

Geise v. Phoenix Company of Chicago, Inc., 246 Ill.App.3d 441, 615 N.E.2d 1179 (2d Dist., 1993), Reversed on other grounds, 159 Ill.2d 507, 639 N.E.2d 1273 (1994)

Gibson v. Brewer, 952 S.W.2d 239 (Mo., Aug 19, 1997) (NO. 79291)

Gladon v. Greater Cleveland Regional Transit Auth., 75 Ohio St.3d 312, 662 N.E.2d 287 (Ohio, Mar 06, 1996) (NO. 94-1063, 9117)

Goss v. American Cyanamid, Co., 278 N.J.Super. 227, 650 A.2d 1001 (N.J.Super.A.D., Dec 22, 1994) (NO. A-4558-92T2

Grable v. Brown, 1997 WL 405210 (E.D.Mich., May 23, 1997) (NO. 97-CV-70218-DT)

Grain Dealers Mut. Ins. Co. v. Pat's Rentals, Inc., S.E.2d, 1997 WL 603787, 97 FCDR 3611 (Ga.App., Oct 02, 1997) (NO. A97A2122)

Gray v. Ward, 950 S.W.2d 232 (Mo., Aug 19, 1997) (NO. 79299)

Green v. City of Paterson, 971 F.Supp. 891 (D.N.J., Jun 26, 1997) (NO. CIV.A. 95-6028 (AJL)

Griffin v. AAA Club South, Inc., 221 Ga.App. 1, 470 S.E.2d 474 (Ga.App., Mar 15, 1996) (NO. A95A1878, A95A2854)

Griffin v. V. & J. Foods, Inc., 199 Wis.2d 524, 546 N.W.2d, N.W.2d 579 (Table, Text in WESTLAW), Uuupublished Disposition, 1996 WL 12826 (Wis.App., Jan 16, 1996) (NO. 95-1335-FT)

Haavistola v. Delta Airlines, Inc., 1997 WL 128097 (Del.Super., Feb 28, 1997) (NO. 96C-06-047-JOH)

Hall v. SSF, Inc., 112 Nev. 1384, 930 F.2d 94, 143 Lab.Cas. P 58,266 (Nev., Dec 20, 1996) (NO. 27067)

Hartsell v. Duplex Products, Inc., 123 F.3d 766, 74 Fair Empl.Prac.Cas. (BNA) 1495, 71 Empl.Prac. Dec. P 44,943 (4th Cir.(N.C.). Aug 25, 1997) (NO.97-1114)

Haynal v. Target Stores, 1996 WL 806706, 6 A.D. Cases 512 (S.D.Cal., Dec 19, 1996) (NO. 96-1599-K(RBB))

Herstrum v. Hodge, 121 F.3d 708 (Table, Test in Westlaw), Unpublished Disposition, 1997 WL 452299 (6th Cir.(Ohio), Aug 07, 1997) (NO. 95-3522)

Hertog v. City of Seattle, 943 P.2d 1153 (Wash.App. Div. 1, Sep 22, 1997) (NO. 37291-2-I)

Heyen v. Coastal Mart, Inc., 1996 WL 563834 (D.Kan., Sep19, 1996) (NO. 95-1284-JTM)

Holbrook v. City of Alpharetta, Ga., 112 F.3d 1522, 65 USLW 2767, 6 A.D. Cases 1409, 10 NDLR P 48, 10 Fla. L. Weekly Fed. C 957 (11th Cir.(Ga.) May 22, 1997) (NO. 95-8691)

Angela N. Holder, f/k/a Angela N. Hamilton v. Mellon Mortgage Company, City of Houston, 1997 WL 461982 (Tex.App.-Hous. (14 Dist.). Aug 14, 1997) (NO. 14-96-00043-CV)

Ible v. MCC Behavioral Care, Inc., 1997 WL 471317 (Minn.App., Aug 19, 1997) (NO. CO-97-170)

In re O'Neill, 1997 WL615661 (Bankr.D.Minn., Oct 02, 1997) (NO.4-95-1477, 4-97-001)

In re Prudential Ins. Co. of America Sales Practices Litigation, 962 F.Supp. 450 (D.N.J., Mar 17, 1997) (NO. MDL 1061, CIV.A. 95-4704)

In re Van Dresser Corp., F.3d, 1997 WL 66873 (6th Cir.(Mich.), Oct 27, 1997) (NO. 96-2179)

Johnson v. Hondo, Inc., 125 F.3d 408, 74 Fair Empl.Prac.Cas. (BNA) 1398, 71 Empl. Prac. Dec. P 44,957 (7th Cir. (Wis.), Jul 28, 1997)

Johnson v. Sawyer, 120 F.3d 1307, 97-2 USTC P 50,616 (5th Cir.(Tex.), Aug 21, 1997) (NO. 96-20667)

Jones v. General Bd. of Global Ministries of United Methodist Church 1997 WL 458790 (S.D.N.Y., Aug 11, 1997) (NO. 96 CIV. 5462 (HB))

Jones v. Johnson & Johnson, 1997 WL 549995 (E.D.Pa., Aug 22, 1997) (NO. CIV. A. 94-7473)

Jordan v. Playtex Family Products, Inc., 116 F.3d 465 (Table, Text in WESTLAW), Unpublished Disposition, 1997 WL 314799 (2nd Cir.(Conn.)., Jun 12, 1997) (NO. 96-7431)

Killinger v. Samford University, 113 F.3d 196, 73 Fair Empl.Prac.Cas. (BNA) 1533, 70 Empl.Prac. Dec. P 44,692, 118 Ed.Law Rep. 48, 10 Fla. L. Weekly Fed. C 955 (11th Cir. (Ala.), May 22, 1997) (NO. 96-6238)

King v Network Management Group, Inc., 1995 WL 234603, No Publication, (Tex.App.-Dallas, Apr 20, 1995) (NO. 05-94-00166-CV)

Kirk v. Consolidated Freightways Corp. of Delaware, 1997 WL 289056 (N.D. Tex., May 21, 1997) (NO. CIV.A. 3:96-CV-1024G)

Ladapo v. City of Dallas, 1997 WL 600696 (N.D. Tex., Sep 19, 1997) (NO. CIV. A. 3:96-CV-1791-G)

LaLone v. Smith, 39 Wash.2d 167, 234 Pa.2d 893 (1951)

Landess v. North American Van Lines, Inc., F.Supp., 1997 WL 566871 (E.D.Tex., Sep 08, 1997) (NO. 1:96-CV-685)

L.B. Russell Chemicals, Inc. v. Liberty Mut. Ins. Co., 116 F.3d 484 (Table, Text in WEST-LAW), Unpublished Disposition, 1997 WL 341761 (9th Cir. (Cal.), Jun 19, 1997) (NO. 94-55407)

Lepera v. ITT Corp., 1997 WL 535165 (E.D.Pa., Aug 12, 1997) (NO. CIV. 97-1461)

Leslie G. v. Perry & Associates, 43 Cal.App.4th 472, 50 Cal.Rptr.2d 785, 96 Cal. Daily Op. Serv. 1731, 96 Daily Journal D.A.R. 2898 (Cal.App. 2 Dist., Mar 11, 1996) (NO. B091339)

Logan v. West Coast Benson Hotel, F.Supp., 1997 WL 677608 (D.Or., Sep 09, 1997) (NO. 96-966-JO)

Ludwig v. Johnson, 243 Ky 533, 49 S.W. 2d 347 (1932)

Maksimovic v. Tsogalis, 1997 WL 638782, 13 IER Cases 653 (Ill., Oct 17, 1997) (NO.81493)

Mallory v. O'Neil, 69 So.2d 313 (Fla. 1954)

Matter of Anis, 126 N.J. 448, 599 A2d 1265, 60 USLW 2465 (N.J., Jan 10, 1992) (NO D-28)

McAllister v. Greyhound LInes, Inc., 1997 WL 642994 (D.N.J., Oct 07, 1997) (NO. CIV. A. 2225)

McGee v. Goodyear Atomic Corp., 103 Ohio App.3d 235, 659 N.E.2d 317 (Ohio App. 4 Dist., Pike County, Mar 23, 1995) (NO. A-4558-92T2)

McKinney-Vareschi v. Paley, 42 Mass.App.Ct. 953, 680 N.E.2d 116 (Mass.App.Ct., Jun 06, 1997) (NO. 96-P-1117)

McLaughlin v. Vinios, 39 Mass.App.Ct. 5, 653 N.E.2d 189 (Mass.App.Ct., Jul 28, 1995) (NO. 94-P-955)

Meadows v. S.E.C., 119 F.3d 1219, Fed. Sec. L. Rep. P 99,520 (5th Cir., Aug 22, 1997) (NO. 96-60328)

Middlemist v. BDO Seldman, LLP, 1997 WL 603886, 97 CJ C.A.R. 2117 (Colo.App., Oct 02, 1997) (NO. 96CA1649)

Mills v. Tippin's Restaurant, Inc., 1996 WL 722072 (N.D.Tex., Dec 06, 1996) (NO. CIV.A. 3:96-CV-2724D)

Missouri, Kansas, & Texas Railway Company v. Texas & Dade, 104 Tex. 237, 136 S.W. 435 (1911)

Moe v. REO Plastics, Inc., 1997 WL 613656 (Minn.App., Oct 07, 1997) (NO. C7-97-814)

Moodie v. Federal Reserve Bank of New York, 862 F.Supp. 59, 65 Fair Empl.Prac.Cas. (BNA) 1245 (S.D.N.Y., Aug 26, 1994) (NO. 91 CIV. 6629 (MEL))

Moore v. First Federal Sav. and Loan Ass'n of Rochester, 654 N.Y.S.2d 900 (N.Y.A.D. 4 Dept., Mar 14, 1997) (NO. 0213)

Moore v. RLCC Technologies, Inc., 95-2621 (La. 2/28/96), 668 So.2d 1135 (La., Feb 28, 1996) (NO. 95-CA-2521)

Moore v. YWCA, NO. 84-CA-1508-MR, (Ky.App. 1985)

Mount Vernon Fire Ins. Co. v. Creative Housing Ltd., 70 F.3d 720 (2nd Cir.(N.Y.), Nov 15, 1995) (NO. 95-7248)

Mulinix v. Mulinix, 1997 WL 585775 (Minn.App., Sep 22, 1997) (NO. C2-97-297)

Mullins v. Hinkle, 953 F.Supp. 744 (S.D.W.Va., Feb 05, 1997) (NO. CIV. A. 5:96-0634)

Murray v. Bryant, 1997 WL 607518 (Tenn.App., Oct 03, 1997) (NO. 01A01-9704-CV-00146)

Myers v. Goodwill Industries of Akron, Inc., 1997 WL 460151 (Ohio App. 9 Dist., Aug 06, 1997) (NO. 18085)

Nadeau v. Thomas, 1997 WL 542708 (N.D.Cal., Aug 21, 1997) (NO. 96-20383 SW)

National Convenience Stores, Inc., v. Stop N Go Markets of Texas, Inc., 1995 WL 144715 (Tex.App.-San Antonio) (APP. NO. 04-94-00065-CV)

Olson v. City of Lakeville, 1997 WL 561254 (Minn.App., Sep 09, 1997) (NO. C3-97-390)

Owens-Corning Fiberglas Corp. v. Martin, 942 S.W.2d 712 (Tex.App.-Dallas Mar 14, 1997) (NO. 05-95-00059-CV)

Parr v. Champion Intern. Corp., 667 So.2d 36 (Ala., May 26, 1995) (NO. 1931040)

Pegump v. Rockwell Intern. Corp., 963 F.Supp. 1518 (S.D. Iowa, Mar 07, 1996) (NO. CIV. 3-94-CV-30187)

Perkins v. Spivey, 911 F.2d 22

Pfau v. Reed, F.3d, 1997 WL 626994 (5th Cir.(Tex.), Oct 27, 1997) (NO. 96-50916)

Pickett v. Yellow Cab Co., 182 Ill.App.3d 62, 537 N.E.2d 933, 130 Ill.Dec. 604 (Ill.App. 1 Dist., Apr 04, 1989) (NO. 1-88-0381)

Pizza Hut of America, Inc., v. Keefe, 900 P.2d 97 (Colo., Jun 30, 1995) (NO. 93SC251)

Pobieglo v. Monsanto Co., 402 Mass. 112, 521 N.E.2d 728 (Mass. Apr 11, 1988) (NO. S-4492)

Pollard v. State, 173 A.D.2d 906, 569 N.Y.S.2d 770 (N.Y.A.D. 3 Dept., May 02, 1991) (NO. 61322)

Poole v. Consolidated Rail Corp., N.Y.S.2d, 1997 WL 604178, 199 N.Y. Slip Op. 7984 (N.Y.A.D. 4 Dept., Sep 30, 1997) (NO. 1214)

Powell v. Johns-Manville Corp., 342 Pa.Super. 544, 493 A.2d724 (Pa.Super., May 24, 1985) (NO. 01077 PITTS. 1982, J. 70030/84)

Preferred Mut. Ins. Co. v. Andrade, 3 Mass. L.Rptr. 505, 1995 WL 809559 (Mass.Super., Mar 03, 1995) (NO. CA93CO1513)

Pugh v. See's Candies, Inc., 116 Cal.App.3rd 311, 171 Cal.Rptr. 917 (1981)

Raimi v. Furlong, So.2d, 1997 WL 574655, 22 Fla. L. Weekly D2184 (Fla.App. 3 Dist., Sep 17, 1997) (NO. 96-954, 96-998, 96-1002, 96-1011, 96-1012)

Ransburg Industries v. Brown, 659 N.E.2d 1081 (Ind.App., Dec 21, 1995) (NO. 76A03-9507-CV-238)

Redeemer Covenant Church of Brooklyn Park v. Church Mut. Ins. Co., 567 N.W.2d 71 (Minn.App., Jul 29, 1997) (NO. C5-96-2616)

Reyes-Lopez v. Misener Marine Const. Co., 854 F.2d 529 (1st Cir. (Puerto Rico), Aug 17, 1988) (NO. 87-1816)

Reynolds v. L & L Management, Inc., S.E.2d, 1997 WL 599380, 97 FCDR 3550 (Ga.App. Sept. 29, 1997) (NO. A97A1474

Roark v. Walker Mfg. Co., 1996 WL 663134 (N.D. Ind., Sep 04, 1996) (NO. 3:95-CV-235RM)

Roberts v. Air Capitol Plating, Inc., 1997 WL 446266 (D. Kan., Jul 22, 1997) (NO. 95-1348-JTM)

Rosenbloom v. Senior Resource, Inc., 974 F.Supp. 738, 74 Fair Empl.Prac.Cas. (BNA) 1345 (D.Minn., Aug 08, 1997) (NO. 3-96 CIV. 520)

Salois v. Dime Sav. Bank of New York, F.3d, 1997 WL 671998 (1st Cir.(Mass.), Nov 03, 1997) (NO. 97-1049, 97-1050)

Saudi Arabia v. Nelson, 507 U.S. 349, 113 S.Ct. 1471, 123 L.Ed.2d 47, 61 USLW 4253 (U.S.Fla., Mar 23, 1993) (NO. 91-522)

Saylor v. Hall, 497 S.W.2d 218 (Ky 1973)

Scott Fetzer Co. v. Read, 945 S.W.2d 854 (Tex.App.-Austin, May 01, 1997) (NO. 03-95-00544-CV)

Schrader v. McGonigal, NO.90-CI-2584 (Ohio App., Mar. 20, 1990)

S.E.C. v. Sayegh, 906 F.Supp. 939 (S.D.N.Y., Nov 21, 1995) (NO. 89 CIV, 0572 (JFK))

Selby v. Revlon Consumer Products Corp., 1997 WL 587472 (N.D.Tex., Sep 17, 1997) (NO. 3-96-CV-2864-D)

Senger v. U.S., 103 F.3d 1437, 65 USLW 2504, 12 IER Cases 598, 96 Cal. Daily Op. Serv 9466, 96 Daily Journal D.A.R. 15, 573 (9th Cir.(Or.), Dec 27, 1996) (NO. 94-35688)

427

Sheehy v. Town of Plymouth, 948 F. Supp. 119 (D.Mass., Dec 12, 1996)
Sheridan v. U.S., 487 U.S. 392, 108 S.Ct. 2449, 101 L.Ed.2d 352, 56 USLW 4761 (U.S.Md., Jun 24, 1988) (NO. 87-626)
Sims v. General Telephone & Electronics, 107 Nev. 516, 815 P.2d 151 (Nev., Jul 12, 1991) (NO. 20727)
Simms v. Village of Alvion, N.Y., 115 F.3d 1098 (2nd Cir.(N.Y.), Jun 16, 1997) (NO. 396, 95-9191)
Skees v. U.S. By and Through Dept. of Army, 107 F.3d 421 (6th Cir, (Ky.), Feb 21, 1997) (NO. 96-5115)
Smith v. Foodmaker, Inc., 928 S.W.2d 683 (Tex.App.-Fort Worth, Aug 08, 1996) (NO. 2-95-282-CV)
Smithson v. V.M.S. Realty, Inc., 536 So.2d 260, 13 Fla. L.Weekly 2459 (Fla.App. 3 Dist., Nov 08, 1988) (NO. 87-25)
Snyder v. Michael's Stores, Inc., 57 Cal.Rptr.2d 105, 49 Cal.App.4th 709, (Cal. Const. art. 5, s 12; Cal. Rules of Court, Rules 28, 976, 977, 979), 65 USLW 2238, 61 Cal. Comp Cases 1019, 96 Cal. Daily Op.Serv. 7141, 96 Daily Journal D.A.R. 11, 633 (Cal.App. 5 Dist., Sep 23, 1996) (NO. F024076)
Stagl v. Delta Air Lines, Inc., 117 F.3d 76 (2nd Cir.(N.Y.), Jul 07, 1997) (NO. 96-9087)
Steffensen v. Smith's Management Corp., 820 P.2d 482 (Utah App., Oct 29, 1991) (NO. 910210-CA)
St. Paul Fire & Marine Insurance v. Knight, 764 S.W.2d 601 (Ark. 1989)
Stratman v. Brent, 291 Ill.App.3d 123, 683 N.E.2d 951, 225 Ill. Dec. 448 (Ill.App 2 Dist., Aug06, 1997) (NO. 2-96-1306)
Suerae Robertson v. Church of God, International, 1997 WL 555626 (Tex.App.-Tyler, Aug 29, 1997) (NO. 12-96-00083-CV)
Sullivan v. Lake Region Yacht and Country Club, Inc., 1997 WL 689799 (M.D.Fla., Oct 21, 1997) (NO. 97-1454-CIV-T-17A)
Tameny v. Altantic Richfield Company, 27 Cal.3rd 167, 610 P.2d 1330, 164 Cal.Rptr.839 (1980)
Taylor v. District of Columbia, 691 A.2d 121 (D.C.App., Mar 20, 1997) (NO. 94-CV-1171)
TeBockhorst v. Bank United of Texas, 1997 WL 471320 (Minn.App., Aug 19, 1997) (NO. C6-97-206)
Texas Workers' Compensation Com'n v. Garcia, 862 S.W.2d 61 (Tex.App.-San Antonio, Aug 11, 1993) (NO. 04-91-00565-CV)
Thomas v. Chicago Housing Authority, 919 F.Supp. 1159 (N.D.111., Mar 15, 1996) (NO. 95 C 4782)
Thompson v. Chapel, S.E.2d, 1997 WL 671986 (Ga.App., Oct 29, 1997) (NO. A97A1333)
Thomas v. Cook Drilling Corp., 79 Ohio St.3d 547, 684 N.E.2d 75 (Ohio, Oct 01, 1997) (NO. 96-0873, 9753)
Thompson v. Olsten Kimberly Quality-Care, Inc., F.Supp, 1997 WL 655922 (D.Minn., Sep 19, 1997) (NO. CIV. 97-11(JRT/RLE))
Tilley v. Franklin Life Ins. Co., 1997 WL 597524 (Mo.App. E.D., Sep 30, 1997) (NO.71615)
Tillman v. Washington Metropolitan Area Transit Authority, 695 A.2d 94 (D.C.App., Apr 16, 1997) (NO. 96-CV-173)
Tranchina v. Howard, Weil, Labouisse, Friedrichs, Inc., 1996 WL 392172 (E.D. La., Jul 11, 1996) (NO. CIV. A. 95-2886)
TransAmerica Ins. Co. v. South, 125 F.3d 392 (7th Cir.(Ill.), Sep 22, 1997) (NO.95-2224)
Transouth Financial Corp. v. Bell, 975 F.Supp. 1305 (M.D.Ala., Aug 25, 1997) (NO. CIV A. 96-T-1747-N)

Travelers Indem. Co. v. AMR Services Corp., 921 F.Supp. 176 (S.D.N.Y., Mar 14, 1996) (NO. 91 CIV 1211 (BN))

U.S. v. Gaubert, 499 U.S. 315, 111 S.Ct. 1267, 113 L.Ed.2d 335, 59 USLW 4244 (U.S.Tex., Mar 26, 1991) (NO. 89-1793)

U.S. v. Sharpe, 470 U.S. 675, 105 S.Ct. 1568, 84 L.Ed.2d 605, 53 USLW 4346 (U.S.S.C., Mar 20, 1985) (NO. 83-529)

U.S. v. Shearer, 473 U.S. 392, 105 S.Ct. 3039, 87 L.Ed.2d 38 (U.S. Pa., Jun 27, 1985) (NO. 84-194)

Van Houten-Maynard v. ANR Pipeline Co., 1995 WL 317056 (N.D.Ill., May 23, 1995) (NO. 89 C 0377)

Vicarelli v. Business Intern., Inc., 973 F.Supp. 241 (D.Mass., Aug 22, 1997) (NO. CIV. A. 95-12401-RCL)

Walker v. Northern San Diego County Hospital District, 135 Cal.App.3d 896, 185 Cal.Rptr 617 (1982)

Wake County Hosp. System, Inc. v. Safety Nat. Cas. Corp., 487 S.E.2d 789 (N.C.App., Aug 05, 1997) (NO. COA96-1038)

Ward v. Bechtel Corp., 102 F.3d 199, 72 Fair Empl. Prac.Cas. (BNA) 1373, 59 Empl. Prac. Dec. P 44,457 (5th Cir. (Tex.) Jan 02, 1997)

Ward v. Bechtel Corp., 1996 WL 763303 (S.D.Tex., May 01, 1996) (NO. CIV. A. H-94-2813)

Wartell v. U.S., 124 F.3d 215 (Table, Text in WESTLAW), Unpublished Disposition, 1997 WL 599980 (9th Cir.(Ariz.), Sep 24,1997) (NO.96-16547)

West By and Through Norris v. Waymire, 114 F.3d 646 (7th Cir.(Ind.), May 21, 1997) (NO. 96-3675)

Wheeler v. Commissioner of Social Services of City of New York, 662 N.Y. S.2d 550, 1997 N.Y. Slip Op. 7720 (N.Y.A.D. 2 Dept., Sep 22, 1997) (NO. 96-05040, 96-10352)

White v. Ransmeier & Spellman, 950 F.Supp. 39, 12 IER Cases 376 (D.N.H., Oct 10, 1996) (NO. CIV. 95-626-JD)

Whitely v. Food Giant, Inc., 693 So.2d 502 (Ala.Civ.App., Mar 28, 1997) (NO. 2960036)

Wilkinson v. Food Service Management Systems of Florida, 639 So.2d 1125, 19 Fla. L. Weekly D1586 (Fla.App. 3 Dist., Jul 26, 1994) (NO. 94-00981)

Wilson v. City & County of San Francisco, 124 F.3d 215 (Table, Text in WESTLAW), Unpublished Disposition, 1997 WL 577157 (9th Cir. (Cal.), Sep 12, 1997) (NO. 96-15692)

Williams v. Chicago Bd. of Educ., 1997 WL 467289, 10 NDLR P 267 (N.D. Ill., Aug 13, 1997) (NO. 97 C 1063)

Williams v. District of Columbia, 1997 WL 31095 (D.D.C., Jan 17, 1997) (NO. 96-0200-LFO)

Williams v. Marriott Corp., 1997 WL 457820 (N.D.Ill., Aug 05, 1997) (NO. 97 C 0091)

Worth v. Tyler, 1997 WL 570762 (N.D.Ill., Sep 09, 1997) (NO. 96 C 3539)

Yunker v. Honeywell, Inc., 496 N.W.2d 419

Zeiger v. State, 58 Cal.App.4th 532, 68 Cal.Rptr.2d 39, 97 Cal. Daily Op. Serv. 7925, 97 Daily Journal D.A.R. 12,725 (Cal.App. 3 Dist., Sep 25, 1997) (NO. 3 CIV. CO19828)

Zelda, Inc. v. Northland Ins. Co., 56 Cal.App.4th 1252, 66 Cal.Rptr.2d 356, 97 Cal. Daily Op. Serv. 6276, 97 Daily Journal D.A.R. 10, 193 (Cal.App. 2 Dist., Aug 05, 1997) (NO. B099018)

Zerangue v. Delta Towers, Ltd., 820 F.2d 130 (5th Cir.(La.), Jun 25, 1987) (NO. 86-3357)

Zoglauer v. City of Wheaton, 1997 WL 519833 (N.D. Ill., Sep 29, 1997) (NO. 96-C 7533)

Zueger v. Carlson, 542 N.W.2d 92 (N.D., Jan 11, 1996) (NO. CIV. 950205)

Appendix B

May 1997

RISK FACTORS FOR INJURY
IN ROBBERIES OF CONVENIENCE STORES
EXAMINED IN NIOSH STUDY

Convenience store owners, operators, and workers are among the occupational groups at highest risk for workplace homicide. A new study by the National Institute for Occupational Safety and Health (NIOSH), "Convenience Store Robberies in Selected Metropolitan Areas: Risk Factors for Employee Injury," published in the May 1997 issue of the Journal of Occupational and Environmental Medicine, notes several factors related to the risk of employee injury in convenience store robberies.

Because of limitations in the data analyzed, the study is most useful as a feasibility study that suggests focal points for further research. It cannot answer key questions about the risk of job-related violence in the convenience store setting -- for example, the question of whether the presence of two or more employees in a store at the same time will deter robbery, because information was available only for robbed stores.

NIOSH initiated the project to determine the feasibility of studying robbery-related injuries in convenience stores, using state criminal justice data. Although a number of studies have addressed the various strategies used in the convenience store setting to prevent robbery and robbery-related injuries, the evidence as to the effectiveness of specific design features is inconsistent. Data were collected from police departments in seven states that had the capacity to identify convenience store robberies and had the highest numbers (of those submitting data) of convenience store robberies during 1992.

In the 1,835 convenience store robberies documented, at least 12 employees were killed and 219 sustained nonfatal injuries. Data from four states -- Florida, Massachusetts, Michigan, and Virginia -- contained more complete information about risk factors than the data from the other states. NIOSH analyzed these data to estimate the risk of employee injury in a robbery, according to various risk factors.

Although the overall risk of employee injury in a robbery and the deterrent effect of any particular strategy could not be estimated because information was available only for robbed stores, with no comparable information for stores that were not robbed, the study found from 758 robberies in these four states that:

• there were 5 homicides and 88 nonfatal assaults of convenience store workers;

• employee probability of injury was significantly lower when the perpetrator used a firearm, although all of the homicides were firearm-related;

• employee probability of injury was significantly lower in stores which had been robbed multiple times versus only once;

• employee probability of injury was significantly lower when money was stolen than when no money was taken;

• employee probability of injury was significantly lower when customers were present at the time of the robbery; and

• employee risk of injury was not significantly different between one and multiple-employee stores.

NIOSH has conducted pioneering research on workplace violence, including a landmark June 1996 report that analyzed national data on assaults and homicides in different industries and identified major risk factors and potential prevention strategies.

For more information on this study or for information on other workplace health and safety concerns, contact NIOSH toll: 1-800-35-NIOSH (1-800-356-4674) or visit the NIOSH site on the World Wide Web at: http://www.cdc.gov.niosh/homepage/html.

This page was last updated on June 17, 1997

NIOSH PUBLICATIONS

You can reach the NIOSH Publications office by:

- FAX 513-533-8573
- Automated router system ▓▓▓▓▓▓▓▓ (1-800-356-4674).
- Telephone number outside the U.S. 1-513-533-8471
- Internet e-mail address: Pubstaff@cdc.gov

Mail: NIOSH Publications
4676 Columbia Parkway, Mail Stop C-13
Cincinnati, OH 45226-1998

- Some files are in .PDF and are identified ▓. Adobe Acrobat Reader is needed to view these files. You can download this program, free of charge, at Adobe's homepage.

NIOSH Publications List

 Alerts

Common Information Service System (CISS) - Mining

 Criteria Document Listing

 Current Intelligence Bulletins Listing

 Fact Sheets

 Federal Register Notices

 Hazard Controls

NIOSH PUBLICATIONS

 Hazard IDs

 Health Hazard Evaluation Program

 Manual of Analytical Methods (NMAM)

 Patents

 Press Releases

 Reports To Congress

 Respirator Users' Notices

 State Profiles

 Technical Summary: Respiratory Protective Devices, 42 CFR Part 84

 Updates

▇Other Publications

This page was last updated : August 18, 1997

VIOLENCE IN THE WORKPLACE

June 1997

The National Institute for Occupational Safety and Health (NIOSH) has found that an average of 20 workers are murdered each week in the United States. In addition, an estimated 1 million workers -18,000 per week - are victims of nonfatal workplace assaults each year.

Homicide is the second leading cause of death on the job, second only to motor vehicle crashes. Homicide is the leading cause of workplace death among females. However, men are at three times higher risk of becoming victims of workplace homicides than women. Homicide is also the leading cause of death for workers under 18 years of age. The majority of workplace homicides are robberyrelated crimes (71%) with only 9% committed by coworkers or former coworkers. Additionally, 76% of all workplace homicides are committed with a firearm.

Most nonfatal workplace assaults occur in service settings such as hospitals, nursing homes, and social service agencies. Forty eight percent of nonfatal assaults in the workplace are committed by a health care patient. Nonfatal workplace assaults result in more than 876,000 lost workdays and $16 million in lost wages. Nonfatal assaults occur among men and women at an almost equal rates.

The circumstances of workplace violence differ significantly from those of other types of homicides. While most workplace homicides are robbery-related, less than 10% of homicides in the general population occur during a robbery. Additionally, in the general population about 50% of all murder victims were related to their assailants whereas the majority of workplace homicides are believed to occur among people who do not know each other. These differences call for unique prevention measures targeted specifically to the workplace.

Risk Factors

Factors that place workers at risk for violence in the workplace include interacting with the public, exchanging money, delivering services or goods, working late at night or during early morning hours, working alone, guarding valuable goods or property, and dealing with violent people or volatile situations.

Anyone can become the victim of a workplace assault, but the risks are much greater in certain industries and occupations. For workplace homicides, the taxicab **industry** has the highest risk at 41.4/100,000, nearly 60 times the national average rate (0.70/100,000). The taxicab industry is followed by liquor stores (7.5) detective/protective services (7.0), gas service stations (4.8) and jewelry stores (4.7). The **occupations** with the highest homicide rates are taxicab drivers/chauffeurs (22.7), sheriffs/bailiffs (10.7), police and detectivespublic service (6.1), gas station/garage workers (5.9), and security guards (5.5). The majority of nonfatal assaults occurred in the service (64%) and retail trade (21%) industries. Specifically, 27% occurred in nursing homes, 13% in social services, 11% in hospitals, 6% in grocery stores, and another 5% occurred in eating and drinking places.

Prevention

A number of environmental, administrative, and behavioral strategies have the potential for reducing the risk of workplace violence. Examples of prevention strategies include (but are not limited to) good visibility within and outside the workplace, cashhandling policies, physical separation of workers from customers or clients, good lighting, security devices, escort services, and employee training. No single strategy is appropriate for all workplaces, but all workers and employers should assess the risk of violence in their workplaces and take appropriate action to reduce those risks. A workplace violence prevention program should include a system for documenting incidents, procedures to be taken in the event of incidents, and open communication between employers and workers.

Additional Information

NIOSH has published *Current Intelligence Bulletin 57: Violence in the WorkplaceRisk Factors and Prevention Strategies* (DHHS [NIOSH] Publication No. 96-100). Copies can be obtained free-of-charge from the NIOSH Publications Office while supplies last:

telephone 1-800-35-NIOSH (1-800-356-4674)

fax 513-533-8573

 e-mail pubstaff@cdc.gov

For a complete listing of documents available on the **CDC Fax Information Service** call **1-888-CDC-FAXX (1-888-232-3299)** and request document #000006. This information is also available on the Internet at CDC's web site.

Document #705002

THIS PAGE WAS LAST UPDATED ON August 12, 1997

National Crime Victim's Rights Week - Resource Guide

WORKPLACE VIOLENCE AND CRIME

Each year nearly one million individuals become victims of violent crime while working or on duty. Although men were more likely to be attacked at work by a stranger, women were more likely to be attacked by someone they knew. *(Bachman, Ronet, Ph.D., July 1994, "Violence and Theft in the Workplace." Crime Data Brief: National Crime Victimization Survey, pg. 1. U.S. Department of Justice, Bureau of Justice Statistics, Washington, D.C.)*

Overall, one out of every six violent crimes experienced by U.S. residents age 12 or older happens at work. *(Ibid. pg. 1)*

According to the latest Bureau of Justice Statistics' annual crime survey, an estimated eight percent of rapes, seven percent of robberies, and 16 percent of all assaults occurred while victims were working or on duty. *(Ibid., pg.1)*

In 1994, 1,071 Americans were victims of workplace homicide, (includes murders by co-workers, personal acquaintances or by persons in the commission of other crimes). Of this number, 887 victims were male and 184 were female; 64 percent of victims were white. *(Sourcebook of Criminal Justice Statistics, 1995, U.S. Department of Justice, Bureau of Justice Statistics,Washington, D.C., pg. 364.)*

Guns were the primary weapon in 86 percent of workplace homicides that took place in 1994, followed by stabbing and beating. *(Ibid., pg. 364.)*

The U.S. Department of Labor reports that homicide was the second leading cause of death of workers killed on the job in 1993. *(Hutchinson, Ty, "Vetoing Violence in the Workplace". Solutions Magazine, April 1995 edition, pg. 41.)*

One-sixth of all workplace homicides of women are committed by a spouse, ex-spouse, boyfriend or ex-boyfriend. *(Windau, J. & Toscano, G., "Workplace Homicides in 1992. Fatal Workplace Injuries in 1992: A Collection of Data and Analysis", May 1994, pg. 1. U.S. Department of Labor, Bureau of Labor Statistics, Washington, D.C.)*

Boyfriends and husbands, both current and former, commit more than 13,000 acts of violence against women in the workplace every year. *(Anfuso, Dawn. "Deflecting Workplace Violence". Personnel Journal (73) 10:66, 1994)*

Note: OVC makes no representation concerning the accuracy of data from non-Department of Justice sources.

Presented as a Public Service by

Back to Victim Statistics from OVC

Appendix C

OSHA Strategic Plan FY1997-FY2002

- The Assistant Secretary
- Information about OSHA
- What's New
- Media Releases
- Publications
- Programs & Services
- OSHA Software/Advisors
- Office Directory
- Statistics & Data

- Compliance Assistance
- Ergonomics
- Federal Register Notices
- Frequently Asked Questions
- Standards
- Other OSHA Documents
- Technical Information
- Vanguard & Customer Service
- US Government Internet Sites
- Safety & Health Internet Sites

Charles Jeffress of North Carolina Sworn In As Head Of OSHA
Biography | News Release | Speech

[Webmaster | **OSHA Home Page** | OSHA-OCIS | US DOL Web Site | Disclaimer]

▶Violence in the Workplace

Needed for PDF Files ➡

Recognition

- OSHA Summary Sheet
- Workplace Violence Initiative; OSHA; 1996
- Long Island Coalition for Workplace Violence Awareness & Prevention

Control

- Guidelines for Preventing Workplace Violence for Health Care and Social Service Workers-OSHA 3148-1996 - **4.67 MB PDF File**
- Guidelines for Workplace Violence Prevention Programs for Night Retail Establishments; OSHA

Training

- Workplace Violence Training and Outreach Materials.

[**Revision Date: 28-August-97** | Disclaimer]

http://www.osha-slc.gov/SLTC/WorkplaceViolence/index.html

Workplace Violence Memo   Page 1 of 2

U.S. Department of Labor
Occupational Safety & Health Administration

March 20, 1997

MEMORANDUM FOR: OSHA Field Staff

FROM: Cathleen Cronin, Chief

Division of Training and Educational

Development

SUBJECT: Workplace Violence Training and Outreach Materials

Attached for your use is a training and outreach package on Workplace Violence. This package is based on the March 1996 "Guidelines for Preventing Workplace Violence for Health Care and Social Service Workers." The package consists of two volumes of materials and includes the following:

- Volume I consists of a script with 52 slides that provides an overview of the Guidelines for Preventing Workplace Violence for Health Care and Social Service Workers. It also includes a hard copy of the text of the slides for overhead transparencies.

- Volume II consists of the guidelines, OSHA Publication 3148-1996; the March 1996 News Release announcing these guidelines; a fact sheet, Protecting Community Workers Against Violence (OSHA Fact sheet 96-53); and a NIOSH document, Violence in the Workplace - Risk Factors and Prevention Strategies (Current Intelligence Bulletin 57, June 1996).

These materials can be used for staff training or as outreach presentations to others. If there are questions concerning the materials, please direct them to me at (847) 297- 4810, extension 125.

Attachments

OSHA's Workplace Violence Prevention Program Volume 1	1,237 KB	PDF Adobe Acrobat
• Slide Presentation	154 KB	PPT (Microsoft PowerPoint)
• Overhead	1,762 KB	PDF Adobe Acrobat
OSHA's Workplace Violence Prevention Program Volume 2		
• OSHA 3148 - Guidelines for Preventing Workplace Violence for Health Care and Social Service Workers	4.9 MB	PDF Adobe Acrobat
• March 14, 1996 News Release	HTML	Web Page
• Fact Sheet 96-53 - Protecting community workers against violence.	HTML	Web Page
• Violence in the Workplace - Risk Factors and Prevention Strategies - (Current Intelligence Bulletin 57, June 1996)	HTML	CDC Web Site
• Violence in the Workplace - Risk Factors and Prevention Strategies - (Current Intelligence Bulletin 57, June 1996)	HTML	CDC Web Site

Appendix D

U.S. Office of Personnel Management
Dealing with Workplace Violence: A Guide for Agency Planners

- Introduction
- The Handbook

Introduction

This handbook, developed by the Office of Personnel Management and the Interagency Working Group on Violence in the Workplace, is the result of a cooperative effort of many Federal agencies sharing their expertise in preventing and dealing with workplace violence. It is intended to assist those who are responsible for establishing workplace violence initiatives at their agencies. However, we anticipate that its usefulness will extend well beyond the planning phase since many of the chapters provide information that can be helpful for managers and specialists as they deal with difficult workplace violence situations.

The Handbook

This handbook is available for downloading as individual chapters in both Adobe Acrobat Portable document Format (PDF) and WordPerfect 5.1 format, and as a single PDF file for the convenience of those persons who would like to print out the entire document.

Full Handbook (PDF Document) -- 346K File

Individual Chapters:

Chapter	Title	PDF Format	WP 5.1 Format
	Contents	PDF	WP
	Foreward	PDF	WP
1	Introduction	PDF	WP
2	Program Development	PDF	WP
3	Deveopment of Written Policy Statement	PDF	WP
4	Prevention	PDF	WP
5	Fact Finding/Investigating	PDF	WP
6	Threat Assessment	PDF	WP
7	Employee Relations Considerations	PDF	WP
8	Employee Assistance Program Considerations	PDF	WP
9	Workplace Security	PDF	WP

Appendix D   *443*

10	Organizational Recovery After An Incident	PDF	WP
11	Case Studies	PDF	WP
12	Resources	PDF	WP

- To <u>Website Index</u>
- To <u>OPM Home Page</u>

Page created 24 October 1997

◈U.S. Department of Labor

Employment Laws Assistance for Workers and Small Businesses

The Department of Labor has developed several *elaws* advisors to provide information individually tailored to specific audiences, such as employees, employers, and DOL compliance officers and policy officials. Each *elaws* advisor is a computer program that imitates the interaction an individual might have with a human expert. Based on an individual's response to questions, the *elaws* advisors provide user specific information on the subject.

Some *elaws* advisors must be run interactively on the Internet. These are identified as being online. Other *elaws* advisors must be downloaded off the Internet, and then installed and run on a PC. These are identified as a download.

Employment Standards Administration (ESA)

- ESA/Office of Federal Contract Compliance Programs' Interactive Compliance Assistance Advisor -- provides Federal contractors and subcontractors with compliance information on the three equal employment opportunity laws administered by OFCCP. online

- ESA/Wage and Hour Division's Family and Medical Leave Act Advisor -- answers a variety of commonly asked questions about FMLA including employee eligibility, valid reasons for leave, employee/employer notification responsibilities, and employee rights/benefits. online

Mine Safety and Health Administration (MSHA)

- MSHA's Form 7000-2 Advisor -- permits mine operators and contractors to electronically file the Quarterly Mine Employment and Coal Production Report Form 7000-2. online

Occupational Safety and Health Administration (OSHA)

- OSHA's Asbestos Advisor -- helps building owners, managers and lessees, as well as some contractors, write a report describing their responsibilities under OSHA's Asbestos rules. download

- OSHA's Confined Spaces Advisor -- assists in the identification of confined spaces and provides information on how to deal with permit required confined spaces. online download

- OSHA's Fire Safety Advisor -- provides interactive expert help to apply OSHA's Fire Safety related standards. download

- OSHA's <u>GOCAD</u> -- assists in the evaluation of laboratory results (of required medical surveillance) under the Cadmium Standard and the preparation of (required and optional) reports, letters, checklists, and statistics. ▪Download▪

Veterans' Employment and Training Service (VETS)

- VETS <u>Uniformed Services Employment and Reemployment Rights Act Advisor</u> -- answers questions regarding employee eligibility, employee job entitlements, employer obligations, benefits, and remedies under USERRA. ▪On-line▪

- VETS <u>Veterans' Preference System</u> -- helps veterans receive the preferences to which they may be entitled in hiring for Federal jobs. ▪On-line▪ ▪Download▪

▪DOL Homepage ▪*elaws* media page ▪Summary of the major laws administered by DOL

▪▪▪U.S. Department of Labor ▪▪▪

CDC Search Results List

Your Query **"WORKPLACE VIOLENCE"** matched 24 documents out of 19966.
24 documents displayed.

0.99 Page Title
   Excerpt: . PURPOSE AND SCOPE . The purpose of this document is to
   review what is known about fatal and nonfatal violence in the workplace
   to determine the focus needed for prevention and research efforts. This...

0.99 Page Title
   Excerpt: . DEVELOPING AND IMPLEMENTING A WORKPLACE
   VIOLENCE PREVENTION PROGRAM AND POLICY . The first
   priority in developing a workplace violence prevention policy is to
   establish a system for documenting viol...

0.92 NIOSH Press Release: Workplace Violence
   Excerpt: HS PRESS RELEASEMonday, July 8, 1996 . .NIOSH
   REPORT ADDRESSES PROBLEM OF WORKPLACE VIOLENCE,
   SUGGESTS STRATEGIES FOR PREVENTING RISKS. Some 1
   million workers are assaulted and more than 1,000 are mur...

0.92 NIOSH Other WWW sites/VIOLENCE (occupational)
   Excerpt: NIOSH Other WWW Sites . .VIOLENCE (occupational).
   Assault Prevention Information Network Index .Employer's Guide to
   Teen Worker Safety (DOL) .Federal Protective Service .Governors Call
   For Banning Fir...

0.92 Page Title
   Excerpt: . ABSTRACT . This document reviews what is known about
   fatal and nonfatal violence in the workplace to determine the focus
   needed for prevention and research. The document also summarizes
   issues to be...

0.90 Page Title
   Excerpt: . RISK FACTORS AND PREVENTION STRATEGIES . Risk
   Factors .A number of factors may increase a worker's risk for workplace
   assault, and they have been described in previous research [Collins and
   Cox 1987...

0.90 Page Title
   Excerpt: . CURRENT EFFORTS AND FUTURE DIRECTIONS:
   RESEARCH AND PREVENTION . Although we are beginning to have
   descriptive information about workplace violence, a number of research
   questions remain: .What are ...

0.87 <u>Page Title</u>
Excerpt: . RELATED READING . AFSCME (American Federation of State, County, and Municipal Employees) and AFL-CIO [1995]. Hidden violence against women at work. Women in Public Service 5(fall):106 .Alexander BH,...

0.87 <u>Page Title</u>
Excerpt: . PUBLIC HEALTH SUMMARY .Violence in the Workplace . What are the hazards? .An average of 20 workers are murdered each week in the United States. The majority of these murders are robbery-related crim...

0.87 <u>Page Title</u>
Excerpt: . FOREWARD . The purpose of the Occupational Safety and Health Act of 1970 (Public Law 910596) is to assure safe and healthful working conditions for every working person and to preserve our human res...

0.87 <u>niosh/fact sheet/violence in the workplace</u>
Excerpt: . .VIOLENCE IN THE WORKPLACE . June 1997 .The National Institute for Occupational Safety and Health (NIOSH) has found that an average of 20 workers are murdered each week in the United States. In addi...

0.80 <u>whatspub</u>

0.80 <u>Family and Intimate Partner Violence Prevention Cooperative Agreements</u>
Excerpt: . NCIPC Cooperative Agreements . Family and Intimate Partner Violence Prevention . COORDINATED COMMUNITY RESPONSE Part 1 projects:. The Chatham Hospital (CH), Family Violence and Rap...

0.80 <u>FAMILY AND INTIMATE VIOLENCE PREVENTION PROGRAM COORINATED COMMUNITY RESPONSE PROJECTS</u>
Excerpt: FAMILY AND INTIMATE VIOLENCE PREVENTION PROGRAM. COORDINATED COMMUNITY RESPONSE . Part 1 projects:. The Chatham Hospital (CH), Family Violence and R...

0.80 <u>whatspub</u>

0.80 <u>schedule</u>

0.80 <u>NIOSH/ Scientists Convene For NIOSH National Symposium On Occupational Injury Prevention</u>

0.80 <u>NIOSH/SHR/child labor research needs/chapter 2-bkg</u>
Excerpt: 2 BACKGROUND. Historical Issues .Historically, regulation

http://www2.cdc.gov/search97cgi/vtopic.exe?ServerKey=Primary&Theme=&Company=Cente112/10/97isease+

of child labor has been motivated by both economic issues and concerns for the safety, health, and normal development of children. In the earl...

. 0.80 NIOSH UPDATE/STOUT
Excerpt: .February 24, 1997 . .NANCY A. STOUT NAMED DIRECTOR OF NIOSH SAFETY RESEARCH DIVISION . .Nancy A. Stout, Ed.D., was named director of the Division of Safety Research in the National Institute fo...

0.80 niosh/update/ RISK FACTORS FOR INJURY IN ROBBERIES OF CONVENIENCE STORES
Excerpt: May 1997. RISK FACTORS FOR INJURY IN ROBBERIES OF CONVENIENCE STORES EXAMINED IN NIOSH STUDY. Convenience store owners, operators, and workers are among the occupational groups at ...

0.80 Violence in the Workplace
Excerpt: CURRENT INTELLIGENCE BULLETIN 57 . Risk Factors and Prevention Strategies . U.S. DEPARTMENT OF HEALTH AND HUMAN SERVICES. Public Health Service. Centers for Disease Control and Prevention. National In...

0.80 Page Title
Excerpt: . NONFATAL ASSAULTS IN THE WORKPLACE . Victimization Studies .Limited information is available in the criminal justice and public health literature regarding the nature and magnitude of nonfatal workp...

0.80 NIOSH/HHS Press Releases

0.80 NASD Safety Resource Directory: Mary Lou Bean
Excerpt: . . . . Mary Lou Bean. Title:. Employee Relations Specialist .Department:. Human Resources-Employee Relations .Institution:. The Ohio State University .Contact Address:. Archer House Columbu...

Friday
January 09, 1998

What's New

Search CIC

Pueblo 81009

Sight & Sound

Links

Special Stuff

Consumer News

Order a Catalog

Download Catalog

Feedback

— Click here! —
1997
Consumer's
Resource
Handbook

Welcome! We have full text versions of hundreds of the best federal consumer publications available. All are FREE! Select a category below to see the publications in our latest Catalog, check out Special Stuff to see other great information that isn't currently in our Catalog, or use our Search to zero in on just what you're looking for. Take a moment and look around. We hope your visit is helpful and thanks for stopping by. (Text version of this page)

This service is provided by the Consumer Information Center of the U.S. General Services Administration. If you have a comment or question, e-mail us at catalog.pueblo@gsa.gov

http://www.pueblo.gsa.gov/

Appendix E

Employee Workplace Rights

Contents
Introduction

OSHA Standards and Workplace Hazards
        Right to Know
        Access to Exposure and Medical Records
        Cooperative Efforts to Reduce Hazards
        OSHA State Consultation Service
OSHA Inspections
        Employee Representative
                Helping the Compliance Officer
                Observing Monitoring
                Reviewing OSHA Form 200
After an Inspection
                Challenging Abatement Period
                Variances
                Confidentiality
                Review If No Inspection Is Made
                Discrimination for Using Rights
Employee Responsibilities
Contacting NIOSH

Other Sources of OSHA Assistance

Introduction

The Occupational Safety and Health (OSH) Act of 1970 created the
Occupational Safety and Health Administration (OSHA) within the
Department of Labor and encouraged employers and employees to reduce
workplace hazards and to implement safety and health programs.

In so doing, this gave employees many new rights and responsibilities,
including the right to do the following:

Review copies of appropriate standards, rules, regulations, and
requirements that the employer should have available at the workplace.
Request information from the employer on safety and health hazards in
the workplace, precautions that may be taken, and procedures to be
followed if the employee is involved in an accident or is exposed to
toxic substances. Have access to relevant employee exposure and medical
records. Request the OSHA area director to conduct an inspection if they
believe hazardous conditions or violations of standards exist in the
workplace. Have an authorized employee representative accompany the OSHA
compliance officer during the inspection tour. Respond to questions from
the OSHA compliance officer, particularly if there is no authorized
employee representative accompanying the compliance officer on the
inspection "walkaround." Observe any monitoring or measuring of
hazardous materials and see the resulting records, as specified under
the Act, and as required by OSHA standards. Have an authorized
representative, or themselves, review the Log and Summary of
Occupational Injuries (OSHA No. 200) at a reasonable time and in a
reasonable manner. Object to the abatement period set by OSHA for
correcting any violation in the citation issued to the employer by
writing to the OSHA area director within 15 working days from the date
the employer receives the citation. Submit a written request to the
National Institute for Occupational Safety and Health (NIOSH) for
information on whether any substance in the workplace has potentially
toxic effects in the concentration being used, and have their names
withheld from the employer, if so requested. Be notified by the employer
if the employer applies for a variance from an OSHA standard, and

testify at a variance hearing, and appeal the final decision. Have their
names withheld from employer, upon request to OSHA, if a written and
signed complaint is filed. Be advised of OSHA actions regarding a
complaint and request an informal review of any decision not to inspect
or to issue a citation. File a Section 11(c) discrimination complaint if
punished for exercising the above rights or for refusing to work when
faced with an imminent danger of death or serious injury and there is
insufficient time for OSHA to inspect; or file a Section 405 reprisal
complaint (under the Surface Transportation Assistance Act (STAA)).

Pursuant to Section 18 of the Act, states can develop and operate their
own occupational safety and health programs under state plans approved
and monitored by Federal OSHA. States that assume responsibility for
their own occupational safety and health program must have provisions at
least as effective as those of Federal OSHA, including the protection of
employee rights. There are currently 25 state plans. Twenty-one states
and two territories administer plans covering both private and state and
local government employment; and two states cover only the public
sector. All the rights and responsibilities described in this booklet
are similarly provided by state programs. (See list of these states at
the end of this booklet.)

Any interested person or groups of persons, including employees, who
have a complaint concerning the operation or administration of a state
plan may submit a Complaint About State Program Administration (CASPA)
to the appropriate OSHA regional administrator (see lists at the end of
this booklet.) Under CASPA procedures, the OSHA regional administrator
investigates these complaints and informs the State and the complainant
of these findings. Corrective action is recommended when required. OSHA
Standards and Workplace Hazards

Before OSHA issues, amends or deletes regulations, the agency publishes
them in the Federal Register so that interested persons or groups may
comment.

The employer has a legal obligation to inform employees of OSHA safety
and health standards that apply to their workplace. Upon request, the
employer must make available copies of those standards and the OSHA law
itself. If more information is needed about workplace hazards than the
employer can supply, it can be obtained from the nearest OSHA area
office.

Under the Act, employers have a general duty to provide work and a
workplace free from recognized hazards. Citations may be issued by OSHA
when violations of standards are found and for violations of the general
duty clause, even if no OSHA standard applies to the particular hazard.

The employer also must display in a prominent place the official OSHA
poster that describes rights and responsibilities under OSHA's law.

Right to Know

Employers must establish a written, comprehensive hazard communication
program that includes provisions for container labeling, material safety
data sheets, and an employee training program. The program must include
a list of the hazardous chemicals in each work area, the means the
employer uses to inform employees of the hazards of non-routine tasks
(for example, the cleaning of reactor vessels), hazards associated with
chemicals in unlabeled pipes, and the way the employer will inform other
employers of the hazards to which their employees may be exposed.

Access to Exposure and Medical Records

Employers must inform employees of the existence, location, and

availability of their medical and exposure records when employees first
begin employment and at least annually thereafter. Employers also must
provide these records to employees or their designated representatives,
upon request. Whenever an employer plans to stop doing business and
there is no successor employer to receive and maintain these records,
the employer must notify employees of their right of access to records
at least 3 months· before the employer ceases to do business. OSHA
standards require the employer to measure exposure to harmful
substances, the employee (or representative) has the right to observe
the testing and to examine the records of the results. If the exposure
levels are above the limit set by the standard, the employer must tell

employees what will be done to reduce the exposure.

Cooperative Efforts to Reduce Hazards

OSHA encourages employers and employees to work together to reduce
hazards. Employees should discuss safety and health problems with the
employer, other workers, and union representatives (if there is a
union). Information on OSHA requirements can be obtained from the OSHA
area office. If there is a state occupational safety and health
program, similar information can be obtained from the state.

OSHA State Consultation Service

If an employer, with the cooperation of employees, is unable to find
acceptable corrections for hazards in the workplace, or if assistance is
needed to identify hazards, employees should be sure the employer is
aware of the OSHA-sponsored, state-delivered, free consultation service.
This service is intended primarily for small employers in high hazard
industries. Employers can request a limited or comprehensive
consultation visit by a consultant from the appropriate state
consultation service.

A list of telephone numbers for contacting state consultation projects
is provided at the end of this booklet.

OSHA Inspections

If a hazard is not being corrected, an employee should contact the OSHA
area office (or state program office) having jurisdiction. If the
employee submits a written complaint and the OSHA area or state office
determines that there are reasonable grounds for believing that a
violation or danger exists, the office conducts an inspection.

Employee Representative

Under Section 8(e) of the Act, the workers' representative has a right
to accompany an OSHA compliance officer (also referred to as a
compliance safety and health officer, CSHO, or inspector) during an
inspection. The representative must be chosen by the union (if there is
one) or by the employees. Under no circumstances may the employer
choose the workers' representative.

If employees are represented by more than one union, each union may
choose a representative. Normally, the representative of each union
will not accompany the inspector for the entire inspection, but will
join the inspection only when it reaches the area where those union
members work.

An OSHA inspector may conduct a comprehensive inspection of the entire
workplace or a partial inspection limited to certain areas or aspects of
the operation.

Helping the Compliance Officer

Workers have a right to talk privately to the compliance officer on a
confidential basis whether or not a workers' representative has been
chosen.

Workers are encouraged to point out hazards, describe accidents or
illnesses that resulted from those hazards, describe past worker
complaints about hazards, and inform the inspector if working conditions
are not normal during the inspection.

Observing Monitoring

If health hazards are present in the workplace, a special OSHA health
inspection may be conducted by an "industrial hygienist." This OSHA
inspector may take samples to measure levels of dust, noise, fumes, or
other hazardous materials.

OSHA will inform the employee representative as to whether the employer
is in compliance. The inspector also will gather detailed information
about the employer's efforts to control health hazards, including
results of tests the employer may have conducted.

Reviewing OSHA Form 200

If the employer has more than 10 employees, the employer must maintain
records of all work-related injuries and illnesses, and the employees or
their representative have the right to review those records. Some
industries with very low injury rates (e.g., insurance and real estate
offices) are exempt from recordkeeping.

Work-related minor injuries must be recorded if they resulted in
restriction of work or motion, loss of consciousness, transfer to
another job, termination of employment, or medical treatment (other than
first-aid). All recognized work-related illnesses and non-minor
injuries also must be recorded.

After an Inspection

At the end of the inspection, the OSHA inspector will meet with the
employer and the employee representatives in a closing conference to
discuss the abatement of hazards that have been found.

If it is not practical to hold a joint conference, separate conferences
will be held, and OSHA will provide written summaries, on request.

During the closing conference, the employee representative may describe,
if not reported already, what hazards exist, what should be done to
correct them, and how long it should take. Other facts about the history
of health and safety conditions at the workplace may also be provided.

Challenging Abatement Period

Whether or not the employer accepts OSHA's actions, the employee (or
representative) has the right to contest the time OSHA allows for
correcting a hazard.

This contest must be filed in writing with the OSHA area director within
15 working days after the citation is issued. The contest will be
decided by the Occupational Safety and Health Review Commission. The
Review Commission is an independent agency and is not part of the
Department of Labor.

Variances

Some employers may not be able to comply fully with a new safety and health standard in the time provided due to shortages of personnel, materials or equipment. In situations like these, employers may apply to OSHA for a temporary variance from the standard. In other cases, employers may be using methods or equipment that differ from those prescribed by OSHA, but which the employer believes are equal to or better than OSHA's requirements, and would qualify for consideration as a permanent variance. Applications for a permanent variance must basically contain the same information as those for temporary variances.

The employer must certify that workers have been informed of the variance application, that a copy has been given to the employee's representative, and that a summary of the application has been posted wherever notices are normally posted in the workplace. Employees also must be informed that they have the right to request a hearing on the application.

Employees, employers, and other interested groups are encouraged to participate in the variance process. Notices of variance application are published in the Federal Register inviting all interested parties to comment on the action.

Confidentiality

OSHA will not tell the employer who requested the inspection unless the complainant indicates that he or she has no objection.

Review If No Inspection Is Made

The OSHA area director evaluates the complaint from the employee or representative and decides whether it is valid. If the area director decides not to inspect the workplace, he or she will send a certified letter to the complainant explaining the decision and the reasons for it. Complainants must be informed that they have the right to request further clarification of the decision from the area director; if still dissatisfied, they can appeal to the OSHA regional administrator for an informal review. Similarly, a decision by an area director not to issue a citation after an inspection is subject to further clarification from the area director and to an informal review by the regional administrator.

Discrimination for Using Rights

Employees have a right to seek safety and health on the job without fear of punishment. That right is spelled out in Section 11(c) of the Act.

The law says the employer "shall not" punish or discriminate against employees for exercising such rights as complaining to the employer, union, OSHA, or any other government agency about job safety and health hazards; or for participating in OSHA inspections, conferences, hearings, or other OSHA-related activities.

Although there is nothing in the OSHA law that specifically gives an employee the right to refuse to perform an unsafe or unhealthful job assignment, OSHA's regulations, which have been upheld by the U.S. Supreme Court, provide that an employee may refuse to work when faced with an imminent danger of death or serious injury. The conditions necessary to justify a work refusal are very stringent, however, and a work refusal should be an action taken only as a last resort. If time permits, the unhealthful or unsafe condition should be reported to OSHA or other appropriate regulatory agency.

A state that is administering its own occupational safety and health enforcement program pursuant to Section 18 of the Act must have provisions as effective as those of Section 11(c) to protect employees from discharge or discrimination. OSHA, however, retains its Section 11(c) authority in all states regardless of the existence of an OSHA-approved state occupational safety and health program.

Workers believing they have been punished for exercising safety and health rights must contact the nearest OSHA office within 30 days of the time they learn of the alleged discrimination. A representative of the employee's choosing can file the 11(c) complaint for the worker. Following a complaint, OSHA will contact the complainant and conduct an indepth interview to determine whether an investigation is necessary.

If evidence supports the conclusion that the employee has been punished for exercising safety and health rights, OSHA will ask the employer to restore that worker's job, earnings, and benefits. If the employer declines to enter into a voluntary settlement, OSHA may take the employer to court. In such cases, an attorney of the Department of Labor will conduct litigation on behalf of the employee to obtain this relief.

Section 405 of the Surface Transportation Assistance Act was enacted on January 6, 1983, and provides protection from reprisal by employers for truckers and certain other employees in the trucking industry involved in activity related to commercial motor vehicle safety and health. Secretary of Labor's Order No. 9-83 (48 Federal Register 35736, August 5, 1983) delegated to the Assistant Secretary of OSHA the authority to investigate and to issue findings and preliminary orders under Section 405.

Employees who believe they have been discriminated against for exercising their rights under Section 405 may file a complaint with OSHA within 180 days of the discrimination. OSHA will then investigate the complaint, and within 60 days after it was filed, issue findings as to whether there is a reason to believe Section 405 has been violated.

If OSHA finds that a complaint has merit, the agency also will issue an order requiring, where appropriate, abatement of the violation, reinstatement with back pay and related compensation, payment of compensatory damages, and the payment of the employee's expenses in bringing the complaint. Either the employee or employer may object to the findings. If no objection is filed within 30 days, the finding and order are final. If a timely filed objection is made, however, the objecting party is entitled to a hearing on the objection before an Administrative Law Judge of the Department of Labor.

Within 120 days of the hearing, the Secretary will issue a final order. A party aggrieved by the final order may seek judicial review in a court of appeals within 60 days of the final order. The following activities of truckers and certain employees involved in commercial motor vehicle operation are protected under Section 405:

Filing of safety or health complaints with OSHA or other regulatory agency relating to a violation of a commercial motor vehicle safety rule, regulation, standard, or order. Instituting or causing to be instituted any proceedings relating to a violation of a commercial motor vehicle safety rule, regulation, standard or order. Testifying in any such proceedings relating to the above items.

Refusing to operate a vehicle when such operation constitutes a violation of any Federal rules, regulations, standards or orders applicable to commercial motor vehicle safety or health; or because of the employee's reasonable apprehension of serious injury to himself or the public due to the unsafe condition of the equipment. Complaining directly to management, coworkers, or others about job safety or health conditions relating to commercial motor vehicle operation.

Complaints under Section 405 are filed in the same manner as complaints under 11(c). The filing period for Section 405 is 180 days from the alleged discrimination, rather than 30 days as under Section 11(c).

In addition, Section 211 of the Asbestos Hazard Emergency Response Act provides employee protection from discrimination by school officials in retaliation for complaints about asbestos hazards in primary and secondary schools.

The protection and procedures are similar to those used under Section 11(c) of the OSH Act. Section 211 complaints must be filed within 90 days of the alleged discrimination.

Finally, Section 7 of the International Safe Container Act also provides employee protection from discrimination in retaliation for safety or health complaints about intermodal cargo containers designed to be transported interchangeably by sea and land carriers. The protection and procedures are similar to those used under Section 11(c) of the OSH Act. Section 7 complaints must be filed within 60 days of the alleged discrimination.

Employee Responsibilities

Although OSHA does not cite employees for violations of their responsibilities, each employee "shall comply with all occupational safety and health standards and all rules, regulations, and orders issued under the Act" that are applicable. Employee responsibilities and rights in states with their own occupational safety and health programs are generally the same as for workers in states covered by Federal OSHA. An employee should do the following:

Read the OSHA Poster at the jobsite. Comply with all applicable OSHA standards. Follow all lawful employer safety and health rules and regulations, and wear or use prescribed protective equipment while working. Report hazardous conditions to the supervisor. Report any job-related injury or illness to the employer, and seek treatment promptly. Cooperate with the OSHA compliance officer conducting an inspection if he or she inquires about safety and health conditions in the workplace. Exercise rights under the Act in a responsible manner.

Contacting NIOSH

NIOSH can provide free information on the potential dangers of substances in the workplace. In some cases, NIOSH may visit a jobsite to evaluate possible health hazards. The address is as follows:

National Institute for Occupational Safety and
Health Centers for Disease Control 1600 Clifton Road
Atlanta, Georgia 30333 Telephone: 404-639-3061

NIOSH will keep confidential the name of the person who asked for help if requested to do so.

Other Sources of OSHA Assistance

Safety and Health Management Guidelines

Effective management of worker safety and health protection is a decisive factor in reducing the extent and severity of work-related injuries and illnesses and their related costs. To assist employers and employees in developing effective safety and health programs, OSHA published recommended Safety and Health Management Program Guidelines (Federal Register 54(18): 3908-3916, January 26, 1989). These voluntary

Appendix E 459

guidelines apply to all places of employment covered by OSHA.

The guidelines identify four general elements that are critical to the development of a successful safety and health management program:

Management commitment and employee involvement, Worksite analysis, Hazard prevention and control, and Safety and health training. The guidelines recommend specific actions, under each of these general elements, to achieve an effective safety and health program. A single free copy of the guidelines can be obtained from the OSHA Publications Office, U.S. Department of Labor, OSHA/OSHA Publications, P.O. Box 37535, Washington, DC 20013-7535, by sending a self-addressed mail label with your request.

Appendix F

MODIFIED FOR THE PURPOSES OF THIS TEXT

Guidelines for Workplace Violence Prevention Programs for Night Retail Establishments

(Draft — June 28, 1996)

Important Note

This is a working draft document for discussion and comment only. Do not cite or quote as though it is a final document. The document does not represent OSHA's final position with respect to the matters discussed therein. It will be revised based upon comments received from stakeholders. The final document will be advisory in nature and informational in content for employers seeking to provide workplace violence prevention strategies. The issuance of these guidelines is not intended as an enhancement of the general duty clause, nor does it reflect any change in OSHA enforcement policy.

Contents

Notice
Acknowledgments
1. Introduction
    OSHA's commitment
2. Extent of the problem
    Fatalities
    Who are the victims?
    Nonfatal assaults
    Risk factors/summary of research
    Preventive strategies
    Overview of guidelines

3. Violence prevention program elements
    Management commitment and employee involvement
    Written program
4. Worksite analysis
    Records analysis and tracking
    Monitoring trends and analyzing incidents
    Screening surveys
    Workplace security analysis
5. Hazard prevention and control
    Engineering controls and workplace adaptation
    Administrative and work practice controls
    Post-incident response
6. Training and education
    General training
    Training for supervisors, managers, and security personnel
7. Recordkeeping and evaluation
    Recordkeeping
    Evaluation
    Sources of assistance
8. Conclusions
    References
    Appendices
        A: Workplace Violence Checklist
        B: Sources of OSHA Assistance
        C: Suggested Readings

1. Introduction

Workplace violence has emerged as a critical safety and health hazard. According to Bureau of Labor statistics, homicide is the second leading cause of death to American workers, claiming the lives of 1,071 workers in 1994 and accounting for 16% of the 6,588 fatal work injuries in the United States (BLS, 1995). Research indicates that measures can be taken to reduce the risks of workplace violence to American workers. OSHA's violence prevention guidelines provide the agency's recommendations for reducing workplace violence developed following a careful review of workplace violence studies, public and private violence prevention programs, and consultation with and input from stakeholders. OSHA encourages employers to establish violence prevention programs and to track their progress in reducing work-related assaults. Although not every incident can be prevented, many can; and the severity of injuries sustained by employees reduced. Adopting practical measures such as those outlined here can significantly reduce this serious threat to worker safety.

OSHA's commitment

The publication and distribution of these guidelines is OSHA's first step in assisting the night retail industry in preventing workplace violence. OSHA plans to conduct a coordinated effort consisting of research, information, training, cooperative programs, and appropriate enforcement to accomplish this goal. The guidelines are not a new standard or regulation. They are advisory in nature, informational in content, and intended for use by employers in providing a safe and healthful workplace through effective violence prevention programs, adapted to the needs and resources of each place of employment.

2. Extent of problem

Fatalities

Workers in retail establishments face an above-average risk for violence. Of the 1,071 deaths due to workplace violence in 1994, as in prior years, half occurred during robberies of small retail establishments, including grocery or convenience stores, restaurants and bars, liquor stores, fast-food restaurants, and gas stations (BLS, 1995). During 1980–89, there were 2,787 homicides in small retail establishments — the highest number of occupational homicides (National Institute for Occupational Safety and Health: NIOSH, 1993). Shootings accounted for four-fifths of fatal workplace assaults, with robbery being the motive in 75% of the homicides (BLS, 1995).

Who are the victims?

At 82%, men account for the majority of all homicide victims. Homicide, however, is the leading way in which women are killed in the workplace, accounting for 40% of such deaths. Women accounted for 53% of the homicides reported in the retail trades. African Americans, Asian Americans, and other minority races comprise an eighth of the workforce but constitute a fourth of all workplace homicide victims. This higher homicide risk is due, in part, to their constituting a disproportionate share of the work force in these occupations (BLS, 1994, 1995).

Nonfatal assaults

Workplace homicides are only part of the problem. According to BLS statistics, about 21,300 workers were injured in nonfatal assaults in the workplace in 1993, women as victims in 56% of these assaults. (BLS Survey of Occupational Injuries and Illnesses, based upon a random sample of establishments, 1992, 1993) The Department of Justice's National Crime Victimization Survey (NCVS) also reported that between 1987 and 1992, approximately 1 million persons annually were assaulted at work. These data include four categories: 615,160 simple assaults, 264,174 aggravated assaults, 79,109 robberies, and 13,068 rapes. Such intentional injuries to workers occur much more frequently than occupational homicides, but efforts to prevent homicides also may reduce the incidences of these nonfatal assaults.

Risk factors/summary of research

In 1992, the Centers for Disease Control (CDC) declared workplace homicide a serious public health problem. NIOSH identified six factors that increase the risks of homicide in the workplace. (CDC/NIOSH Alert, 1993) All of these factors are usually present in night retail establishments:

1. Exchange of money with the public
2. Working alone or in small numbers
3. Working late-night or early-morning hours
4. Working in high-crime areas
5. Guarding valuable property or possessions
6. Working in community settings

Several studies have examined risk factors for robbery. The groundbreaking experiment conducted by Crow and Bull (1975), at the request of the Southland Corporation (owner of the 7-Eleven convenience store chain), for example, enlisted the aid of ex-convicts to determine which stores were most "attractive" to robbers. The study confirmed that robberies are not randomly distributed, but robbers are selective in their targets. Stores most vulnerable had large amounts of cash on hand, an obstructed view of counters with inattentive clerks, poor outdoor lighting, and easy escape routes. Studies conducted by the State of Florida's Attorney General's office and the State of Virginia's Crime Commission have both corroborated these results and added an additional risk factor: clerks working alone at night. According to these studies, the top five "attractiveness" features for choosing a store to rob are: remote area, only one clerk, no customers, easy access/getaway, and lots of money on hand. The five most "unattractive" (deterrent) features were: many customers, heavy traffic in store front, two clerks, a visible back room where another employee might be, and a male clerk. It was determined that the primary deterrent was having two clerks on duty. Other deterrents significantly reducing the potential for robbery included: security cameras, time-release safes, a location near other 24-hour businesses, and a midnight store closing time. Retail robberies occur in the late evening and early morning hours more often than during daylight hours. Rates run from 65% from 11:00 p.m. to 6:00 a.m. (Erickson, 1995), 65% from 9:00 p.m. to 3:00 a.m. (Kraus, et al., 1995), to 69% for the time frame 9:00 p.m. to 5:00 a.m. (State of Virginia, 1993). Generally, a lone clerk is working at night when an attack occurs. According to studies of robberies in Gainesville, Florida from 1981 to 1986, 92% occurred when one clerk was working. In 85% of the cases, the clerk was completely alone — no customers were present in the store.

Preventive Strategies

"Situational crime prevention" reduces opportunities for crime and increases its risks. The situational approach is based on the presumption that perpetrators of crime "rationally" select their targets and that the chance of victimization can be reduced by: increasing the effort (target hardening, controlling access, deflecting offenders, and controlling facilitators); increasing the risks (entry/exit screening, formal surveillance, surveillance by employees and others); and reducing the rewards (removing the target, identifying property, removing inducements, and setting rules). Physical and behavioral changes at a site can substantially reduce robberies. A test group of stores eliminated or reduced the risk factors and subsequently experienced a 30% drop in robberies compared to a control group, thus supporting the hypotheses that robbers select their targets and that physical change to the establishment can reduce the incidence or change the location of robberies. The target-hardening efforts, including a basic robbery deterrence package, were implemented in 7-Eleven stores nationwide in 1976. The 7-Eleven store robberies decreased by 65% from 1976 to 1986 and have held at a 50% reduction since that time. The National Association of Convenience Stores (NACS) adopted the program for use nationwide in 1987 (Erickson, 1995). The rationale of the NACS program is to make the target less attractive by reducing the cash, maximizing the take/risk ratio, and training employees. The program includes the following:

7. Clearing windows for increased visibility; improving lighting
8. Maintaining low cash in register
9. Installing time-controlled drop safes
10. Posting signs about low cash
11. Altering escape routes
12. Training employees in not resisting (Erickson, 1995)

Employing two clerks is a form of "target hardening" because it may make a robbery more difficult to complete and, therefore, more unsuitable. Having two clerks may also increase the likelihood that they will be effective witnesses in a police investigation. Since the enactment of the Gainesville ordinance requiring two clerks, robbery was reduced by 92% between the hours of 8:00 p.m. and 4:00 a.m. (Florida Office of Attorney General, 1991).

Overview of guidelines

In January 1989, OSHA published voluntary, generic safety and health program management guidelines for all employers to use as a foundation for their safety and health programs, including a workplace violence prevention program. OSHA's violence prevention guidelines build on the 1989 generic guidelines by identifying common risk factors and describing some feasible solutions. Although not exhaustive, the new workplace violence guidelines include policy recommendations and practical corrective methods to help prevent and mitigate the effects of workplace violence. The goal is to eliminate or reduce worker exposure to conditions that lead to death or injury from violence by implementing effective security devices and administrative work practices, among other control measures. These guidelines are intended to cover a broad spectrum of workers in retail trades who provide services during evening and night hours. They are particularly appropriate for workers in convenience stores, liquor stores, and gasoline stations with grocery stores providing services late at night. It is anticipated, however, that other types of establishments — such as drug stores, grocery stores, supermarkets, and eating and drinking establishments — may find these recommendations helpful.

3. *Violence prevention program elements*

There are four main components to any effective safety and health program that also apply to preventing workplace violence:

1. Management commitment and employee involvement
2. Worksite analysis
3. Hazard prevention and control
4. Safety and health training

Management commitment and employee involvement

Management commitment and employee involvement are complementary and essential elements of an effective safety and health program. To ensure an effective program, management and front-line employees must work together, perhaps through a team or committee approach. If employers opt for this strategy, they must be careful to comply with the applicable provisions of the National Labor Relations Act which prohibits unfair labor practices. Management commitment, including the endorsement and visible involvement of top management, provides the motivation and resources to deal effectively with workplace violence, and should include the following:

5. Demonstrated organizational concern for employee emotional and physical safety and health
6. Equal commitment to worker safety and health and customer safety
7. Assigned responsibility for the various aspects of the workplace violence prevention program to ensure that all managers, supervisors, and employees understand their obligations
8. Appropriate allocation of authority and resources to all responsible parties
9. A system of accountability for involved managers, supervisors, and employees

10. A comprehensive program of medical and psychological counseling and debriefing for employees experiencing or witnessing assaults and other violent incidents
11. Commitment to support and implement appropriate recommendations from safety and health committees

Employee involvement and feedback enable workers to develop and express their own commitment to safety and health and provide useful information to design, implement, and evaluate the program. Employee involvement should include the following:

12. Understanding and complying with the workplace violence prevention program and other safety and security measures
13. Participation in an employee complaint or suggestion procedure covering safety and security concerns
14. Prompt and accurate reporting of violent incidents
15. Participation on safety and health committees or teams that receive reports of violent incidents or security problems, make facility inspections, and respond with recommendations for corrective strategies
16. Taking part in a continuing education program that covers techniques to recognize escalating agitation, assaultive behavior, or criminal intent, and discusses appropriate responses

Written program

A written program for job safety and security, incorporated into the organization's overall safety and health program, offers an effective approach for larger organizations. In smaller establishments, the program need not be written or heavily documented to be satisfactory. What is needed are clear goals and objectives to prevent workplace violence suitable for the size and complexity of the workplace operation and adaptable to specific situations in each establishment. The prevention program and startup date must be communicated to all employees. At a minimum, workplace violence prevention programs should do the following:

17. Create and disseminate a clear policy of zero-tolerance for workplace violence, verbal and nonverbal threats, and related actions. Managers, supervisors, co-workers, and customers must be advised of this policy.
18. Ensure that no reprisals are taken against an employee who reports or experiences workplace violence.
19. Encourage employees to report incidents promptly and to suggest ways to reduce or eliminate risks. Require records of incidents to assess risk and to measure progress.
20. Outline a comprehensive plan for maintaining security in the workplace, which includes establishing a liaison with law enforcement

representatives and others who can help identify ways to prevent and mitigate workplace violence.

21. Assign responsibility and authority for the program to individuals or teams with appropriate training and skills. The written plan should ensure that there are adequate resources available for this effort and that the team or responsible individuals develop expertise on workplace violence prevention in night retail establishments.

22. Affirm management commitment to a worker-supportive environment that places as much importance on employee safety and health as on serving the customer.

23. Set up a company briefing as part of the initial effort to address such issues as preserving safety, supporting affected employees, and facilitating recovery.

4. Worksite analysis

Worksite analysis involves a step-by-step, common-sense look at the workplace to find existing or potential hazards for workplace violence. This entails reviewing specific procedures or operations that contribute to hazards and specific locales where hazards may develop. A "Threat Assessment Team," similar task force, or coordinator may assess the vulnerability to workplace violence and determine the appropriate preventive actions to be taken. Implementing the workplace violence prevention program then may be assigned to this group. The team should include representatives from senior management, operations, employee assistance, security, occupational safety and health, legal, and human resources staff. The team or coordinator can review injury and illness records and workers' compensation claims to identify patterns of assaults that could be prevented by workplace adaptation, procedural changes, or employee training. As the team or coordinator identifies appropriate controls, these should be instituted. The recommended program for worksite analysis includes, but is not limited to, analyzing and tracking records; monitoring trends and analyzing incidents; screening surveys; and analyzing workplace security.

Records analysis and tracking

This activity should include reviewing medical, safety, workers' compensation, and insurance records — including the OSHA 200 log, if required — to pinpoint instances of workplace violence. Scan unit logs and employee and police reports of incidents or near-incidents of assaultive behavior to identify and analyze trends in assaults relative to particular departments, units, job titles, unit activities, workstations, and/or time of day. Tabulate these data to target the frequency and severity of incidents to establish a baseline for measuring improvement.

Monitoring trends and analyzing incidents

Contacting similar local businesses, trade associations, and community and civic groups is one way to learn about their experiences with workplace violence and to help identify trends. Use several years of data, if possible, to trace trends of injuries and incidents of actual or potential workplace violence. Identify and analyze any apparent trends in injuries or incidents relating to a particular worksite, job title, activity, or time of day or week. The following questions may be helpful in ascertaining trends:

1. How many times has your establishment or similar company establishments been robbed in the past three years? Have other violent incidents occurred? Analyze incidents, noting characteristics of assailants and victims, brief account of what happened before and during the incident, noting relevant details of the situation, and its outcome.
2. How many injuries occurred during those robberies or other incidents?
3. How many times was a firearm used?
4. How many times was a firearm discharged?
5. How many times was the threat of a firearm used?
6. How many times were other weapons used?
7. What job or workstation was involved in the robbery/incident? Identify, based upon the risk factors identified in these guidelines, those work positions in which staff are at risk of assaultive behavior.
8. What specific tasks were employees performing prior to the robbery/incident? Identify processes and procedures which put employees at risk of assault. When do these occur (e.g., on all shifts)?
9. What time was the robbery/incident committed?
10. How many other incidents of violent attack or threats of violence on employees have been reported?
11. How many times have police been called to your establishment to arrest a perpetrator, deter what you considered criminal intentions, or protect property? When possible, obtain reports of police who investigated the incident, and their recommendations.

Screening surveys

One important screening tool is to give employees a questionnaire or survey to get their ideas on the potential for violent incidents and to identify or confirm the need for improved security measures. Detailed baseline screening surveys can help pinpoint tasks that put employees at risk. Periodic surveys — conducted at least annually or whenever operations change or incidents of workplace violence occur — help identify new or previously unnoticed risk factors and deficiencies or failures in work practices, procedures, or controls. Also, the surveys help assess the effects of changes in the work processes. The periodic review process should also include feedback and follow-up. Independent reviewers, such as safety and health professionals, law enforcement or security specialists, insurance safety auditors, and other qualified persons may offer advice to strengthen programs. These experts also can provide fresh perspectives to improve a violence prevention program.

Workplace security analysis

The team or coordinator should periodically inspect the workplace and evaluate employee tasks to identify hazards, conditions, operations, and situations that could lead to violence. To find areas requiring further evaluation, the team or coordinator should do the following:

12. Analyze incidents, including the characteristics of assailants and victims, and give an account of what happened before and during the incident, and the relevant details of the situation and its outcome. When possible, obtain police reports and recommendations.
13. Identify jobs or locations with the greatest risk of violence as well as processes and procedures that put employees at risk of assault, including how often and when.
14. Note high-risk factors such as: physical risk factors of the building; isolated locations/job activities; lighting problems; lack of phones and other communication devices, areas of easy, unsecured access; and areas with previous security problems.
15. Evaluate the effectiveness of existing security measures, including engineering, administrative, other control measures. Determine if risk factors have been reduced or eliminated, and take appropriate action. The following questions can help assess risks. (See sample checklist for assessing hazards in Appendix).
16. Do employees exchange money with the public or guard valuable property or possessions during evening or late-night hours of operation?
17. Is cash control a key element of the establishment's violence and robbery prevention program?
18. Is there a drop safe to minimize cash on hand?
19. Does the site have a policy to maintain less than $50 in the cash register?
20. Are there signs posted stating that limited cash is on hand?
21. Does the site have a policy limiting the number of cash registers open during late night hours?
22. Do employees receive training in emergency procedures for robberies, in conflict resolution, and in nonviolent response?
23. At sites with a history of robbery or assaults, are employees provided with bullet-proof barriers or enclosures? Is there more than one

employee on duty? After conducting a worksite analysis appropriate for the size and conditions of the workplace, the employer may find that there are no significant hazards related to violence in the establishment. If there are no hazards, the employer need not implement the other program elements recommended by the guidelines. The employer should, however, continue efforts to ensure workplace safety and health, monitoring changes in the workplace that might indicate hazards related to violence.

5. Hazard prevention and control

After hazards of violence are identified through the systematic worksite analysis, the next step is to design measures through engineering or administrative and work practices to prevent or control these hazards. If violence does occur, post-incidence response can be an important tool in preventing future incidents. The workplace violence prevention and control program should indicate specific engineering, administrative and work practice controls, and other appropriate interventions that address specific hazards at the worksite. Presented in the sections that follow are general recommendations for establishments that provide retail services and are open late at night. These illustrative examples cover engineering and workplace adaptation, and administrative and work practice controls as recommendations to help employers prevent workplace violence.

Engineering controls and workplace adaptation

Engineering controls remove the hazard from the workplace or create a barrier between the worker and the hazard. The primary hazard faced by employees working in a late-night retail establishment is assault during an armed robbery. It is essential, therefore, to find ways to protect workers from assaults involving guns or other weapons. Physical changes in the workplace that help eliminate or reduce these hazards might include using some or all of the following measures. The selection of any measure should be based upon the hazards identified in the workplace security analysis of each facility:

1. Physical barriers such as bullet proof enclosures between customers and employees provide the greatest protection for workers. Installing pass-through windows for customer transactions and limiting entry to authorized persons during certain hours of operation also limit risk. Doors used for deliveries should be locked when not in use.
2. Mechanisms that permit employees to have a complete view of their surroundings such as convex mirrors, an elevated vantage point, and placement of the employee/customer service and cash register area so that it is clearly visible outside of the retail establishment serve as deterrents.

3. Video surveillance equipment and closed-circuit TV can increase the possibility of detection and apprehension of the criminal, thus deterring crime.

4. Adequate outside lighting of the parking area and approach to the retail establishment during nighttime hours of operation enhances employee protection. Surveillance lighting to detect and observe pedestrian and vehicular entrances of the retail establishment also helps. Adequate lighting within and outside the establishment (see ANSI/IES RP7-1993) makes the store less appealing to a potential robber by making detection more likely.

5. Speed bumps placed in traffic lanes used to exit drive-up windows can deter would-be criminals by reducing the chance for a quick escape.

6. An unobstructed view to the street from the store, clear of shrubbery, trees, or any form of clutter that a criminal could use to hide can help protect employees.

7. Cash-handling controls, including the use of locked drop-safes and the posting of signs stating that limited cash is on hand, can also deter would-be robbers.

8. Height markers on exit doors should be installed to help provide more complete descriptions of assailants.

9. Strategically placed fences can control access to the store.

10. Garbage areas and external walk-in freezers or refrigerators should be located so as to ensure safety of employees who use them. There should be good visibility with no potential hiding places for assailants near these areas.

Administrative and work practice controls

Administrative and work practice controls affect the way jobs or tasks are performed. An effective program of hazard prevention and control includes safe and proper work procedures that are understood and followed by managers, supervisors, and workers. Key elements for preventing workplace violence include proper work practices, regular monitoring and feedback, modifications, and enforcement of the program. The following examples illustrate work practices and administrative procedures that can help prevent workplace violence incidents:

11. *Proper Work Practices*. A program establishing proper work practices should include appropriate training and practice time for employees. Following are some suggested work practices:

1. Employees should wear conservative clothing (such as a company uniform) and be discouraged from wearing jewelry.

2. Employees should not carry cash while on duty unless it is absolutely necessary.

3. During evening and late-night hours of operation, cash levels should be kept to a minimal amount per cash register ($50 or less) to conduct business. Transactions with large bills (over $20) should be prohibited.

4. Stores should adopt proper emergency procedures for employees to use in case of a robbery or security breach. These should include incident report forms which prompt for perpetrator information (e.g., sex, height, build, age, race) to be completed immediately following a violent event. Emergency telephone numbers should be accessible and the notification policy clearly established. All violent incidents should be reported to local police.

5. Alarm systems, video surveillance equipment, drop-safes or comparable devices, surveillance lighting, or other security devices in the establishment must be used and maintained properly.

6. Any physical barriers and/or pass-through windows must be used correctly to provide effective deterrence to robbers.

7. At sites with a history of robbery or assaults, employees should not be required to work alone.

12. *Monitoring and Feedback.* Regular monitoring helps ensure that employees continue to use proper work practices. Monitoring should include review of the specific procedures in use and their effectiveness, including a determination of whether the procedures actually used are those specified in the hazard prevention and control program. The review should address any deficiencies, changes that have occurred, employee or customer complaints about lack of security, instances of violence or threats of violence, and whether corrective action is necessary. Giving regular, constructive feedback to employees helps to ensure their commitment to the prevention program.

13. *Adjustments and Modification.* Monitoring may show a need to modify administrative and work practice controls. Such adjustments could include:
   1. Limiting or restricting customer access
   2. Increasing staffing levels
   3. Reducing the number of cashier positions
   4. Increasing security or surveillance
   5. Reducing the hours of operation

14. *Enforcement.* For an effective program, the employer should establish employee sanctions for those employees who chronically and/or purposefully violate administrative controls or work practices. An employee who has been properly trained and consulted after such a violation, but who continues to violate established written work practice, should be disciplined accordingly.

Post-incident response

Post-incident response and evaluation are essential to an effective violence prevention program. All workplace violence programs should provide comprehensive treatment for victimized employees and employees who may be traumatized by witnessing a workplace violence incident. Injured staff should receive prompt treatment and psychological evaluation whenever an assault takes place, regardless of severity. Transportation of the injured to medical care should be provided if care is not available on-site. Victims of workplace violence suffer a variety of consequences in addition to their actual physical injuries. These include short- and long-term psychological trauma, fear of returning to work, changes in relationships with co-workers and family, feelings of incompetence, guilt, powerlessness, and fear of criticism by supervisors or managers. Consequently, a strong follow-up program for these employees will not only help them to deal with these problems but also to help prepare them to confront or prevent future incidents of violence (Flannery, 1991, 1993; 1995). There are several types of assistance that can be incorporated into the post-incident response. For example, trauma-crisis counseling, critical incident stress debriefing, or employee assistance programs may be provided to assist victims. Certified employee assistance professionals, psychologists, psychiatrists, clinical nurse specialists, or social workers could provide this counseling, or the employer can refer staff victims to an outside specialist. In addition, an employee counseling service, peer counseling, or support groups may be established. In any case, counselors must be well trained and have a good understanding of the issues and consequences of assaults and other aggressive, violent behavior. Appropriate and promptly rendered post-incident debriefings and counseling reduce acute psychological trauma and general stress levels among victims and witnesses. In addition, such counseling educates staff about workplace violence and positively influences workplace and organizational cultural norms to reduce trauma associated with future incidents.

6. Training and education

Training and education ensure that all staff are aware of potential security hazards and how to protect themselves and their co-workers through established policies and procedures. Training should cover robbery prevention, reaction, and post-robbery strategies, conflict resolution, and personal safety.

General training

A training program should include all employees, especially those engaged in exchanging money with customers, such as sales clerks, waitresses, or bartenders. Employees who may face safety and security hazards should receive formal instruction on the specific hazards associated with the unit or job and facility. This includes information on potential injuries, problems identified in the facility, and the methods to control the specific hazards. Training for employees should be repeated for each employee as necessary, at least annually. In large establishments, refresher programs may be needed more frequently (monthly or quarterly) to effectively reach and inform all employees. New and returning employees should receive appropriate training through the establishment's violence prevention program. Employees should thoroughly understand the preventive measures and safeguards used at the specific worksite. Employees reassigned to jobs or tasks that increase the risk of potential violence (e.g., serving as cashier, making bank deposits, dealing with irate customers) should have specific training that covers these situations. Training should be designed and implemented by qualified persons. Appropriate special training should be provided for personnel responsible for administering the program, and training should be presented in language and comprehension levels appropriate for the individuals being trained. It should provide an overview of the potential risk of assault, the prevention measures used to deter robbery or other assaults, the behavioral skills necessary to reduce the likelihood of a violent outcome, and the appropriate steps to take in case of an emergency. The training program also should include an evaluation component. This might include supervisor and/or employee interviews, testing and observing work practices in use, and reviewing actual incident reports of assaultive behavior.

Training for supervisors, managers, and security personnel

Supervisors and managers are responsible for ensuring that employees follow safe work practices and that they receive appropriate training to accomplish this goal. Therefore, management personnel should undergo training comparable to that of the employees, plus additional training to enable them to recognize, analyze, and establish violence prevention controls. Training for managers should address any specific duties and responsibilities they have that could increase their risk of assault. Security personnel need specific training including the psychological components of handling aggressive and abusive customers and ways to handle aggression and defuse hostile situations. The training program should also include an evaluation. The content, methods, and frequency of training should be reviewed and evaluated annually by the team or coordinator responsible for implementation. Program evaluation may involve supervisor and/or employee interviews, testing and observing, and/or reviewing reports of behavior of individuals in threatening situations.

7. Recordkeeping and evaluation

Recordkeeping and evaluation of the violence prevention program are necessary to determine overall effectiveness and identify any deficiencies or changes that should be made.

Recordkeeping

Recordkeeping is essential to the success of a workplace violence prevention program. Good records help employers determine the severity of the problem, evaluate methods of hazard control, and identify training needs. Records can be especially useful to large organizations and for members of a business group or trade association who "pool" data. Records of injuries, illnesses, accidents, assaults, hazards, corrective actions, and training, among others, can help identify problems and solutions for an effective program. The following records are important:

1. OSHA Log of Injury and Illness (OSHA 200). OSHA regulations require entry on the Injury and Illness Log of any injury that requires more than first aid, is a lost-time injury, requires modified duty, or causes loss of consciousness. (This applies only to establishments required to keep OSHA logs.) Injuries caused by assaults, which are otherwise recordable, also must be entered on the log. A fatality or catastrophe that results in the hospitalization of 3 or more employees must be reported to OSHA within 8 hours. This includes those resulting from workplace violence and applies to all establishments.
2. Medical reports of work injury and supervisors' reports for each recorded assault should be kept. These records should describe the type of assault, i.e., unprovoked sudden attack; who was assaulted; and all other circumstances of the incident. The records should include a description of the environment or location, potential or actual cost, lost time, and the nature of injuries sustained.
3. Incidents of abuse, verbal attacks, or aggressive behavior — which may be threatening to the worker but do not result in injury, such as pushing or shouting and acts of aggression toward customers — should be recorded, perhaps as part of an assaultive incident report. These reports should be evaluated routinely by the affected department.
4. Minutes of safety meetings, records of hazard analyses, and corrective actions recommended and taken should be documented.
5. Records of all training programs, attendees, and qualifications of trainers should be maintained.

Evaluation

As part of their overall program, employers should evaluate their safety and security measures. Top management should review the program regularly, and with each incident, to evaluate program success. Responsible parties (managers, supervisors, and employees) should collectively reevaluate policies and procedures on a regular basis. Deficiencies should be identified and corrective action taken. An evaluation program should involve the following:

6. Establishing a uniform violence reporting system and regular review of reports.
7. Reviewing reports and minutes from staff meetings on safety and security issues.
8. Analyzing trends and rates in illness/injury or fatalities caused by violence relative to initial or "baseline" rates.
9. Measuring improvement based on lowering the frequency and severity of workplace violence.
10. Keeping up-to-date records of administrative and work practice changes to prevent workplace violence to evaluate their effectiveness.
11. Surveying employees before and after making job or worksite changes or installing security measures or new systems to determine their effectiveness.
12. Keeping abreast of new strategies available to deal with violence in the night retail industry as these develop.
13. Surveying employees who experience hostile situations about the medical treatment they received initially and, again, several weeks afterward, and then several months later.
14. Complying with OSHA and state requirements for recording and reporting deaths, injuries, and illnesses.
15. Requesting periodic law enforcement or outside consultant review of the worksite for recommendations on improving employee safety.

Management should share workplace violence prevention program evaluation reports with all employees. Any changes in the program should be discussed at regular meetings of the safety committee, union representatives, or other employee groups.

Sources of assistance

Employers who would like assistance in implementing an appropriate workplace violence prevention program can turn to the OSHA Consultation service provided in their state. Primarily targeted at smaller companies, the consultation service is provided at no charge to the employer and is independent of OSHA's enforcement activity. (See Appendix.)

OSHA's efforts to assist employers combat workplace violence are complemented by those of NIOSH (1-800-35-NIOSH) and public safety officials, trade associations, unions, insurers, human resource and employee assistance professionals, as well as other interested groups. Employers and employees may contact these groups for additional advice and information.

8. Conclusions

OSHA recognizes the importance of effective safety and health program management in providing safe and healthful workplaces. In fact, OSHA's consultation services help employers establish and maintain safe and healthful workplaces, and the agency's Voluntary Protection Programs was specifically established to recognize worksites with exemplary safety and health programs. (See Appendix) Effective programs are known to improve both morale and productivity and reduce workers' compensation costs. OSHA's violence prevention guidelines are an essential component to workplace safety and health programs. OSHA believes that the performance-oriented approach of the guidelines provide employers with flexibility in their efforts to maintain safe and healthful working conditions.

Index

A

Accident
  preventive action, 29
Actions
  disciplinary, 29
  job-related, 19
  preventive, 29
Administrative hearing process, 77
Affirmative duty, 219
Americans with Disabilities Act, 336
*Annual Survey of Occupational Injuries and
      Illnesses*, 10
Assaults
  physical, 10-11
  workplace, 10
Associated Press reports, 2-5
ASOII See Annual Survey of
      Occupational Injuries
      & Illnesses
At Risk, 15, 18-21

B

Beating, 427
Behavior, 451-454
Bensimon, Helen Frank, 1
Bona fide, 329
Brief,
  definition, 16
Briefing cases,
  dissent, 14
  example, 16-17
  holding/decision, 14
  facts, 13
  format, 12-13
  issues, 13

purpose, 12
opinion, 14
underlying policy/reason, 14
Brown v. Board of Education, 14
Farris Bryant, v. Mark Livigni, 136
Bureau of Justice Statistics, 10

C

*Emma Carter, Administrator of the Estate of
      Emma L. Hopkins, v. Skokie Valley
      Detective Agency, LTD.*, 123
Cases,
  cites, 11
  decisions, 11
  finding cases, 11
Case incidents, 425-454
Case law,
  analyzing, 11
  briefing, 11
Case studies,
  basic concepts, 423-424
Causal links, 185
Census
  death-certificate based, 5
  fatal occupational injuries, 1
C.F.R. *See Code of Federal Regulations*
Coal Mine Safety Act of 1952, 39
Code of Federal Regulations, 12
Common myths, 18
Compliance officer, 49-51
Compliant, 66
Components, 24, 27-28
Co-worker, 427
Criminal
  Liability, 47-48
  Penalties, 60

Crisis plan
  agencies/resources, 30-31
  chain of command, 30
  preparation/planning, 30-33
  procedures, 31
  investigation, 31

D

Databases,
  WESTLAW, 12
  LEXIS, 12
Death
  causes, 20
  census, 5
Dedicated employee, 15
Deep pocket, 264
Defamation, 334
Direct threat, 338
Discrimination laws, 13
Doctrine
  respondeat superior, 119
Domestic violence, 243, 432
Drug-free workplace, 20
Dunlop v. Hanover Shoe Farms, 65

E

EAP *See Employee Assistance Program*
E-mail, 332
Edwards v. Robinson-Humphrey Company,
  120
Employee
  at risk, 15, 18-21
  dedicated, 15
  loner, 16
  violent, 15-16
Employee Assistance Program, 33
Employer rights, 62

F

False light, 331
Fatal occupational injuries, 1
FBI data, 9
*The Federal Occupational Safety & Health
  Act of 1970*, 38
Federal Tort Claim Act, 71
Filing, 66

Forms
  incident report, 29
  workplace privacy, 327
The Four W's, 28
Freedom of Information Act, 67
*Kenneth D. Fund, Administrator, v. Hotel
  Lenox of Boston, Inc.*, 248

G

General Duty Clause, 37, 43-45
*Griffin, v. AAA Auto Club South, Inc., v. All
  South Security, Inc.*, 253

H

History, 1
Homicide, 5

I

Incident, 11, 19
  report form, 29
  response/recovery process,
    409-411
Individual profile, 12
Industrial revolution, 12
Information laws, 13
Injuries
  fatal/occupational, 1
Intervention, 31-32
Interviewer signals, 29
Intimidation, 447-448
Introduction, 19
Intrusion, 328
Investigation
  The Four W's, 28

J

*James Robert Johnson, Personal
  Representative of the Estate of Robert
  Wayne Johnson, v. Richard E. Thoni*,
  245
Judgment proof, 264

K

Kurlan, Warren M., 11

L

*Labor Management Relations Act
    (Taft-Hartley Act) of 1947*, 38-39
Laws
  discrimination, 13
  information, 13
  privacy, 13
Legal considerations, 327
Legal dictionaries
  Black's Law Dictionary, 15
  Ballantine Law Dictionary, 15
  Gilmer's Law Dictionary, 15
Legislative
  history, 39
Level of security, 244
Liability
  potential, 243-244
Litigation, 67
Loner, 16

M

Management team, 23
*Gail Mandy, v. Minnesota Mining &
    Manufacturing, a/k/a 3M*, 220
*McNamara-O'Hara Public Service Contract
    Act of 1965*, 39
Miranda v. Arizona, 15
Monetary
  fines/penalties, 45-46
  liabilities, 47
Myths, 18

N

NCVS *See National Crime Victimization
    Survey*
NIOSH See National Institute for
    Occupational Safety and Health
Narrow tunnel, 20-21
National Crime Victimization Survey,
    10
National Foundation of Arts &
    Humanities Act, 40
National Institute for Occupational
    Safety and Health, 1, 37
National Safe Workplace Institute, 1
National Traumatic Occupational
    Fatalities, 5–6

Negligence
  defined by *Black's Law Dictionary*, 68
  four basic elements, 133
Negligent
  actions, 68
  basic elements, 217-218
  causal links, 185
  hiring, 117
  retention, 133
  security, 241
  supervision, 185
  theories, 12, 185
  training, 217
Negligent Entrustment, 186
Newspaper Articles, 2-5
NTOF See National Traumatic
    Occupational Fatalities

O

Occupational Safety and Health
    Administration, 12, 37
Oklahoma City, 10
OSH Act
  coverage, 40
  jurisdiction, 40
  private litigation, 67
  rights/responsibilities, 49
OSHA See Occupational Safety and
    Health Administration
  defense, 70
  enforcement, 45
  inspection, 49-53
  potential liabilities, 37
  standards, 43
Overview/history, 1

P

Parallel action, 186
*Melody Perkins v. Thomas S. Spivey, v.
    General Motors Corp.*, 156
*Pizza Hut of America, Inc., v. Ray Keefe and
    Paula Keefe*, 80
Ponticas, v. K.M.S. Investments, 120
Postal service, 11
Postal workers, 11
Preplanning, 23
Prevention, 31-32
Prevention and Response Program, 23

Prima facie, 133
  elements, 335
Primary federal groups
  National Institute of Safety and Health
    (NIOSH), 37
  Occupational Safety and Health
    Administration (OSHA), 37
  United States Department of Labor, 37
Primary theories, 78
Privacy
  issues, 330-331
  drug testing, 332
  laws, 13
Probability, 19
Profile
  general, 12
  individuals, 12, 15, 17-19
Cornelius Pollard, v. State of New York, 258
Promulgation of standards, 43-44
Protected activities, 64-65
Publicity, 330
Publicized incidents, 11

Q

Questions
  basic, 19

R

Rate, 19
reasonable accommodation, 338
Research
  selected readings/studies, 417
Respondeat superior doctrine, 119
Response, 31-32
Risks, 19-20
Rosen-House v. Sloan's Supermarket, 120
Rule of law, 15

S

Safeguard Computer Systems, 33
*Barbara G. Santiago, etc., v. Edward Allen
    and Dorothy Allen*, 281
Screening applicants, 27
Security
  basic elements, 242
  development, 30
  improvement, 30

level, 244
physical, 32-33
pivotal issues, 241
private/public sector, 241-242
Seclusion, 328
*Larry L. Sharp, Administrator of the Estate
    of David C. Sharp et al., v. Norbert
    E. Mitchell, Sr., et al.*, 267
Shooting, 425
Stalking, 430
Standard of care, 219
*State of Nevada, Department of Human
    Resources, Division of Mental
    Hygiene & Mental Retardation, v.
    Julie Jimenez, Guardian Ad Litem*,
    188
State safety plans, 41-42
Statistics, 1
Statutory discrimination, 186
Suicide, 429
Surveillance system, 5
Sympathy factor, 266

T

Taft-Hartley Act, of 1947, 38-39
Threats
  coping, 34-35
  direct, 338
Threats of Violence, 31-35, 434, 437,
    440-445
Theory
  negligent, 12

U

Unabomber, 10
U.S. Bureau of Labor Statistics, 1
U.S. Department of Labor, 37
U.S. Postal
  services, 11
  workers, 11

V

Violations
  de minimis, 53
  failure to abate, 58-59
  non-serious, 54
  penalty schedule, 46

repeat, 58-59
serious, 55
willful, 57
Violation notices, 60
Violence
  development, 1, 27
  domestic, 243
  training, 1, 27
  workplace, 1, 27
Violence in the workplace, 1, 27
Violent employee, 15-16

W

Waiver of rights, 66
*Kimberly M. Walsh, v. City School District
  of Albany*, 212
Walsh-Healey Public Contracts Act of 1936,
  38
Walton v. Potlatch Corp., 69
Warning Signs, 15, 17-19
Whirlpool Corp. v. Marshall, 67
Workers' compensation
  administrative hearing process, 77
  features, 74
  liability, 73
  state systems, 73
Workplace
  death, 20
  development, 1, 27
  drug-free, 20
  homicides, 5
  privacy, 327
  training, 1, 27
  violence, 1, 27
Workplace homicides
  age group/sex, 7
  bureau of census/geographic region, 7
  high-risk industries, 9
  industry/sex, 6, 8

methods, 8
NIOSH data, 6, 7, 9
race, 7
Workplace violence
  definition, 11
  emergency/disaster program, 412
  issues, 37
  prevention policy, 26
  prevention and response program, 23,
    412
World Trade Center, 10
Wounding, 427
Written plan
  areas to include, 27
Written policy statement
  developing considerations, 25
  inclusions, 24-25
  issuing advantages, 24
  statement sample, 26
  workplace violence prevention policy,
    26
Wrongful death
  actions, 263
  damages, 254-265
  defenses, 265-266
  elements, 264-265
  immunity protection, 266
  statutes, 263
Wrongful discharge, 186

Y

*Jean Marie Yunker, as Trustee for the heirs
  and next of kin of Kathleen M. Nesser,
  v. Honeywell, Inc.*, 283

Z

Perry Zeiger, v. State of California et al.,
  94
Zero tolerance/policy, 25, 30-32

For Product Safety Concerns and Information please contact our EU
representative GPSR@taylorandfrancis.com
Taylor & Francis Verlag GmbH, Kaufingerstraße 24, 80331 München, Germany

www.ingramcontent.com/pod-product-compliance
Ingram Content Group UK Ltd.
Pitfield, Milton Keynes, MK11 3LW, UK
UKHW021624240425
457818UK00018B/724